SYSTEM DYNAMICS:
A UNIFIED APPROACH

SYSTEM DYNAMICS:
A UNIFIED APPROACH

DEAN KARNOPP

Department of Mechanical Engineering
University of California, Davis Campus

RONALD ROSENBERG

Department of Mechanical Engineering
Michigan State University

A WILEY–INTERSCIENCE PUBLICATION

JOHN WILEY & SONS, New York • Chichester • Brisbane • Toronto

Library of Congress Cataloging in Publication Data:

Karnopp, Dean.
  System dynamics.

  "A Wiley-Interscience publication."
  Includes bibliographical references.
  1. Systems engineering. 2. System analysis.
3. Bond graphs. I. Rosenberg, Ronald C., joint
author. II. Title.

TA168.K362   620'.72   74-22466
ISBN 0-471-45940-2

Printed in the United States of America

10 9 8 7 6 5

PREFACE

The principal contribution of this book is its unified approach to the modeling and manipulation of dynamic engineering systems, an approach made possible by the use of bond graphs. The basic physics involved in the description of mechanical, electrical, hydraulic, thermal, magnetic, and fluid dynamic systems is phrased in terms of four types of generalized variables—effort, flow, displacement, and momentum. Bond-graph models can be manipulated systematically to yield state-space equations of standard form. Furthermore, the ENPORT computer program can provide dynamic responses directly from a suitable bond-graph model and specified excitations, without requiring the explicit prior formulation of state equations. We believe that bond-graph methods offer the most unified and understandable way to proceed from the basic physical modeling of components, devices, and their connections to analytical and computational results for complex systems involving a variety of types of energy flow.

This book was written with several goals in mind. First, despite the range of coverage, we desired the book to be reasonably compact. Second, we wanted to enable a student to use part of the material to begin a study of the simpler mechanical, electrical, hydraulic, thermal, and transducer systems, but in a style that would generalize suitably if he or she should continue more deeply into the areas of system dynamics, control, and computation. Third, we wished to provide ample advanced material to demonstrate the power of the approach for complex systems including transduction and modulation, and for large-scale systems. We hope the inclusion of this material will stimulate interest in the earlier parts of the book and, at the same time, allow the book to be used for more advanced courses.

Naturally, nothing is without its price, and the most difficult decision for us has been which mathematical methods to include and which to

v

omit. The book falls naturally into two parts. In the first part, Chapters 1 through 6, the level of mathematical sophistication has been kept low. In fact, the first part can be studied by a student who is taking a first course in differential equations concurrently. All formulations are organized in a state-space-compatible form, but the actual formalism of vectors and matrices has been omitted. In the second part of the book, which deals with large-scale and nonlinear systems and with a richer variety of physical systems, matrix notation is used, and some familarity with differential equations is assumed.

The existence of various excellent computer programs for performing many phases of systems analysis once a model is made has enabled us to count on the student being able to examine the dynamics of a variety of systems without having to program at a language level like FORTRAN. This generally has increased the student's incentive and interest and has permitted us to consider a greater variety of engineering examples than otherwise might have been possible. We strongly recommend the use of automated computational aids in conjunction with this book.

The first part has been used in a number of engineering schools across the country in a preliminary version for several years, in courses at both the under- and upper-division levels. The second part has been used in senior-graduate courses, where the students have had some prior systems experience, whether bond-graph style or not. Various instructors have chosen to shape their own special variations by the material they add and omit. At some point, students must become conversant with block diagrams, sinusoidal frequency response, transfer functions, Laplace transforms, and other topics that we have omitted or treated in a somewhat abbreviated fashion. Experience has shown that additional topics may be fitted in very nicely to the flow of the book. In that sense the book serves as a unifying basis on which to build one's understanding of system dynamics, rather than a complete statement of all the techniques useful in attaining this understanding.

Because this is the first system dynamics textbook to adopt the bond-graph approach as its basis, we would like to express our appreciation to those who have helped us in our efforts to bring it to publication. To Professor Henry Paynter, who started the whole thing in the late 1950s, and to the many intrepid investigators who were willing to use novel methods in a hard-headed business like engineering, go our thanks for advancing the state-of-the-art to its present condition. To the many engineering instructors who found themselves explaining subleties of a method they had only recently learned themselves go our thanks for their professorial courage. To the College of Engineering at Michigan State University, which made available the preliminary edition of the material, on which so many of us have relied for so long, go our prepublication thanks. And to our wives, who have been with us somewhat longer than bond graphs, go our thanks and probably our royalties. As a final note we would like to add that flashes of insight and brilliance detectable in the text should be credited to the authors, while errors of substance should be attributed to faulty interpretation of our intended meaning.

DEAN C. KARNOPP
RONALD C. ROSENBERG

Davis, California
East Lansing, Michigan
July, 1974

CONTENTS

SYSTEM DYNAMICS:
A UNIFIED APPROACH

1

INTRODUCTION

This book is concerned with the development of an understanding of the dynamic physical systems that engineers are called upon to design. Methods for modeling real systems will be presented, ways of analyzing systems in order to shed light on system behavior will be shown, and techniques for using computers to simulate the dynamic response of systems to external stimuli will be developed. Before beginning the study of physical systems it is worthwhile to reflect a moment on the nature of the discipline that is usually called "system dynamics" in engineering.

The word "system" is used so often and so loosely to describe a variety of concepts that it is hard to give a meaningful definition of the word or even to see the basic concept that unites its diverse meanings. When the word "system" is used in this book, two basic assumptions are being made: (1) A system is assumed to be an entity separable from the rest of the universe (the environment of the system) by means of a physical or conceptual boundary. An animal, for example, can be thought of as a system that reacts to its environment (the temperature of the air, for example) and that interchanges energy and information with its environment. In this case the boundary is physical or spatial. An air traffic control system, on the other hand, is a complex, man-made system, the environment of which is not only the physical surroundings but also the fluctuating demands for air traffic which ultimately come from human decisions about travel and the shipping of goods. The unifying element in these two disparate systems is the ability to decide what belongs in the system and what represents an external disturbance or command originating from outside the system. (2) A system is composed of interacting parts. In an animal we recognize organs with specific functions, nerves that transmit information, and so on. The air traffic control system is

1

composed of men and machines with communication links between them. Clearly the *reticulation* of a system into its component parts is something that requires skill and art since most systems could be broken up into so many parts that any analysis would be swamped with largely irrelevant detail.

These two aspects of systems can be recognized in everyday situations as well as in the more specific and technical applications that form the subject matter of most of this book. For example, when one hears a complaint that the transportation system in this country does not work well, one may see that there is some logic in using the word system. First of all, the transportation system is roughly identifiable as an entity. It consists of air, land, and sea vehicles and the men, machines, and decision rules by which they are operated. In addition, many parts of the system can be identified—cars, planes, ships, baggage handling equipment, computers, and the like. Each part of the transportation system could be further reticulated into parts (i.e., each component part is itself a system), but for obvious reasons we must exercise restraint in this division process.

The essence of what may be called the "systems viewpoint" is to concern oneself with the operation of a complete system rather than with just the operation of the component parts. Complaints about the transportation system are often real "system" complaints. It is possible to start a trip in a private car that functions just as its designers had hoped it would, transfer to an airplane that can fly at its design speed with no failures, and end in a taxi that does what a taxi is supposed to do and yet have a terrible trip because of traffic jams, air traffic delays, and the like. Perfectly good components can be assembled into an unsatisfactory system.

In engineering, as indeed in virtually all other types of human endeavor, tasks associated with the design or operation of a system are broken up into parts which can be worked on in isolation to some extent. In a power plant, for example, the generator, turbine, boiler, and feed water pumps typically will be designed by separate groups. Furthermore, heat transfer, stress analysis, fluid dynamic, and electrical studies will be undertaken by subsets of these groups. In the same way, the bureaucracy of the federal government represents a splitting up of the various functions of government. All the separate groups working on an overall task must interact in some manner to make sure that not only will the parts of the system work but also the system as a whole will perform its intended function. Many times, however, oversimplified assumptions about how the system will operate are made by those working on a small part of the system. When this happens the results can be disappointing. The power plant may undergo damage during a full load rejection or the economy of a country may collapse because of the unfavorable interaction of segments of government each of which assiduously pursues seemingly reasonable policies.

In this book, the main emphasis will be on studying system aspects of behavior as distinct from component aspects. This requires a knowledge of the component parts of the systems of interest and hence some knowledge in certain areas of engineering that are taught and sometimes even practiced in splendid isolation from other areas. In the engineering systems of primary interest in this book, topics from vibrations, strength of materials, dynamics, fluid mechanics, thermodynamics, automatic control, and electrical circuits will be used. It is possible, and perhaps even common, for an engineer to spend a major part of his professional career in just one of these disciplines, despite the fact that few significant engineering projects concern a single discipline. Systems engineers, on the other hand, must have a reasonable command of several of the engineering sciences as well as knowledge pertinent to the study of systems per se.

Although many systems may be successfully designed by careful attention to static or steady-state operation in which the system variables are assumed to remain constant in time, in this book the main concern will be with *dynamic* systems, that is, those systems whose behavior as a function of time is important. For a transport aircraft that will spend most of its flight time at a nearly steady speed, the fuel economy at constant speed is important. For the same plane, the stress in the wing spars during steady flight is probably less important than the time varying stress during flight through turbulent air, during emergency maneuvers, or during hard landings. In studying the fuel economy of the aircraft, a static system analysis might suffice. For stress prediction, a dynamic system analysis would be required.

Generally, of course, no system can operate in a truly static or steady state, and both slow evolutionary changes in the system and shorter time transient effects associated, for example, with start up and shut down, are important. In this book, despite the importance of steady-state analysis in design studies, the emphasis will be on dynamic systems. Dynamic system analysis is more complex than static analysis but is extremely important since decisions based on static analyses can be misleading. Systems may never actually achieve a possible steady state, because of external disturbances or instabilities that appear when the system dynamics are taken into account. Also, systems of all kinds can exhibit counter-intuitive behavior when considered statically. A change in a system or a control policy may appear beneficial in the short run based on static considerations, but may have long-run repercussions opposite to the initial effect. The history of social systems abounds with sometimes tragic examples and there is hope that dynamic system analysis can help avoid some of the errors in "static thinking" [1]. However, even in engineering with rather simple systems, one must have some understanding of the dynamic response of a system before one can reasonably study the system on a static basis.

A simple example of a counter-intuitive system in engineering is the case of a hydraulic power generating plant. In order to reduce power, wicket gates just before the turbine are moved toward the closed position. Temporarily, however, the power actually increases as the inertia of the water in the penstock forces the flow through the gates to remain almost constant, resulting in a higher velocity of flow through the smaller gate area. Ultimately, the water in the penstock slows down and power is reduced. Without an understanding of the dynamics of this system, one would be led to open the gates to *reduce* power. If this were done, the immediate result would be a gratifying decrease in power followed by a surprising and inevitable increase in power. Clearly a good understanding of dynamic response is crucial to the design of a control policy for dynamic systems.

1.1 MODELS OF SYSTEMS

A central idea involved in the study of the dynamics of real systems is the idea of a *model* of the system. Models of systems are simplified, abstracted constructs used to predict the behavior of systems of interest. Scaled, physical models are well known in engineering. In this category fall the wind tunnel models of aircraft, ship hull models used in towing tanks, structural models used in civil engineering, plastic models of metal parts used in photoelastic stress analysis, and the "breadboard" models used in the design of electric circuits.

The characteristic feature of these models is that some, but not all, of the features of the real system are reflected in the model. In the wind tunnel aircraft model, for example, no attempt is made to reproduce the color or interior seating arrangement of the real aircraft. Aeronautical engineers assume that some aspects of a real craft are unimportant in determining the aerodynamic forces on the craft and thus the model contains only those aspects of the real system that are supposed to be important to the characteristics of the system under study.

In this book, another type of model, often called a *mathematical model*, is considered. Although this type of model may seem much more abstract than the physical model, there are strong similarities between physical and mathematical models. The mathematical model is used to predict only certain aspects of the system response to inputs. For example, a mathematical model might be used to predict how a proposed aircraft would respond to pilot input command signals during test maneuvers. But such a model would not have the capability of predicting every aspect of the real aircraft response. The model might not contain any information on changes in aerodynamic heating during maneuvers nor about high-frequency vibrations of the aircraft structure, for example.

Because a model must be a simplification of reality, there is a great deal of art in the construction of models. An overly complex and detailed model may contain parameters virtually impossible to estimate, may be practically impossible to analyze, and may cloud important results in a welter of irrelevant detail if it can be analyzed. An overly simplified model will not be capable of exhibiting important effects. It is important, then, to realize that *no system can be modeled exactly* and that any competent system designer needs to have a procedure for constructing a variety of system models of varying complexity so that he can find the simplest model capable of answering the questions he has about the system under study.

The remainder of this book deals with models of systems and with the procedures for constructing models and for extracting system characteristics from models. The models will be mathematical models in the usual meaning of the term even though the models may be represented by stylized graphs and computer printouts rather than the more conventional sets of differential equations.

System models will be constructed using a uniform notation for all types of physical systems. It is a remarkable fact that models based on apparently diverse branches of engineering science all can be express using the notation of *bond graphs* based on energy and information flow. This allows one to study the *structure* of a system model. The nature of the parts of the model and the manner in which the parts interact can be made evident in a graphical format. In this way, analogies between various types of systems are made evident and experience in one field can be extended to other fields.

Using the language of bond graphs, one may construct models of electrical, magnetic, mechanical, hydraulic, pneumatic, thermal, and other systems using only a rather small set of ideal elements. Standard techniques allow the models to be translated into differential equations or computer simulation schemes. Historically, diagrams for representing dynamic system models developed separately for each type of system. For example, parts *a*, *b*, and *c* of Figure 1.1 each represent a diagram of a typical model. Note that in each case the elements in the diagram seem to have evolved from sketches of devices, but in fact a photograph of the

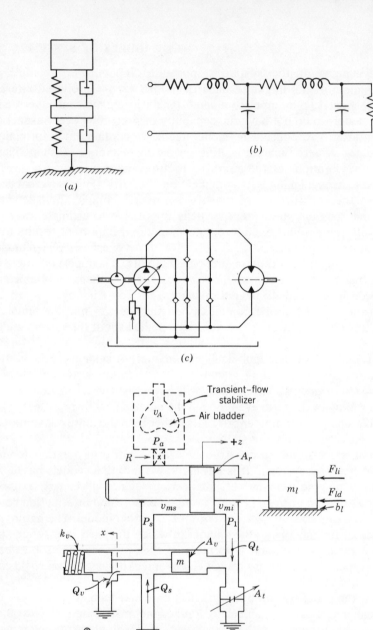

Figure 1.1. (*a*) Typical mechanical schematic diagram; (*b*) typical electric circuit diagram; (*c*) typical hydraulic diagram; (*d*) schematic diagram of system containing mechanical, electrical, and hydraulic components.

6

real system would not resemble the diagram at all. Figure 1.1a, might well represent the dynamics of heave motion of an automobile, but the masses, springs, and dampers of the model are not directly related to the parts of an automobile visible in a photograph. Similarly, symbols for resistors and inductors in diagrams such as Figure 1.1b may not correspond to separate physical elements called resistors and chokes but instead may correspond to the resistance and inductance effects present in a single physical device. Thus, even semipictorial diagrams are often a good deal more abstract than they might at first appear.

When mixed systems such as that shown in Figure 1.1d are to be studied, the conventional means of displaying the system model are less well developed. Indeed, few such diagrams are very explicit about just what effects are to be included in the model. The basic structure of the model may not be evident from the diagram. A bond graph is more abstract than the type of diagrams shown in Figure 1.1, but it is explicit and has the great advantage that all the models shown in Figure 1.1 would be represented using exactly the same set of symbols. For mixed systems such as Figure 1.1d, a universal language such as bond graphs provide is required in order to display the essential structure of the system model.

1.2 SYSTEMS, SUBSYSTEMS, AND COMPONENTS

In order to model a system it is usually necessary to first break up the system into smaller parts that can be modeled and perhaps studied experimentally and then to assemble the system model from the parts. Often, the breaking up of the system is conveniently accomplished in several stages. In this book major parts of a system will be called *subsystems* and primitive parts of subsystems will be called *components*. Of course, the hierarchy of components, subsystems, and systems can never be absolute, since even the most primitive part of a system could be modeled in such detail that it would be a complex subsystem. On the other hand, in many engineering applications, the subsystem and component categories are fairly obvious.

Basically, a subsystem is a part of a system that will be modeled as a system itself; that is, the subsystem will be broken into interacting component parts. A component, on the other hand, is modeled as an entity and is not thought of as composed of simpler parts. One needs to know how the component interacts with other components and one must have a characterization of the component, but otherwise a component is treated as a "black box" without any need to know what caused it to act as it does.

To illustrate these ideas, consider the vibration test system shown in

Figure 1.2. The system is intended to subject a test structure to a vibration environment specified by a signal generator. For example, if the signal generator delivers a random noise signal, it may be desired that the acceleration of the shaker table be a faithful reproduction of the electrical noise signal waveform. In a system that is assembled from physically separate pieces, it is natural to consider the parts that are assembled by connecting wires, hydraulic lines, or by mechanical fasteners as subsystems. Certainly, the electronic boxes labeled signal generator, controller and electrical amplifier are subsystems, as are the electrohydraulic valve, the hydraulic shaker, and the test structure. It may be possible to treat some of these subsystems as components if their interactions with the rest of the system can be specified without knowledge of the internal construction of the subsystem. The electrical amplifier is obviously composed of many components, such as resistors, capacitors, transistors, and the like, but if the amplifier is sized correctly so that it is not overloaded, then it may be possible to treat the amplifier as a component specified by the manufacturers input-output data. Other subsystems may require a subsystem analysis in order to achieve a dynamic description suitable for the overall system study.

Consider, for example, the electrohydraulic valve. A typical servo valve is shown in Figure 1.3. Clearly, the valve is composed of a variety of electrical, mechanical, and hydraulic parts that work together to produce the dynamic response of the valve. For this subsystem the components might be the torque motor, the hydraulic amplifier, mechanical springs, hydraulic passages, and the spool valve. A subsystem dynamic analysis can reveal weaknesses in the subsystem design that may necessitate the substitution of another subsystem or a reconfiguration of the overall system. On the other hand, such an analysis may indicate that, from the

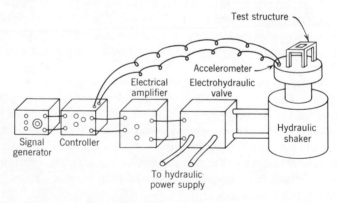

Figure 1.2.   Vibration test system.

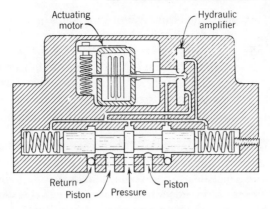

Figure 1.3.   Electrohydraulic valve.

point of view of the overall system, the subsystem may be adequately characterized as a simple component. A skilled and experienced system designer often makes a judgement on the appropriate level of detail for the modeling of a subsystem on an intuitive basis. A major purpose of the methods presented in this book is to show how system models can be assembled conveniently from component models. It is then possible to experiment with subsystem models of varying degrees of sophistication in order to verify or disprove initial modeling decisions.

1.3 STATE-DETERMINED SYSTEMS

The goal of this book is to describe means for setting up mathematical models for systems. The type of model that will be found often is described as a "state-determined system." In mathematical notation, such a system model often is described by a set of ordinary differential equations in terms of so-called *state variables* and a set of algebraic equations that relate other system variables of interest to the state variables. In succeeding chapters an orderly procedure beginning with physical effects to be modeled and ending with state equations will be demonstrated. Even though some techniques of analysis and computer simulation do not require that the state equations be written, from a mathematical point of view all the system models are state-determined systems.

The future of all the variables associated with a state-determined system can be predicted if (1) the state variables are known at some initial time and (2) the future time history of the input quantities from the environment are known.

Such models, which are virtually the only ones used in engineering, have some built-in philosophical implications. For example, events in the future do not affect the present state of the system. This implication is correlated with the assumption that time runs only in one direction—from past to future. That models should have these properties probably seems plausible, if not obvious, yet it is remarkably difficult to conceive of a demonstration that real systems always have these properties.

Clearly, past history can have an effect on a system; yet the influence of the past is exhibited in a special way in state-determined systems. All the past history of a state-determined system is summed up in the present value of its state variables. This means that many past histories could have resulted in the same present value of state variables and hence the same future behavior of the system. It also means that if one can condition the system to bring the state variables to some particular values, then the future system response is determined by the future inputs and nothing is important about the past except that the state variables were brought to particular values.

Scientific experiments are run as if the systems to be studied were state determined. The system is always started from controlled conditions that are expressed in terms of carefully monitored variables. If the experiment is repeatable, then the assumption is that the state variables are properly initialized by the operations used to set up the experiment. If the experiment is not repeatable, then the assumption is that some important influence has not been controlled. This influence can be either a state variable that was not monitored and initialized properly or an unrecognized input quantity through which the environment influences the system.

State-determined system models have proved useful over centuries of scientific and technical work. For the usual macroscopic systems encountered in engineering, state-determined system models are nearly universal, and there is continuing interest in developing such models for social and economic systems. This book can be regarded as a textbook devoted to the establishment and study of state-determined system models using well-defined physical systems of interest to engineers as examples.

1.4 USES OF DYNAMIC MODELS

In Figure 1.4 a general dynamic system model is shown schematically. The system, S, is characterized by a set of state variables indicated by X that are influenced by a set of input variables, U, that represent the action of the system's environment on the system. The set of output variables,

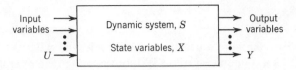

Figure 1.4. General dynamic system model.

Y, are observable aspects of the system's response or back effects from the system onto the environment. This type of dynamic system model may be used in three quite distinct ways:

1. Analysis. Given U for the future, X at the present, and the model, S, predict the future of Y. Assuming that the system model is an accurate representation of the real system, analysis techniques allow one to predict system behavior.

2. Identification. Given time histories of U and Y, usually by experimentation on real systems, find a model, S, and state variables, X, which are consistent with U and Y. This is the essence of scientific experimentation. Clearly, a "good" model is one that is consistent with a great variety of U and Y sets.

3. Synthesis. Given U and some desired Y, find S such that U acting on S will produce Y. Most of engineering deals with synthesis, but only in limited contexts are there direct synthesis methods. Often we must be content to accomplish synthesis of systems via a trial and error process of repetitive analysis of a series of candidate systems. In this regard, dynamic models play a vital role since progress would be slow indeed if one had to construct each candidate system "in the metal" in order to discover its properties.

In this book we will concentrate on setting up system models and predicting the behavior of the systems using analytical or computational techniques. Thus we will concentrate on analysis, but it is important to remember that the techniques are useful for identification problems and that the major challenge to a systems engineer is to synthesize desirable systems. It may not be too much to say that analysis, except in the service of synthesis, is a rather sterile pursuit for an engineer.

REFERENCE

1. J. W. Forrester, *Urban Dynamics*, Cambridge: M.I.T. Press, 1969.

PROBLEMS

1-1 Suppose you were a heating engineer and you wished to consider a house as a dynamic system. Without a heater the average temperature in the house would clearly vary over a 24-hr period. What might you consider for inputs, outputs, and state variables for a simple dynamic model? How would you expand your model so that it would predict temperatures in several rooms of the house? How does the installation of a thermostatically-controlled heater change your model?

1-2 For a particular car operated on a level road at steady conditions there is a relation between throttle position and speed. Sketch the general shape you would expect for this curve. If recordings were made of instantaneous speed and throttle position while the car was driven normally for several miles on ordinary roads, do you think that the instantaneous values of speed and throttle position would fall on the steady-state curve? What inputs, outputs, and state variable might prove useful in trying to find a dynamic model useful in predicting dynamic speed variations?

1-3 A car is driven over a curb twice—once very slowly and once quite rapidly. What would you need to know about the car in the second case that you did not need to know in the first case if you were required to find the tire force which resulted from going over the curb?

1-4 In the steady state a good weather vane points into the wind, but when the wind shifts, the vane cannot always be trusted to be pointing into the wind. Identify inputs, outputs, and the parameters of the weather vane system which affect its response to the wind. Sketch your idea of how the position of the vane would change in time if the wind suddenly shifted 10°.

1-5 The height of water in a reservoir fluctuates in time. If you had to construct a dynamic system model to help water resource planners predict variations in the height, what input quantities would you consider? How many state variables do you think you would need for your model?

2

MULTIPORT SYSTEMS AND BOND GRAPHS

In this chapter the first steps are taken toward the development of system-modeling techniques for the kinds of systems encountered in engineering. First, major subsystems are identified, and the means by which the subsystems are interconnected are studied. The fact that interacting physical systems must transmit power then is used to unify the description of interconnected subsystems. A uniform classification of the variables associated with power and energy is established, and bond graphs showing the interconnection of subsystems are introduced. Finally, the notions of inputs, outputs, and pure signal flows are discussed.

2.1 ENGINEERING MULTIPORTS

In Figure 2.1 a random collection of sybsystems or components of engineering systems is shown. Although the subsystems sketched are quite elementary, they will serve to introduce the concept of an engineering multiport. "Engineering" is used to imply that the devices are used to build up a systems such as automobiles, television sets, machine tools, or electric power plants that are designed to accomplish some specific objectives. "Multiport" refers to a point of view taken in the description of the subsystems.

Inspection of Figure 2.1 reveals that a number of variables have been labeled on the subsystems. These variables are torques, angular speeds, forces, velocities, voltages, currents, pressures, and volume flow rates. The variables occur in pairs associated with points at which the subsystems could be connected with other subsystems to form a system. It

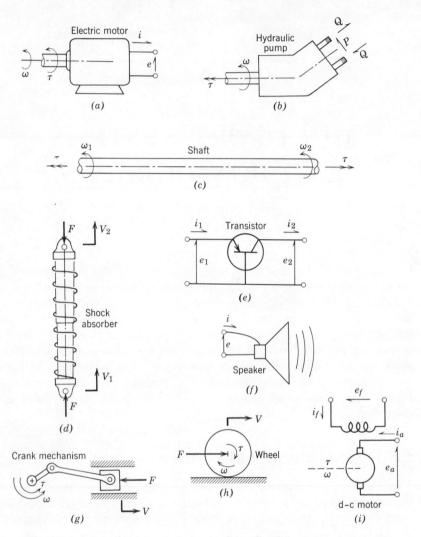

Figure 2.1. A collection of engineering multiports. (*a*) Electric motor; torque τ, angular speed ω, voltage e, current i; (*b*) hydraulic pump; torque τ, angular speed ω, pressure P, volume flow rate Q; (*c*) drive shaft; torque τ, angular speeds ω_1 and ω_2; (*d*) spring shock absorber unit; force F, velocities V_1 and V_2; (*e*) transistor; voltages e_1 and e_2, currents i_1 and i_2; (*f*) loudspeaker; voltage e, current i; (*g*) crank and slider mechanism; torque τ, angular speed ω, force F, velocity V; (*h*) wheel; force F, velocity V, torque τ, angular speed ω; (*i*) separately excited d-c motor; torque τ, angular speed ω, voltages e_a and e_f, currents i_a and i_f.

14

would be possible, for example, to couple the shaft of the electric motor (*a*) to one end of the drive shaft (*c*), and the hydraulic motor shaft (*b*) to the other end of the drive shaft. After the coupling the motor torque and speed would be identical to the torque and speed of one end of the drive shaft. Similarly, the torque and speed of the other end of the drive shaft would be identical after coupling to the torque and speed of the hydraulic pump. (If the drive shaft were not rigid, then the two angular speeds, ω_1 and ω_2, on the drive shaft ends would not necessarily be equal at all times.) Similarly, one could connect the two terminals of the transistor (*e*) associated with e_2 and i_2 to the terminals of the loudspeaker (*f*). After the connection the voltage and current associated with one terminal pair of the transistor would be identical to the voltage and current associated with the loudspeaker terminals. Generally, when two subsystems or components are joined together physically, two complementary variables are simultaneously constrained to be equal for the two subsystems.

Places at which subsystems can be interconnected are places at which power can flow between the subsystems. Such places are called *ports* and physical subsystems with one or more ports are called *multiports*. A system with a single port is called a *1-port*, a system with two ports is called a *2-port*, and so on. The multiports in Figure 2.1*a* through *h* are shown as 2-ports. Figure 2.1*f* is a 1-port as long as it is considered only as an electrical element and not as an element coupling electrical and acoustic subsystems. Figure 2.1*i* is shown as a 3-port.

The variables listed for the multiports in Figure 2.1 and the variables that are forced to be identical when two multiports are connected are called *power variables* because the product of the two variables considered as functions of time is the instantaneous power flowing between the two multiports. For example, if the electrical motor, Figure 2.1*a*, were coupled to the hydraulic pump, Figure 2.1*b*, the power flowing from the motor to the pump would be given by the product of the angular speed and the torque. Since power could flow in either direction, a sign convention for the power variables will be established. Similarly, power can be expressed as the product of a force and a velocity for a multiport in which mechanical translation is involved, as the product of voltage and current for an electrical port, and as the product of pressure and volume flow rate for a port at which hydraulic power is interchanged.

Since power interactions are always present when two multiports are connected, it is useful to classify the various power variables in a universal scheme and to describe all types of multiports in a common language. In this book all power variables are called either *effort* or *flow*. Table 2.1 shows effort and flow variables for several types of power interchange.

TABLE 2.1.   Some effort and flow quantities

Domain	Effort, $e(t)$	Flow, $f(t)$
Mechanical translation	Force component, $F(t)$	Velocity component, $V(t)$
Mechanical rotation	Torque component, $\tau(t)$	Angular velocity component, $\omega(t)$
Hydraulic	Pressure, $P(t)$	Volume flow rate, $Q(t)$
Electric	Voltage, $e(t)$	Current, $i(t)$

As Table 2.1 indicates, in general discussions the symbols $e(t)$ and $f(t)$ are used to denote effort and flow quantities as functions of time. For specific applications, more traditional notation suggestive of the physical variable involved may be used. A curse of system analysis that becomes evident as soon as problems involving several energy domains are studied is that it is hard to establish notation that does not conflict with conventional usage. In Table 2.1, for example, a force is an effort quantity, $e(t)$, even though the common use of the letter F to stand for a force might be confused with the $f(t)$, which stands for a flow quantity. These notational difficulties are bothersome but not fundamental, and cannot be avoided except by using entirely new notation. For example, the letter Q has been used for charge in electric circuits, volume flow rate in hydraulics, and heat in thermodynamics. In this book, both the generalized notation e and f and the physical notation in Figure 2.1 and Table 2.1 are used. The context in which the symbols are used will resolve any possible ambiguities in meaning.

The power, $P(t)$, flowing into or out of a port can be expressed as the product of an effort and a flow variable, and thus in general notation is given by the following expression:

$$P(t) = e(t)f(t). \tag{2.1}$$

In a dynamic system the effort and the flow variables, and hence the power, fluctuate in time. Two other types of variables turn out to be important in describing dynamic systems. These variables, sometimes called *energy variables* for reasons which will become clearer later, are called *momentum*, $p(t)$, and *displacement*, $q(t)$, in generalized notation.

The momentum is defined as the time integral of an effort. That is,

$$p(t) \equiv \int^{t} e(t)\,dt = p_0 + \int_{t_0}^{t} e(t)\,dt, \tag{2.2}$$

in which either the indefinite time integral can be used, or one may define p_0 to be the initial momentum at time t_0 and use the definite integral from t_0 to t. In the same way a displacement variable is the time integral of a flow variable.

$$q(t) \equiv \int^t f(t)\, dt = q_0 + \int_{t_0}^t f(t)\, dt. \tag{2.3}$$

Again, the second integral expression in Eq. (2.3) indicates that at time t_0 the displacement is q_0.

Other ways of writing the definitions in Eqs. (2.2) and (2.3) follow by considering the differential rather than the integral forms:

$$\frac{dp(t)}{dt} = e(t); \qquad dp = e\, dt. \tag{2.2a}$$

$$\frac{dq(t)}{dt} = f(t); \qquad dq = f\, dt. \tag{2.3a}$$

The energy, $E(t)$, which has passed into or out of a port is the time integral of the power, $P(t)$. Thus

$$E(t) \equiv \int^t P(t)\, dt = \int^t e(t) \cdot f(t)\, dt. \tag{2.4}$$

The reason p and q are sometimes called energy variables is that in Eq. (2.4) one may be able to write $e\, dt$ as dp or $f\, dt$ as dq by using Eq. (2.2a) or Eq. (2.3a). Alternate expressions for E then follow:

$$E(t) = \int^t e(t)\, dq(t) = \int^t f(t)\, dp(t). \tag{2.5}$$

In the next chapter, cases will be encountered in which an effort is a function of a displacement or a flow is a function of a momentum. Then the energy can be expressed not only as a function of time but also as a function of one of the energy variables; thus,

$$E(q) = \int^q e(q)\, dq, \tag{2.5a}$$

or

$$E(p) = \int^p f(p)\, dp. \tag{2.5b}$$

This provides the motivation for calling p and q energy variables in distinction to the power variables e and f.

In Figure 2.2 a mnemonic device fancifully called the "tetrahedron of state" is shown. The four variable types, e, f, p, and q, are associated with the four vertices of a tetrahedron. Along two of the edges of the tetrahedron are indicated the relationships between e and p and f and q. In Chapter 3 the same figure will be augmented to display the variables related by certain basic multiport elements.

It is an interesting fact that the only types of variables that will be needed to model physical systems are represented by the power and energy variables, e, f, p, and q. In order to make this statement more plausible, let us study the variables in several energy domains in more detail.

Table 2.2 displays power and energy variables for mechanical translational ports. Since the power variables force and velocity are considered primitive, the units of the remaining variables follow from a choice in units for the power variables. Units are the shoals on which many a system analysis has foundered, and it is worthwhile to begin a study of the problem of conversion from one system of units to another. In the English system, engineers typically use pounds for a force unit and feet per second for a velocity unit. The mass unit then turns out to be a slug or a $(\text{lb-sec}^2)/\text{ft}$, as we shall see in Chapter 3. Two sets of metric units are shown in Table 2.2. The first column of metric units is based on the mks system in which the $m$eter is the length unit, the $k$ilogram is the mass unit, and the $s$econd is the time unit. The newton is a derived force unit which follows from Newton's law; a newton is the force that gives a mass of one kilogram an acceleration of one meter per second squared. The mass unit

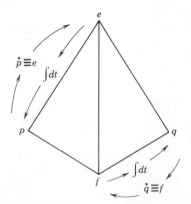

Figure 2.2.   The tetrahedron of state.

TABLE 2.2. Power and energy variables for mechanical translational systems

Generalized Variables	Mechanical Translation	English Units	Metric Units	
Effort, e	Force, F	pounds force [lbf]	newtons [N]	kilograms force [kgf]
Flow, f	Velocity, V	feet/second [ft/sec]	meters/sec [m/sec]	meters/sec [m/sec]
Momentum, p	Momentum, P	[lbf-sec]	[N-sec]	[kgf-sec]
Displacement, q	Displacement, X	[ft]	[m]	[m]
Power, P	$F(t) \cdot V(t)$	[(ft-lbf)/sec]	[(N-m)/sec]	[(kgf-m)/sec]
Energy, E	$\int^{x} F\,dx, \int^{p} V\,dP$	[ft-lbf]	[N-m]	[kgf-m]

in the mks system is the kilogram or, equivalently, a $(\text{N-sec}^2)/\text{m}$ if expressed in terms of the force unit.

The last column in Table 2.2 gives units related to the mks system, but in which the weight of the kilogram in a standard gravitational field is used as a force unit rather than the newton. The force unit is the kilogram-force, abbreviated kgf or kp. Quite often engineering data are presented in this system of units since the kgf is more closely related to everyday experience than the newton. (Household scales are often calibrated in kilograms, for example.) It is important to remember, however, that if kgf and m/sec are used for force and velocity units, the mass unit will not be the kilogram as in the mks system but rather an undesignated unit equivalent to a $\text{kgf-sec}^2/\text{m}$ or 9.80665 kilograms mass. The relation between newtons and kilograms of force is readily computed using Newton's law and the standardized sea level acceleration of gravity, 9.80665 m/sec^2. A newton gives one kilogram one m/sec^2 acceleration, a kgf gives one kilogram 9.80665 m/sec^2. Therefore,

$$1.0 \text{ kgf} = 9.80665 \text{ N}. \tag{2.6}$$

Where confusion may result, kgm may be used to indicate that kilogram is being used as a mass unit rather than a force unit. The standardized sea level acceleration in English units is 32.1740 ft/sec^2.

Table 2.3 gives power and energy variables for ports involving mechanical rotation. The shafts of motors, pumps, gears, and many other useful devices represent such ports.

The entries in Table 2.4 for hydraulic power again are related to the variables used in solid mechanics, but some unusual quantities are defined. The momentum quantity is defined according to Eq. (2.2) as the integral of the effort, or in this case, the pressure. Not only is the pressure

TABLE 2.3. Power and energy variables for mechanical rotational ports

Generalized Variables	Mechanical Rotation	English Units	Metric Units	
Effort, e	Torque, τ	foot-pounds [ft-lbf]	newton-meters [N-m]	kilogram-meters [kgf-m]
Flow, f	Angular velocity ω	radians/second [rad/sec][a]	radians/second [rad/sec][a]	radians/second [rad/sec][a]
Momentum, p	Angular momentum, p_τ	[ft-lbf-sec]	[N-m-sec]	[kgf-m-sec]
Displacement, q	Angle, θ	[rad][a]	[rad][a]	[rad][a]
Power, **P**	$\tau(t) \cdot \omega(t)$	[(ft-lbf)/sec]	[(N-m)/sec]	[(kgf-m)/sec]
Energy, **E**	$\int^\theta \tau\, d\theta, \int^{p_\tau} \omega\, dp_\tau$	[ft-lbf]	[N-m]	[kgf-m]

[a] Radians and other angular measures are dimensionless, but there are scale factors between, say, radians, revolutions, and degrees which can cause errors not discoverable by dimensional analysis. The formulas used in this book all are based on the radian as the unit of angular measure.

momentum a quantity not often encountered in conventional fluid mechanics, but it is a quantity without an obvious symbol. The symbol p_p is meant to indicate a momentum quantity that is the integral of $P(t)$ just as in Table 2.3, p_τ was a momentum quantity defined as the time integral of $\tau(t)$. Fortunately, the lack of a commonly accepted symbol for certain variables is not a serious handicap. When some facility in system modeling has been developed, the generalized variables, e, f, p, and q, can be used for variables in all the energy domains, if desired.

Finally, Table 2.5 gives power and energy variables for electrical ports. The units for electrical quantities are the same for both the English and the metric systems and are directly related to the mks system. The only

TABLE 2.4. Power and energy variables for hydraulic ports

Generalized Variables	Hydraulic Variables	English Units	Metric Units	
Effort, e	Pressure, P	pounds/foot^2 [lbf/ft^2]	newtons/meter^2 [N/m^2]	kilograms/meter^2 [kgf/m^2]
Flow, f	Volume flow rate, Q	feet^3/second [ft^3/sec]	meters^3/second [m^3/sec]	meters^3/second [m^3/sec]
Momentum, p	Pressure momentum, p_p	[(lbf-sec)/ft^2]	[(N-sec)/m^2]	[(kgf-sec)/m^2]
Displacement, q	Volume, V	[ft^3]	[m^3]	[m^3]
Power, **P**	$P(t) \cdot Q(t)$	[(ft-lbf)/sec]	[(N-m)/sec]	[(kgf-m)/sec]
Energy, **E**	$\int^V P\, dV, \int^{p_p} Q\, dp_p$	[ft-lbf]	[N-m]	[kgf-m]

TABLE 2.5.   Power and energy variables for electrical ports

Generalized Variable	Electrical Variables	Units
Effort, e	Voltage, e	volt = newton-meter/coulomb [V] = [(N-m)/C]
Flow, f	Current, i	ampere = coulomb/second [A] = [C/sec]
Momentum, p	Flux linkage variable, λ	[V-sec]
Displacement, q	Charge, q	[C] = [A-sec]
Power, P	$e(t) \cdot i(t)$	[V-A] = [W]^a = [(N-m)/sec]
Energy, E	$\int^q e\,dq, \int^\lambda i\,d\lambda$	[V-A-sec] = [W-sec] = [N-m]

^{a} Capital W stands for watts.

new quantity that needs to be defined is the unit of electrical charge, the coulomb. In engineering, however, it is more common to use volts and amperes for units of voltage and current rather than their equivalents in coulombs and mks units. Most of the variables in Table 2.5 should be familiar, with the possible exception of the momentum or flux linkage variable, λ. The usefulness of this variable will become evident when inductors are studied in Chapter 3.

The tables of variables presented above give at least preliminary evidence that the variables associated with a variety of physical systems can be fit into the scheme of Figure 2.2. The usefulness of this view point will become increasingly evident as systems are modeled in detail. In the next sections the ways in which subsystem interconnections can be indicated graphically using the e, f, p, q classification will be shown.

2.2 PORTS, BONDS, AND POWER

The devices sketched in Figure 2.1 all can be treated as multiport elements with ports that can be connected to other multiports to form systems. Further, when two multiports are connected, power can flow through the connected ports and the power can be expressed as the product of an effort and a flow quantity as given in Tables 2.2–2.5. We now develop a universal way to represent multiports and systems of interconnected multiports based on the variable classifications in the tables.

Consider the separately excited d-c motor shown in Figure 2.3. Physically, such motors have three obvious ports. The two electrical ports are represented by armature and field terminal pairs, and the shaft is a rotary mechanical port as sketched in Figure 2.3a. Figure 2.3b is a conventional schematic diagram in which the mechanical shaft is represented by a dashed line, the field coils are represented by a symbol similar to the circuit symbol for an inductance, and the armature is represented by highly schematic sketch of a commutator and brushes. Note that the

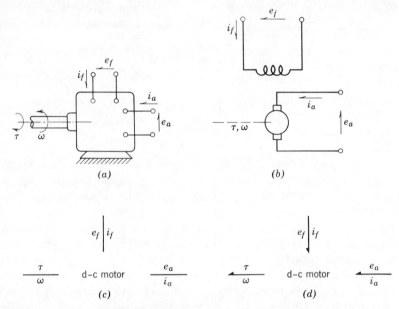

Figure 2.3.   Separately excited d-c motor. (a) Sketch of motor; (b) conventional schematic diagram; (c) multiport representation; (d) multiport representation with sign convention for power.

schematic diagram does not indicate what the detailed internal model of this subsystem or component will be. To write down equations describing the motor, an analyst must decide how detailed a model is necessary.

Figure 2.3c represents a further step in simplifying the representation of this engineering multiport. The name, d-c motor, is used to stand for the device, and the ports are simply indicated by single lines emanating from the word representing the device. As a convenience, the effort and flow variables are written next to the lines representing the ports. Whenever the port lines are either horizontal or vertical, it is useful to use the following convention:

> Efforts are placed either *above* or to the *left* of the port lines, and Flows are placed either *below* or to the *right* of the port lines.

Note that Figures 2.3*a*, *b*, and *c* all contain the same information, namely, that the d-c motor is a 3-port with power variables τ, ω, e_f, i_f, e_a, and i_a. In Figure 2.3*d*, a sign convention has been added: The *half arrow* on a port line indicates the direction of power flow at any instant of time when the effort and flow variables both happen to be positive.

For example, if ω is positive in the direction shown in Figure 2.3*a*, and if τ is interpreted to be the torque on the motor shaft resulting from a connection to some other multiport and is positive in the direction shown in Figure 2.3*a*, then when τ and ω are both positive (or for that matter, both negative), the product $\tau\omega$ is positive and represents power flowing *from* the motor to some other multiport coupled to the motor shaft. Thus, the half arrow in Figure 2.3*d* points *away from* d-c motor. Similarly, when e_f, i_f, e_a, and i_a are positive, power flows *to* the motor from whatever other multiports are connected to the field and armature terminals. Hence, the half arrows associated with the field and armature ports point *toward* the motor.

Any time one desires to be specific about the characteristics of a multiport, for instance in equation form or in the form of tabulated data, then a sign convention is necessary. The establishment of sign conventions is fairly straightforward for electric circuits or for the circuit-like parts of representations of multiports such as those of Figures 2.3*a* and *b*. Anyone who has struggled with the definition of forces and moments on interconnected rigid bodies using "free body diagrams," however, knows that the establishment of sign conventions in mechanical systems is not trivial. The problem is that the action and reaction forces show up as oppositely directed in most representations. Thus, in Figure 2.3*a*, one must decide whether τ represents the torque *on* the motor shaft or *from* the motor onto some other multiport. On diagrams such as Figure 2.3*b*, the mechanical signs are often not indicated at all, and it is up to the analyst to insert plus or minus signs in his equations without much help from the schematic diagram of the system.

When two multiports are coupled together so that the effort and flow variables become identical, the two multiports are said to have a common *bond* in analogy to the bonds between component parts of molecules. Figure 2.4 shows part of a system consisting of three multiports bonded together. The motor and pump have a common angular speed, ω, and torque at the coupling, τ. The battery and the motor have a common voltage and current defined at the terminals at which the battery leads connect to the motor armature. To represent this type of subsystem interconnection in the manner of Figures 2.3*c* or *d* is very straightforward: the joined ports are represented by a single line or bond between the multiports. This has been done in Figure 2.5. The line between the

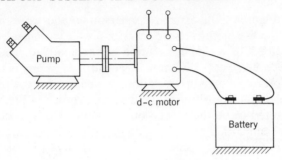

Figure 2.4.   Partially assembled system.

pump and motor in Figure 2.5 implies that a port of the motor and a port of the pump have been connected, and, hence, a single torque and a single angular speed pertain to both the pump and the motor. The half arrow on the bond means that the torque and the speed are defined in such a way that when the product, $\tau\omega$, is positive, power is flowing from the motor to the pump. Thus, lines associated with isolated multiports indicate ports or potential bonds. For interconnected multiports, a line represents the conjunction of two ports, that is, a bond.

2.3 BOND GRAPHS

The mechanism for studying dynamic systems to be used subsequently in this book is the *bond graph*. A bond graph simply consists of subsystems linked together by lines representing power bonds as in Figure 2.5. When major subsystems are represented by words, as in Figure 2.5, then the graph is called a *word bond graph*. Such a bond graph establishes multiport subsystems, the way in which the subsystems are bonded together, effort and flow variables at the ports of the subsystems, and sign conventions for power interchanges.

Since the word bond graph serves to make some initial decisions about the representation of dynamic systems, it is worthwhile to consider some

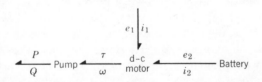

Figure 2.5.   Word bond graph for system of Figure 2.4.

example systems even before the details of dynamic systems have been presented. In Figure 2.6 part of a positioning system for a radar antenna is shown. The word bond graph indicates the major subsystems to be considered, and the bonds with the effort and flow variables indicated introduce some variables which will be useful in characterizing the subsystems at a later stage in the analysis. You should be able to associate all the efforts and flows on the bond graph with physical quantities associated with the physical system being modeled. Try it.

In Figure 2.7 another example system is shown. Again, it is instructive to try to understand the effort and flow quantities associated with the bonds in the word bond graph. For example, what are the three efforts and three flows associated with the 3-port –Diff–? Can you see that –Wheel– is a 2-port that relates a torque and an angular speed to a force and a

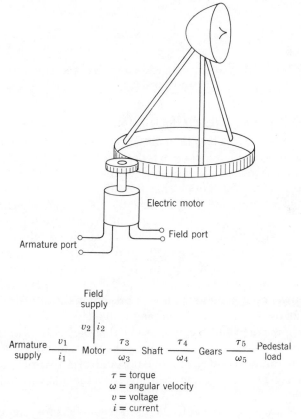

τ = torque
ω = angular velocity
v = voltage
i = current

Figure 2.6. Schematic diagram and word bond graph for radar antenna pedestal drive system.

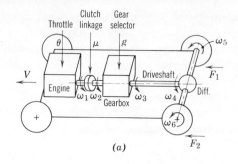

(a)

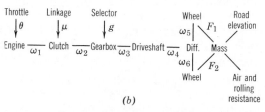

(b)

Figure 2.7.   Automotive drive train example. (a) Schematic diagram; (b) word bond graph.

velocity? Do not be surprised if the construction of a word bond graph for a dynamic system seems less than obvious at this stage. As you progress in the details of system modeling, it will become easier to recognize ports and bonds and, hence, multiport subsystems.

In Figure 2.7 you will notice that the influences of throttle position, clutch linkage position, and the gear selector position are indicated using a bond with a full arrowhead. This notation, which is discussed in more detail in the next section, indicates that an influence on the system from its environment occurs at essentially zero power flow. In the present example, the driver of the car is part of the environment of the car, and he can control the car using the accelerator pedal, clutch pedal, and gear shift using low power compared to the power present in the drive train. A bond with a full arrow is an *active bond*, and it indicates a signal flow at very low power. In the present case, we assume that the controls of the car can be moved by the driver at will, and our dynamic model need not concern itself with the forces required to move the controls. A word bond graph is useful for sorting true power interactions from the one-way influences of active bonds.

Bond graphs will subsequently be used to model subsystems in detail internally. For this purpose, a set of basic multiport elements denoted not by words but by letters and numbers will be developed in the next chapter. Ultimately, detailed bond graphs must be substituted for the multiports designated by words in a word bond graph. From a sufficiently detailed bond graph, state equations may be derived using standard techniques, or computer simulations of the system can be made. The ENPORT computer programs* will accept a wide variety of bond graphs directly and produce either state equations for subsequent analysis or system response predictions. In addition, some types of analysis can be performed on a bond graph without either writing the state equations or using a computer.

2.4 INPUTS, OUTPUTS, AND SIGNALS

Multiport subsystem characteristics typically are determined by a combination of experimental and theoretical methods. It might be fairly easy to compute the moment of inertia of a rotor, for example, merely by knowing the density of the material of which the rotor was made and having a drawing of the part, but to predict the port characteristics of a fan in great detail by theoretical means would be much more difficult than by measuring the characteristics. In performing experiments on a subsystem the notions of *input* and *output* or, equivalently, *excitation* and *response* arise. The same concepts will carry over when "mathematical" models of subsystems are assembled into a system model.

In performing experiments on a multiport, one must make a decision about what is to be done at the ports. At each port, both an effort and a flow variable exist, and one can control either one but not both of these variables simultaneously. As an example, consider the problem of determining the steady-state characteristics of a d-c motor such as the one shown in Figures 2.3, 2.4, and 2.5.

Figure 2.8*a* shows a sketch of equipment that could be used in experimenting on the motor. The dynamometer is supposed to be capable of setting the speed of the motor regardless of the torque delivered by the motor. This speed, ω, is then an *input variable* to the motor. The torque being delivered by the motor is then measured by means of a torque gage. The torque is thus an output variable of the motor. Note that, in general, it is not possible to adjust the dynamometer for both torque and speed. The nature of the experiment is to discover what the motor torque is at a given speed.

Similarly, if voltages are supplied to the two electrical ports, that is, if voltages are input variables, then the motor responds with measurable

---

*The programs are described in R. Rosenberg, *A Users Guide to ENPORT-4*, J. Wiley & Sons, N.Y., 1974.

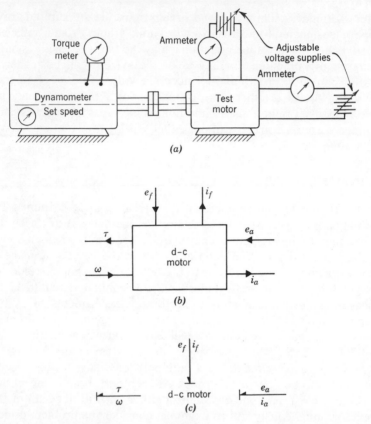

Figure 2.8. Experimental testing of a d-c motor. (*a*) Sketch of the test apparatus; (*b*) block diagram showing input and input signal flow; (*c*) causal strokes added to multiport representation.

currents that are output variables of the motor. Figure 2.8*b* is an attempt to use lines and arrows to show which quantities are inputs to the motor and which are outputs. Figure 2.8*b* is a simple example of a "*block diagram*," in which lines with arrows indicate the direction of flow of *signals*. For multiports each port or bond has both an effort and a flow, and when these two types of variables are represented as paired signals, it is only possible for one of these signals to be an input and the other to be an output.

To know which of the effort and flow signals at a port is the input of the multiport, only one piece of information must be supplied to Figures 2.3*c*, 2.3*d*, or 2.5. This is because if one of the effort and flow variables is an input, the other is an output. In bond graphs the way in which inputs and outputs are specified is by means of the *causal stroke*. The causal stroke is a short, perpendicular line made at one end of a bond or port line. The causal stroke indicates the direction in which the effort signal is directed. (By implication, the end of a bond that does not have a causal stroke is the end toward which the flow signal arrow points.) In Figure 2.8c causal strokes have been added to the multiport representation of Figure 2.3d. By comparing Figures 2.8a, b, and c, all of which contain the same information regarding input and output variables, the meaning of causal strokes may be appreciated. The meaning of the causal stroke is summarized in Figure 2.9, in which both bond graphs and block diagrams are

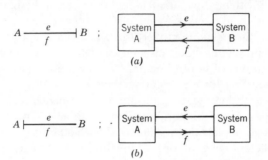

Figure 2.9. The meaning of causal strokes. (*a*) Effort is output of *A*, input to *B*; flow is output of *B*, input to *A*; (*b*) effort is output of *B*, input to *A*; flow is output of *A*, input to *B*.

shown. Note that the half arrow sign convention for power flow and the causal stroke are completely independent. Thus, using *A* and *B* to stand for subsystems as in Figure 2.9, all the following combinations of sign convention and causal strokes are possible: $A \rightharpoondown|B$, $A \leftharpoondown|B$, $A| \rightharpoondown B$, and $A| \leftharpoondown B$. The study of input-output *causality*, which is a uniquely useful feature of bond graphs, will be dealt with at length in succeeding chapters.

Finally, we come to the question of pure signal flow, or the transfer of information with negligible power flow, which we already encountered in the example of Figure 2.7. Multiports in principle all transmit finite power when interconnected. This is correlated with the fact that both an effort and a flow variable exist when multiports are coupled. Thus, systems are interconnected by the matching of a *pair* of signals representing the power variables.

In many important cases, however, systems are so designed that only one of the power variables is important, that is, so that a single signal is transmitted between two subsystems. For example, an electronic amplifier may be designed so that the voltage from a circuit influences the amplifier, but the current drawn by the amplifier has virtually no effect on the circuit. Essentially, the amplifier reacts to a voltage, but extracts negligible power in doing so compared to the rest of the power levels in the circuit. No information can really be transmitted at zero power, but, practically speaking, information can be transmitted at power levels that are negligible compared to other system power levels. Every instrument is designed to extract information about some system variable without seriously disturbing the system to which the instrument is attached. An ideal ammeter indicates current but introduces no voltage drop, an ideal voltmeter reads a voltage while passing no current, an ideal pressure gage reads pressure with no flow, an ideal tachometer reads angular speed with no added torque, and the like. When an instrument reads an effort or flow variable, but with negligible power, there is a signal connection between subsystems without the back effect associated with power interaction. The block diagrams of control engineering or the signal flow graphs that were developed first for electrical systems ideally show signal coupling. As Figure 2.8b shows, when multiports are considered, power interactions require a pair of bilaterally oriented signals. The bond graph, in which each bond implies the existence of both an effort and a flow signal, is a more efficient way of describing multiports than by using block diagrams or signal flow graphs. On the other hand, when the system is dominated by signal interactions due to the presence of instruments, isolating amplifiers, and the like, then either an effort or a flow signal may be suppressed at many interconnection points. In such a case, a bond degenerates to a single signal and may be shown as an *active bond*. The notation for an active bond is identical to that for a signal in a block diagram, for example, $A \overset{e}{\rightarrow} B$ indicates that effort, e, is determined by subsystem A and is an input to subsystem B. Normally, this situation would be indicated by $A \overset{e}{\underset{f}{\rightharpoondown}} B$ in which the flow, f, is determined by B and is an input to A. When e is shown as a signal (by means of the full arrow on the bond) or, in other words, an activated bond, the implication is that the flow, f, has a negligible effect on B. When automatic control systems are added to physical systems, the control systems usually receive signals by means of nearly ideal instruments and affect the systems through nearly ideal amplifiers. The use of active bonds for such cases simplifies the analysis of the systems. Notice that in using bond graphs, one always assumes that multiports are coupled with both forward and backward effects unless a specific modeling decision has been made that a back effect is negligible.

In the following chapter we begin the detailed modeling of susbystems by considering a basic idealized set of multiports that can be assembled to model the pertinent physical effects in a subsystem. At this detailed level physical parameters must be estimated and the rules of causality among ideal multiports must be discovered and obeyed in assembling the subsystem model from elemental multiports. As this process goes on, the notation and concepts briefly introduced in this chapter will become more familiar and useful.

PROBLEMS

2-1 Construct four tetrahedra of state similar to that shown in Figure 2.2 for the four physical domains: mechanical translation, mechanical rotation, hydraulic systems, electrical systems. Replace e, f, p, q with their physical couterpart variables and list the dimensions of each variable.

2-2 For each multiport in Figure 2.1 construct a word bond graph similar to that shown in Figure 2.3. Construct several systems by bonding several multiports together.

2-3 Suppose a pump was tested by running it at various speeds and measuring the volume flow rate and torque for various pressures at the pump outlet. Draw a schematic diagram, block diagram, and bond graph for the pump test analogous to those shown in Figure 2.8 for an electric motor test.

2-4 If the system of Figure 2.5 had the following causality,

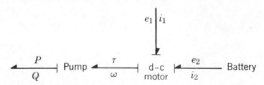

show how the signals flow by using a block diagram of the type used in Figure 2.8*b* for each multiport. See also Figure 2.9.

2-5 Apply causal strokes in an arbitrary manner to each bond in the bond graph of Figure 2.6. Construct an equivalent block diagram for this system using one block for each multiport as in Figure 2.8*b*. Indicate the signal flow directions which correspond to your causal marks as in Figure 2.9.

2-6 Repeat Problem 2-5 for the system of Figure 2.7. (Note that active bonds act just like one-way signal flows in a block diagram but that normal bonds each result in two signal flows for e and f.)

2-7 Consider the system of Problem 2-4. Identify the system input variables which come from the environment of the system, given the causal stroke pattern indicated. What variables are indicated as system outputs (and hence inputs to the environment)?

2-8   What is the maximum number of 100-W light bulbs which a 100-horsepower engine could keep lit if it drove an efficient generator?

2-9   Represent an electric drill as multiport. Consider the switch position influence as occurring on an active bond. Apply causal strokes to your bond graph, assuming that the drill is plugged into a 100-V outlet and that the torque is determined by the material being drilled. Show a block diagram for the drill corresponding to your choice of causality at the ports.

2-10   If a positive displacement hydraulic pump is 100% efficient (so that the mechanical power is always instantaneously equal to the hydraulic power) and if a torque of 5 ft-lb produces a pressure of $15,000 \, lb/ft^2$, what is the relationship between volume flow and angular speed? Convert these relationships to the metric system.

3

BASIC COMPONENT MODELS

In Chapter 2 real devices were considered as subsystems from the point of view of power exchanges and external port variables. In this chapter a basic set of multiports are defined that can be used to model subsystems in detail. These multiports function as components of subsystem and system models and are, in many cases, idealized mathematical versions of real components such as resistors, capacitors, masses, springs, pipes, and so on. In other cases, however, the basic multiports are used to model *physical effects* in a device and cannot be put into a one-to-one correspondence with physical components of the device. For example, one might create a model of an electrical or fluid transmission line using a finite collection of resistance, capacitance, and inertia elements, even though in the real device the effects being modeled are distributed along the transmission line and not concentrated into lumps, as in the model.

Using bond graphs and the classification of power and energy variables presented in the previous chapter, it turns out that only a few basic types of multiport elements are required in order to represent models in a variety of energy domains. The bond-graph notation often allows one to visualize aspects of the system more easily than would be possible with just the state equations or with some other graphical notation designed for a single energy domain or for signal flow rather than power flow. The search for a bond-graph model of a complex system frequently increases one's physical understanding of the system.

3.1 BASIC 1-PORT ELEMENTS

A 1-port element is addressed through a single power port, and at the port a single pair of effort and flow variables exists. Generally, a 1-port can be a very complex subsystem. An ordinary electrical wall outlet can represent the port of a 1-port in a system analysis. The port actually connects to a vast network of power generation and distribution equipment, yet, from the point of view of a system model, a relatively simple characterization of what is behind the wall outlet as a 1-port may suffice. Here we deal with the most primitive 1-ports. We consider, in order, elements which dissipate power, store energy, and supply power.

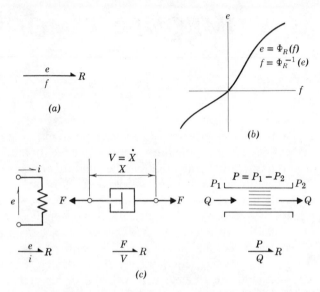

Figure 3.1. The 1-port resistor. (a) Bond-graph symbol; (b) defining relation; (c) representations in several physical domains.

The *1-port resistor* is an element in which the effort and flow variables at the single port are related by a static function. Figure 3.1 shows the bond-graph symbol for the resistor, a typical graph of the constitutive relation between e and f, and sketches of resistors in several energy domains. Usually, resistors dissipate energy. This must be true for simple electrical resistors, mechanical dampers or dashpots, porous plugs in fluid lines, and other analogous passive elements. Noting from Figure 3.1a that power flows *into* the port when the product of e and f is positive according to the sign convention shown, we may deduce that power is always dissipated if the defining constitutive relation between e and f lies in the first and third quadrants of the e–f plane as shown in Figure 3.1b, for then the product, $(e)(f)$, is positive.

When the relation between e and f for a 1-port resistor plots as a curved line as in Figure 3.1*b*, then the resistor is a *nonlinear element*. If the relation is a straight line, then it is a *linear* resistor. In the special case of a linear element, a coefficient, the *resistance*, or its inverse, the *conductance* may be defined. The resistance relationships are summarized in the first lines of Table 3.1. Note that for power-dissipating resistors, with the sign convention shown in Figure 3.1 and Table 3.1, the resistance and

TABLE 3.1. The 1-port resistor, $\dfrac{e}{f} \rightarrow R$

| | General Relation | Linear Relation | Units for Linear Resistance Parameter | | |
| | | | | Metric | |
			English	(a)	(b)
Generalized variables	$e = \Phi_R(f)$ $f = \Phi_R^{-1}(e)$	$e = Rf$ $f = Ge = e/R$	$[R] = [e]/[f]$	$[R] = [e]/[f]$	$[R] = [e]/[f]$
Mechanical translation	$F = \Phi(V)$ $V = \Phi^{-1}(F)$	$F = bV$	$[b] = [\text{lb-sec}]/[\text{ft}]$	$[b] = [\text{N-sec}]/[\text{m}]$	$[b] = [\text{kgf-sec}]/[\text{m}]$
Mechanical rotation	$\tau = \Phi(\omega)$ $\omega = \Phi^{-1}(\tau)$	$\tau = c\omega$	$[c] = [\text{ft-lb-sec}]$	$[c] = [\text{N-m-sec}]$	$[c] = [\text{kgf-m-sec}]$
Hydraulic systems	$P = \Phi(Q)$ $Q = \Phi^{-1}(P)$	$P = RQ$	$[R] = [\text{lb-sec}]/[\text{ft}^5]$	$[R] = [\text{N-sec}]/[\text{m}^5]$	$[R] = [\text{kgf-sec}]/[\text{m}^5]$
Electrical systems	$e = \Phi(i)$ $i = \Phi^{-1}(e)$	$e = Ri$ $i = Ge$		$[R] = [\text{V/A}] = [\Omega] = [\text{ohm}]$	

conductance parameters, R and G, respectively, are positive. For simplicity, we establish the following arbitrary but useful rule:

For passive resistors, establish the power sign convention by means of a half arrow pointing *toward the resistor*. Then linear resistance parameters will be positive and nonlinear relations will fall in the first and third quadrants of the *e*–*f* plane.

Occasionally, $-R$ may be used to model a constitutive relation between *e* and *f* which actually arises from a subsystem that can both supply and dissipate power. For example, the torque-speed characteristics of motors and engines can be plotted as Figure 3.1*b*. With the sign convention, $\rightarrow R$, however, the constitutive relation would not lie in the first and third quadrants. In fact, in most operating regimes, a motor or engine supplies rather than dissipates power. For such nonpassive resistors, one often uses this sign convention: $\leftarrow R$.

Table 3.1 displays resistance relations in terms of the generalized variables, *e* and *f*, and also in terms of the same physical variables used in the tables of Chapter 2. In the first column, Φ stands for a general function relating two variables such as the one plotted in Figure 3.1b. The function might be read two ways: (1) the effort can be found if the flow is given, or (2) the flow can be found if the effort is given. This necessitates the definition of both the function, Φ, and its inverse, Φ^{-1}. For example, if Φ represented a cubic law

$$e = af^3,$$

then the inverse function, Φ^{-1}, would represent a cube root law

$$f = \left(\frac{1}{a}\right)^{1/3} e^{1/3}.$$

In the linear case, of course, the inversion of the function, $e = Rf$, involves a simple inversion of the parameter, R, $f = (1/R)e$.

Since linear models are of great usefulness in certain fields, (vibrations and electric circuits, for example), the linear versions of resistance relations in various energy domains are shown in Table 3.1 with the same notation employed in Chapter 2. The units of the linear resistance parameter are simply the units of effort divided by the units of flow. It is worth studying the units displayed in Table 3.1 since many of the resulting units may not be familiar. The only resistance unit dignified with its own name is the electrical ohm.

Next consider a 1-port device in which a static constitutive relation exists between an effort and a displacement. Such a device stores and gives up energy without loss. In bond-graph terminology, an element that relates e to q is called a *1-port capacitor*. In the physical terms, a capacitor is an idealization of devices called springs, torsion bars, electrical capacitors, gravity tanks, and accumulators. The bond-graph symbol, defining constitutive relation, and some physical examples are shown in Figure 3.2. Note that when a sign convention similar to that used for the resistor, namely, $\rightharpoonup C$, is used for the $-C$ element, then $(e)(f)$ represents power flowing *to* the capacitor and

$$\mathbf{E}(t) = \int_0^t e(t)f(t)\, dt + \mathbf{E}_0 \qquad (3.1)$$

represents the energy stored in the capacitor at any time t. The energy stored initially at $t = 0$ (if any) is called $\mathbf{E}_0$.

Since from Eq. (2.3a) the displacement, q, is defined so that $f\, dt \equiv dq$, and the constitutive relation of a $-C$ implies that e is a function of q,

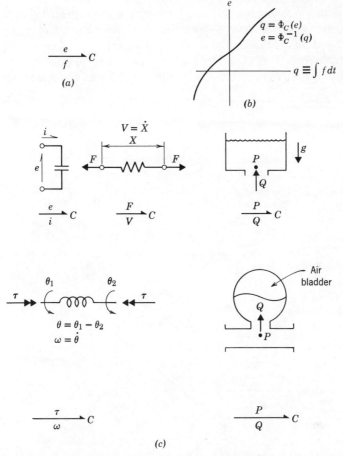

Figure 3.2. The 1-port capacitor. (a) Bond-graph symbol; (b) defining relation; (c) representation in several physical domains.

$e = e(q)$, then Eq. (3.1) can be rewritten as

$$\mathbf{E}(q) = \int_{q_0}^{q} e(q)\, dq + \mathbf{E}_0 \qquad (3.2)$$

where $\mathbf{E}_0$ is the energy stored when $q = q_0$. Usually, it is convenient to define the energy stored to be zero when the effort is zero. Then, if q_0 is that value of q at which $e = 0$, and $\mathbf{E}_0 = 0$, Eq. (3.2) may be written as

$$\mathbf{E}(q) = \int_{q_0}^{q} e(q)\, dq. \qquad (3.2a)$$

The operation indicated in Eq. (3.2a) may be interpreted graphically as shown in Figure 3.3. As q varies, the area under the curve of e versus q varies, and this area is equal to **E**. The *conservation of energy* for $-C$ is almost obvious. If q goes from q_0 to q as in Figure 3.3a, then energy is stored; if q then ever returns to q_0, the shaded area disappears and all the

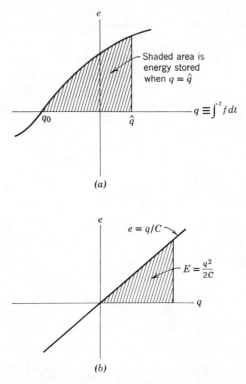

Figure 3.3.   Area interpretation of stored energy for 1-port capacitor. (a) Nonlinear case; (b) linear case.

stored energy disappears. The power flow into the port, which resulted in the storage of energy, reverses and power flows out of the port. During the process, no energy is lost.

Table 3.2 summarizes the relationships characterizing capacitors. The units for linear capacitance parameters are given and again it may be noted that only the electrical unit is given a name, the farad. For mechanical systems, it is common to use the *spring constant, k*, rather than the *compliance, $C \equiv 1/k$*, which is analogous to the electrical capacitance, C, and the parameter, C, in generalized variables. In mixed electrical-mechanical systems, one must simply be careful to note whether a numerical parameter corresponds to C or the inverse of C in a bond graph. Once again, the reader is urged to study the units shown since some units will probably be unfamiliar.

A second energy storing 1-port arises if the momentum, p, is related by a static constitutive law to the flow f. Such an element is called an *inertia* in bond-graph terminology. The bond-graph symbol for an inertia, the constitutive relation, and several physical examples are shown in Figure

TABLE 3.2. The 1-port capacitor, $\dfrac{e}{f=\dot{q}}\!\succ C$

	General Relation	Linear Relation	English	Metric (a)	(b)
				Units for Linear Capacitance Parameters	
Generalized	$q = \Phi_C(e)$ $e = \Phi_C^{-1}(q)$	$q = Ce$ $e = q/C$	$[C] = [q]/[e]$ $[1/C] = [e]/[q]$		
Mechanical translation	$X = \Phi_C(F)$ $F = \Phi_C^{-1}(X)$	$X = CF$ $F = kX$	$[C] = [\text{ft}]/[\text{lb}]$ $[k] = [\text{lb}]/[\text{ft}]$	$[C] = [\text{m/N}]$ $[k] = [\text{N/m}]$	$[C] = [\text{m/kgf}]$ $[k] = [\text{kgf/m}]$
Mechanical rotation	$\theta = \Phi_C(\tau)$ $\tau = \Phi_C^{-1}(\theta)$	$\theta = C\tau$ $\tau = k\theta$	$[C] = [\text{rad/ft-lb}]$ $[k] = [\text{ft-lb/rad}]$	$[C] = [\text{rad/N-m}]$ $[k] = [\text{N-m/rad}]$	$[C] = [\text{rad/kgf-m}]$ $[k] = [\text{kgf-m/rad}]$
Hydraulic systems	$V = \Phi_C(P)$ $P = \Phi_C^{-1}(V)$	$V = CP$ $P = V/C$	$[C] = [\text{ft}^5/\text{lb}]$	$[C] = [\text{m}^5/\text{N}]$	$[C] = [\text{m}^5/\text{kgf}]$
Electrical systems	$q = \Phi_C(e)$ $e = \Phi_C^{-1}(q)$	$q = Ce$ $e = q/C$	$[C] = [c/v] = [\text{farad}] = [f]$		

3.4. The inertia is used to model inductance effects in electrical systems and mass or inertia effects in mechanical or fluid systems.

Using the sign convention, $\succ I$, the power flowing into the inertia is given by the expression in Eq. (3.1). In the present case, Eq. (2.2a) allows us to write $e\,dt \equiv dp$ and, if $f = f(p)$, then Eq. (3.1) can be rewritten thus:

$$\mathbf{E}(p) = \int_{p_0}^{p} f(p)\,dp + \mathbf{E}_0. \tag{3.3}$$

If the energy is defined to vanish when f vanishes and if p_0 corresponds to that point in the function of f versus p at which $f = 0$, then

$$\mathbf{E}(p) = \int_{p_0}^{p} f(p)\,dp. \tag{3.3a}$$

The similarities between Eq. (3.2) and Eq. (3.3) should be noted. Often the

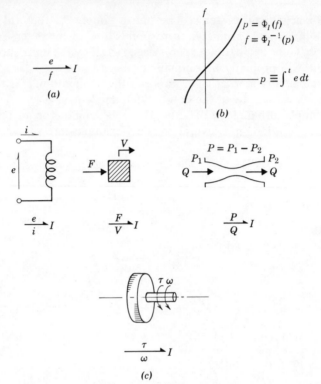

Figure 3.4. The 1-port inertia. (a) Bond-graph symbol; (b) defining relation; (c) representation in several physical domains.

energy associated with a capacitor is called *potential energy*, whereas the energy associated with an inertia is called *kinetic energy*. These names are applied primarily to mechanical systems. In electrical systems, the corresponding two forms of stored energy are sometimes called *electric* and *magnetic* energy.

As in the case of the capacitor, if the constitutive relation for the inertia is plotted, then there is an area interpretation of the stored energy. This interpretation is shown in Figure 3.5, and, again, you should be able to demonstrate that any energy stored in an $-I$ can be recovered without loss.

Table 3.3 shows the constitutive relations for inertias and gives unit for inertance parameters for the linear case. Since most of engineering work is accomplished successfully using Newton's laws rather than the postulates of relativity, the relation between velocity and momentum is linear and the mass or moment of inertia is the inertance parameter. Although it

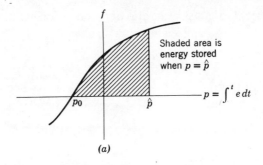

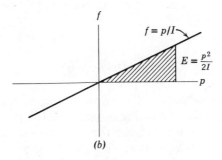

Figure 3.5. Area interpretation of stored energy for 1-port inertia. (*a*) Nonlinear case; (*b*) linear case.

TABLE 3.3. The 1-port inertia, $\xrightarrow{e = \dot{p}}_{f} I$

	General Relation	Linear Relation	Units for Linear Inertance Parameters		
				Metric	
			English	(*a*)	(*b*)
Generalized variables	$p = \Phi_I(f)$ $f = \Phi_I^{-1}(p)$	$p = If$ $f = p/I$	$[I] = [p]/[f]$ $[1/I] = [f]/[p]$		
Mechanical translation	$p = \Phi_I(V)$ $V = \Phi_I^{-1}(p)$	$p = mV$ $V = p/m$	$[m] = [\text{lb-sec}^2]/[\text{ft}]$	$[m] = [\text{N-sec}^2/\text{m}]$	$[m] = [\text{kgf-sec}^2/\text{m}]$
Mechanical rotation	$p_\tau = \Phi_I(\omega)$ $\omega = \Phi_I^{-1}(p_\tau)$	$p_\tau = J\omega$ $\omega = p_\tau/J$	$[J] = [\text{ft-lb-sec}^2]$	$[J] = [\text{N-m-sec}^2]$	$[J] = [\text{kgf-m-sec}^2]$
Hydraulic systems	$p_p = \Phi_I(Q)$ $Q = \Phi_I^{-1}(p_p)$	$p_p = IQ$ $Q = p_p/I$	$[I] = [\text{lb-sec}^2/\text{ft}^5]$	$[I] = [\text{N-sec}^2/\text{m}^5]$	$[I] = [\text{kgf-sec}^2/\text{m}^5]$
Electrical systems	$\lambda = \Phi_I(i)$ $i = \Phi_I^{-1}(\lambda)$	$\lambda = Li$ $i = \lambda/L$	$[L] = [\text{V-sec/A}] = [\text{henrys}] = [\text{H}]$		

is common to think of mass as a ratio of force to acceleration, a, from the equation,

$$F = ma, \qquad a \equiv \dot{V}, \tag{3.4}$$

the table gives the fundamental definition of a mass according to

$$p \equiv mV \tag{3.5}$$

with

$$\dot{p} \equiv F. \tag{3.6}$$

Clearly, when Eq. (3.5) is differentiated with respect to time, and Eq. (3.6) is used, then Eq. (3.4) can be derived. If, on the other hand, Eq. (3.5) is replaced with a nonlinear relation,

$$p = \Phi_I(V) = \frac{mV}{(1 - V^2/c^2)^{1/2}}, \tag{3.7}$$

where m is the rest mass and c is the velocity of light, then Eqs. (3.5) and (3.6) hold for the special theory of relativity. See [1], p. 19, for example. Thus, there is some justification for the general constitutive relations given for mechanical systems, even though engineering is overwhelmingly concerned with the linear case. For electrical systems, however, the relation between the flux linkage variable (the time integral of the voltage) and the current in an inductor is nonlinear in typical cases. The use of the linear parameter, L, is then the result of a modeling decision. It is more satisfactory to generalize $\lambda = Li$ to $\lambda = \Phi(i)$ with $\dot{\lambda} = e$ than to try to generalize $e = L\, di/dt$ to the nonlinear case.

As an aid in remembering which variables the three 1-ports relate, the tetahedron of state introduced in Figure 2.2 may be used. See Figure 3.6. We now know something about five of the six edges of the tetrahedron. The sixth edge, which stretches between the vertices representing p and q, is hidden from view in Figure 3.6. This is just as well since no element will relate p and q.*

Finally, two useful and rather simple 1-ports must be defined: the *effort source* and the *flow source*. The 1-port sources are idealized versions of

---

* One can, in fact, define an element corresponding to the hidden edge, the "memristor." While interesting and occasionally useful, memristors can be represented in terms of other elements to be introduced later, so the memristor will not be considered to be a basic element. See Oster, G. F. and Auslander, D. M., "The Memristor: A New Bond Graph Element," *Trans. ASME, J. Dynamic Systems, Measurement, and Control*, **94**, Ser. G, n. 3 (Sept. 1972), pp. 249–252.

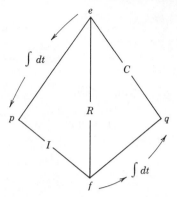

Figure 3.6. The three 1-ports placed on the tetrahedron of state according to the variables they relate.

voltage supplies, pressure sources, vibration shakers, constant flow systems, and the like. In each case, an effort or flow is either maintained sensibly constant independent of the power supplied or absorbed by the source or is constrained to be some particular function of time. As an example of a constant effort source, consider the gravity force on a mass. Near the surface of the earth, this force is essentially constant independent of the velocity of the mass. As an example of a time varying source, the electrical wall outlet will serve. The wall outlet enforces a sinusoidal voltage across the power cord wires of most small appliances. Over a reasonable range of currents, the voltage is independent of fluctuations in the current. Of course the voltage is actually affected by large currents, and a fuze will blow to protect the circuits if very large currents build up, but this simply means that the real outlet is not modeled exactly by an ideal source of effort.

Table 3.4 presents bond-graph symbols and the constitutive relations for sources. Typically, source elements are thought of as supplying power to a system. This accounts for the sign convention half arrow shown which implies that when $e(t)f(t)$ is positive, power flows from the source to whatever system is connected to the source. Since a source maintains one of the power variables constant or a specified function of time no matter how large the other variable may be, a source can supply an indefinitely large amount of power. This is, of course, not a realistic state of affairs, and real devices are not really sources even though they may be modeled approximately by sources. As an example, consider the problem of predicting the current flowing from a 12 V automotive battery into a variable resistor connected to the battery. Figure 3.7 shows a circuit diagram, bond graph, and a plot of voltage versus current. This is a static

TABLE 3.4

	Bond-Graph Symbol	Defining Relation
Generalized variables	$S_e \dashv$ $S_f \vdash$	$e(t)$ given, $f(t)$ arbitrary $f(t)$ given, $e(t)$ arbitrary
Mechanical translation	$S_F \dashv$ $S_V \vdash$	$F(t)$ given, $V(t)$ arbitrary $V(t)$ given, $F(t)$ arbitrary
Mechanical rotation	$S_\tau \dashv$ $S_\omega \vdash$	$\tau(t)$ given, $\omega(t)$ arbitrary $\omega(t)$ given, $\tau(t)$ arbitrary
Hydraulic systems	$S_P \dashv$ $S_Q \vdash$	$P(t)$ given, $Q(t)$ arbitrary $Q(t)$ given, $P(t)$ arbitrary
Electrical systems	$S_e \dashv$ $S_i \vdash$	$e(t)$ given, $i(t)$ arbitrary $i(t)$ given, $e(t)$ arbitrary

(a)   *(b)*

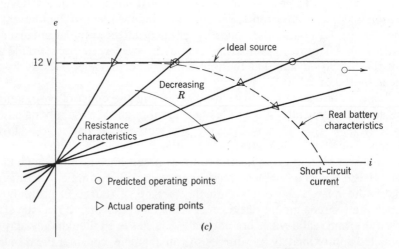

(c)

Figure 3.7. Study of a battery connected to a variable resistance. (*a*) Electric circuit diagram; (*b*) bond graph; (*c*) plot of source, real battery, and resistance characteristics.

system operating at points at which the source characteristic intersects the resistor characteristics. For small currents (or high values of resistance, R), the battery is almost a constant voltage source. When the resistance is lowered toward zero, the predicted current approaches infinity. Actually, when the current gets large, the internal resistance in the battery reduces the voltage below the nominal 12 V. In fact, if the resistance approaches zero as it will when a shorting bar is put across the battery terminals, the battery voltage will approach a finite value, labeled "short-circuit current" in Figure 3.7c. When more basic multiports have been defined, it will be possible to model the battery with an ideal source and a resistor in such a way that the actual characteristic in Figure 3.7 will be reproduced by the model. For now, we simply note that ideal sources are useful in modeling real devices, but should not be expected to be realistic models in all power ranges unless supplemented by other multiports.

A universe made up only of 1-ports would be very simple since bond graphs more complicated than that in Figure 3.7b would be impossible. This leads one to anticipate that 1-ports are not the whole story. A logical next step is to consider 2-ports.

3.2 BASIC 2-PORT ELEMENTS

One might expect that it would be necessary to define more basic types of 2-ports than 1-ports, but, in fact, only two basic types of 2-ports are required. There are, of course, an unlimited number of 2-port subsystems, but we need to discuss here only those which cannot be modeled using the basic 1-ports of the previous section and other elements to be defined later.

The 2-ports to be discussed here are ideal in the specific sense that *power is conserved.* If any 2-port, $—TP—$, has the sign convention,

$$\xrightarrow[f_1]{e_1} TP \xrightarrow[f_2]{e_2} ,$$

then power conservation means that at every instant of time

$$e_1(t)f_1(t) = e_2(t)f_2(t). \tag{3.8}$$

The power sign convention implied in Eq. (3.8) and shown in the bond graph just above is a "through power" sign convention in the sense that power is thought of as flowing through the 2-port. Equation (3.8) states

that whatever power is flowing into one side of the 2-port is simultaneously flowing out of the other side.

One way in which Eq. (3.8) can be satisfied is found in the 2-port known
as a transformer, and given the bond-graph symbol, —TF—. The constitutive laws of the ideal 2-port transformer are

$$e_1 = me_2,$$
$$mf_1 = f_2,$$

$$(3.9)$$

in which the parameter, m, is called the *transformer modulus* and the
subscripts 1 and 2 correspond to the two ports as shown in Figure 3.8a.
Note that Eqs. (3.9) and (3.8) both imply the use of the through sign
convention shown in the figure. Also shown in Figure (3.8) are a number
of devices which in idealized form are modeled by transformers. In no
case is the physical device exactly a transformer. For example, the lever
in Figure 3.8b would only be a —TF— if it were massless, rigid, and

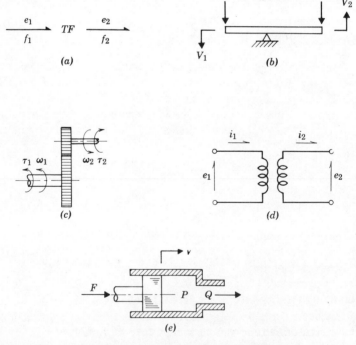

Figure 3.8.   Transformers. (a) Bond graph; (b) ideal rigid lever; (c) gear
pair; (d) electrical transformer; (e) hydraulic ram.

frictionless. Similar restrictions can be made on the validity of the transformer as a model for the other physical devices. Actual models of the devices can be made using the ideal transformer and other multiports to account for nonideal effects if these effects are important to the system under study.

Of particular interest is the ram of Figure 3.8e in which hydraulic power is transduced into mechanical power. The constitutive laws of the ideal version of this device are

$$F = AP,$$
$$AV = Q, \tag{3.10}$$

in which the area of the piston, A, functions as the transformer modulus, m, as it appears in Eq. (3.9). The two equations of (3.10) can be derived separately from physical considerations, or if one is derived, the other follows because of power conservation. The fact that only a single modulus exists serves as a useful check on constitutive equations such as (3.10).

Another way in which the power balance of Eq. (3.8) may be satisfied is embodied in the *gyrator*, which is symbolized thus: —GY—. The constitutive laws of the gyrator are

$$e_1 = rf_2,$$
$$rf_1 = e_2, \tag{3.11}$$

in which r is the *gyrator modulus* and the through sign convention of Figure 3.9a is implied. The modulus is called r because Eq. (3.11) reminds one of the 1-port linear resistance law as shown in Table 3.1. In Eq. (3.11), however, the effort and flow at two *different* ports are statically related. Thus, a 1-port resistor dissipates power, whereas the 2-port gyrator conserves power, as can be seen by multiplying the two equations in Eq. (3.11) together.

Figure 3.9 shows some physical devices that are at least approximately gyrators. The electric circuit symbol is used to represent gyrators in electric network diagrams. Electric gyrators can be made using the Hall effect, and the gyrator is needed to model effects at microwave frequencies even though the gyrator in that case cannot be identified as a separate physical device. Anyone who has played with a toy gyroscope has observed a gyrator. If the rotor of Figure 3.9c spins very rapidly, a gentle push in the direction of F_1 will yield a proportional velocity, V_2. Similarly, a force, F_2, will result in a velocity, V_1. The counter-intuitive behavior of

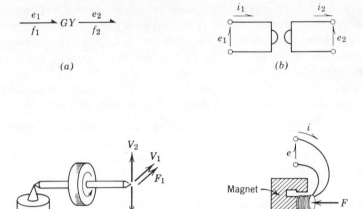

Figure 3.9.   Gyrators. (*a*) Bond graph; (*b*) symbol for electrical gyrator; (*c*) mechanical gyrator; (*d*) voice coil transducer.

the gyroscope is predicted by Eq. (3.11). For example, if the gravity force is in the direction of F_2, then the device precesses in a horizontal path. If a gyro rotates slowly, or if large disturbances are applied, the gyro must be modeled as a rigid body. The bond graph, then, is much more complex than a simple $—GY—$, but it does contain gyrators. For a restricted range of spin speeds and forces, the gyro is approximately a gyrator and gives the gyrator its name.

Figure 3.9*d* shows a useful transducer which is a gyrator if certain nonideal effects may be neglected. It is the *voice coil* used in electrodynamic loud speakers, vibration shakers, seismic mass accelerometers, and many other devices. The constitutive laws of the device are

$$e = TV,$$
$$Ti = F,$$
$$(3.12)$$

in which T plays the role of r in Eq. (3.11). The units of T might be [V/(ft/sec)] from one equation and [lb/A] from the other equation, and one might be tempted to measure the two versions of T which have such different units in two separate experiments. Actually, there is only a single T for a gyrator. The different units merely arise because power is measured in volt-amperes or watts on one side of the transducer and in foot pounds per second on the other. Thus, one may measure T using one of the equations in Eq. (3.12), invoke power conservation using the conversion factor between [W] and [(ft-lb)/sec], and deduce the T in the other equation. The two values of T are numerically different *only* because of the choice of units. The recognition that the device is power conserving and representable as a gyrator helps one avoid the mistake of using numerical values of the two Ts in Eq. (3.12), which would allow the model to create power out of nothing. An important advantage of the metric system is that the numerical values of T in the two parts of Eq. (3.12) are identical.

The gyrator always seems to be a more mysterious element than the transformer. Before the importance of the gyrator was recognized, it was common to make equivalent electrical network diagrams for electromechanical or electrohydraulic systems using only transformers. This is not possible, in general, but in many special cases one may switch the analogy between electrical and mechanical or hydraulic variables until a gyrator is treated as a transformer. In bond-graph terms, the voice coil is a transformer if, for example, we call current an effort and voltage a flow. This switching of the identification of effort-flow variables is entirely unnecessary if one only recognizes that gyrators are really necessary for devices such as those shown in Figures 3.9*b* and *c*, in any case.

In fact, a gyrator is a more fundamental element than a transformer. Two gyrators cascaded are equivalent to a transformer:

$$\xrightarrow{\frac{e_1}{f_1}} GY_1 \xrightarrow{\frac{e_2}{f_2}} GY_2 \xrightarrow{\frac{e_3}{f_3}} = \xrightarrow{\frac{e_1}{f_1}} TF_3 \xrightarrow{\frac{e_3}{f_3}}$$

$$e_1 = r_1 f_2; \qquad r_2 f_2 = e_3; \quad \rightarrow e_1 = (r_1/r_2)e_3;$$

$$r_1 f_1 = e_2; \qquad e_2 = r_2 f_3; \quad \rightarrow (r_1/r_2)f_1 = f_3.$$

In contrast, cascaded transformers are equivalent only to another transformer:

$$\xrightarrow{\frac{e_1}{f_1}} TF_1 \xrightarrow{\frac{e_2}{f_2}} TF_2 \xrightarrow{\frac{e_3}{f_3}} = \xrightarrow{\frac{e_1}{f_1}} TF_3 \xrightarrow{\frac{e_3}{f_3}}$$

$$e_1 = m_1 e_2; \qquad e_2 = m_2 e_3; \quad \rightarrow e_1 = m_1 m_2 e_3;$$

$$m_1 f_1 = f_2; \qquad m_2 f_2 = f_3; \quad \rightarrow m_1 m_2 f_1 = f_3.$$

Thus, one could, in principle, consider every transformer as a cascade combination of two gyrators and dispense with —*TF*— as a basic 2-port. It is more convenient, however, to retain —*TF*— as a basic bond-graph element.

It is important also to realize that the gyrator essentially interchanges the roles of effort and flow. This may be seen by replacing r in Eq. (3.11) by unity. Then the effort at one port of the $-GY-$ is just the flow at the other, and vice versa. Thus, the combination, $-GY-I$, is equivalent to $-C$. To see this, recall that $-I$ relates f to the integral of e, or p. After the gyrator is added, and the roles of e and f are interchanged, the combination relates e and the integral of f or q at the external port. The element relating e to q is $-C$. Similarly, $-GY-C$ is equivalent to $-I$. Thus, one could in principle dispense with either $-C$ or $-I$ as a basic 1-port as long as $-GY-$ is available. Again, it is more convenient and natural to retain both $-C$ and $-I$ as basic 1-ports.

As an example of a deduction about a system based purely on its bond graph representation, consider the following equivalence $I_1-GY-I_2 = I_1-C$, in which $-GY-I_2$ has been replaced by $-C$. An $-I$ bonded to a $-C$ is an oscillator. In physical terms, it could be a mass-spring or inductor-capacitor system. Thus, we see that I_1-GY-I_2 (or, for that matter, C_1-GY-C_2) will act just like an inertia-capacitor system.

Finally, there is a generalization of the transformers and gyrators discussed above based on the curious fact that in both Eq. (3.9) and Eq. (3.11) the power conservation between the two ports is maintained even when the moduli, m and r, are not constant. This gives rise to the *modulated transformer* and the *modulated gyrator* denoted in bond-graph symbolism by

$$\xrightarrow[f_1]{e_1} M\overset{\downarrow m}{T}F \xrightarrow[f_2]{e_2} \quad \text{and} \quad \xrightarrow[f_1]{e_1} M\overset{\downarrow r}{G}Y \xrightarrow[f_2]{e_2}.$$

Note that m and r are shown as *signals* on an *activated bond*. This means that no power is associated with the changes in m and r and $e_1 f_1$ is always exactly equal to $e_2 f_2$, as in the case of the constant modulus, $-TF-$ and $-GY-$.

Many physical devices may be modeled by the modulated 2-ports. For example, the electrical autotransformer contains a mechanical wiper which, when moved, alters the turns ratio between the primary and secondary coils and thus changes the transformer ratio. This alteration takes no power (if we can assume mechanical friction is negligible), and for any wiper position the device essentially conserves electrical power.

Both the gyrators in Figures 3.9c and d have moduli that can be changed without changing the fact that power is conserved at the two ports. For the gyroscope, a motor can be made to change the spin speed of the rotor which changes r. Similarly, for the voice coil, if an electromagnet is substituted for a permanent magnet, then the transduction coefficient, T, in Eqs. (3.12) may be varied. At every instant, the power is conserved at the two ports, but the characteristics of the device change.

In mechanics, the *MTF* is particularly important and may be used to represent geometric transformations or kinematic linkages. As a simple example consider the rotating arm shown in Figure 3.10. The arm is in equilibrium under the action of the torque, τ, and force, F, and it provides a relation between θ and y or $\dot\theta \equiv \omega$ and $\dot y \equiv V_y$. Writing the displacement relation first

$$y = l \sin \theta, \tag{3.13}$$

we can differentiate this to yield a constitutive relation for velocities

$$\dot y = (l \cos \theta)\dot\theta,$$

or

$$V_y = (l \cos \theta)\omega. \tag{3.14}$$

The equilibrium relation between τ and F is

$$(l \cos \theta)F = \tau. \tag{3.15}$$

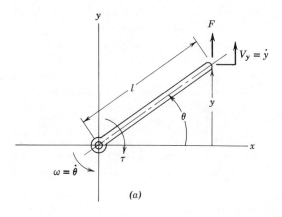

(a)

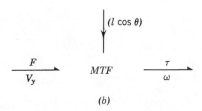

(b)

Figure 3.10. A displacement-modulated transformer. (a) Sketch of rigid, massless rotating arm; (b) bond graph.

The occurrence of $l \cos \theta$ in both Eq. (3.14) and Eq. (3.15) might appear to be coincidental until we remember that the device must conserve power. Equations (3.14) and (3.15) are embodied in the bond graph of Figure 3.10b. The *MTF* is called a *displacement*-modulated transformer since the modulus, r, is a function of a displacement variable, θ. Such transformers allow one to create a bond graph for the extremely complex dynamic systems associated with three dimensional rigid body motion.

Although some useful 2-port components have been defined, still only very simple bond graphs can be assembled from 2-ports and 1-ports. Only chains of 2-ports ended in 1-ports at the ends can be made. In order to create the complex models used in engineering, it turns out that 3-ports are required, but only two basic 3-ports are required to model a very rich variety of systems.

3.3 THE 3-PORT JUNCTION ELEMENTS

We now introduce two 3-port components which, like the 2-ports of the previous section, are power conserving. These 3-ports are called *junctions* since they serve to interconnect other multiports into subsystem or system models. These 3-ports represent one of the most fundamental ideas behind the bond-graph formalism. The idea is to represent in multiport form the two types of connections which, in electrical terms, are called the *series* and *parallel* connections. As we shall see, such connections really occur in all types of systems, even though traditional treatments may not recognize the existence of the junctions as multiports.

First, consider the *flow junction*, *0-junction*, or *common effort junction*. The symbol for this junction is a zero with three bonds emanating from it. (As will become evident, it is easy to extend the definition to a 4-, 5-, or more-port version of this 3-port.)

$$\begin{array}{ccc} \mid \\ \text{---}0\text{---} \end{array}, \qquad \begin{array}{c} \mid 2 \\ \underset{1}{\text{---}}0\underset{3}{\text{---}} \end{array}, \qquad \begin{array}{c} e_2\mid f_2 \\ \overset{e_1}{\underset{f_1}{\diagdown}}0\overset{e_3}{\underset{f_3}{\diagup}} \end{array}$$

Using the inward power sign convention shown in the last version of the junction, the constitutive relations may be written.

$$e_1(t) = e_2(t) = e_3(t), \tag{3.16}$$

$$f_1(t) + f_2(t) + f_3(t) = 0. \tag{3.17}$$

In words, the efforts on all bonds of a 0-junction are always identical, and the algebraic sum of the flows always vanishes. Taken together, the equations imply that power on all the bonds sums to zero.

$$e_1f_1 + e_2f_2 + e_3f_3 = 0; \qquad (3.18)$$

that is, if power is flowing into the 0-junction on two of the ports, it must be flowing out at the third.

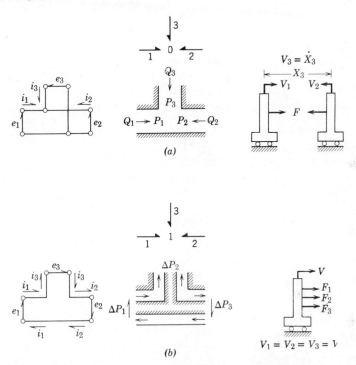

Figure 3.11. Basic 3-ports in various physical domains. (a) 0-junction; (b) 1-junction.

The use of the 0-junction is suggested by Figure 3.11a. The most obvious examples of 0-junctions are the electrical conductors connected as shown to provide three terminal pairs and the pipe "tee" junction that is an idealized version of the hardware-store variety. The mechanical example may seem obscure, and it is contrived. Mechanical 0-junctions are just as necessary as electrical or hydraulic ones, but they do not appear so readily in gadget form. The two carts in the mechanical example of Figure 3.11a are supposed to be rigid and massless. Note that $V_3 = V_1 - V_2$, which conforms with Eq. (3.17). If F is the force across the gap, X_3, then F is the port effort for V_1, V_2, and V_3, in accordance with Eq. (3.16). Such a force would, in fact, exist if F were due to a massless spring connected between the two carts. With the spring connected, a bond graph of the system would be

$$
\begin{array}{c}
C \\
F_3 {\Big|} V_3 \\
\xrightarrow[V_1]{F_1} 0 \xleftarrow[V_2]{F_2} \, .
\end{array}
$$

Before considering more examples in which the 0-junction is used, consider the dual of the 0-junction, that is, a multiport in which the roles of effort and flow are interchanged. Such an element is *effort junction*, *1-junction*, or *common-flow junction*. The symbol for this multiport is a 1 with three bonds.

$$
\underrightarrow{\quad\big|\quad} 1 \underrightarrow{\quad} \,, \quad \underrightarrow{\big|^2}_{1} 1 \underrightarrow{\quad}_{3} \,, \quad \xrightarrow[f_1]{e_1} \overset{e_2 \big| f_2}{1} \xleftarrow[f_3]{e_3}
$$

With the indicated power sign convention, the constitutive relations for this element are

$$
f_1(t) = f_2(t) = f_3(t). \tag{3.19}
$$

and

$$
e_1(t) + e_2(t) + e_3(t) = 0. \tag{3.20}
$$

As with the 0-junction, the constitutive equations for the 1-junction combine to insure power conservation in the form of Eq. (3.18).

The 1-junction has a single flow, and the sum of the effort variables on the bonds vanishes. Figure 3.11b shows some instances in which a 1-junction can be used to model physical situations. Both the electrical conductors and the hydraulic passages are arranged so that if 1-port components are attached to the ports one could describe the resulting connection as a series connection. A single current or volume flow would circulate, and the voltages and pressures at the ports would sum algebraically to zero. In the mechanical example, the three forces all are associated with a common velocity, and the forces must sum to zero since the cart is supposed to be massless.

The understanding of the meaning of the 0- and 1-junctions is important for anyone learning bond-graph techniques, and it may be helpful to give some physical interpretations for these multiports in several physical domains:

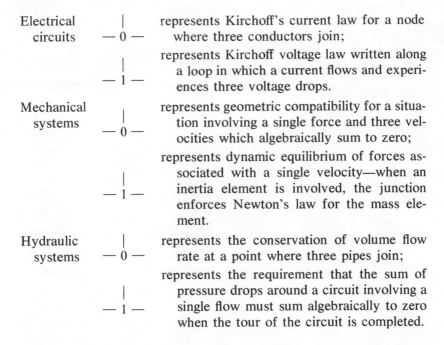

Electrical circuits	— 0 —	represents Kirchoff's current law for a node where three conductors join;
	— 1 —	represents Kirchoff voltage law written along a loop in which a current flows and experiences three voltage drops.
Mechanical systems	— 0 —	represents geometric compatibility for a situation involving a single force and three velocities which algebraically sum to zero;
	— 1 —	represents dynamic equilibrium of forces associated with a single velocity—when an inertia element is involved, the junction enforces Newton's law for the mass element.
Hydraulic systems	— 0 —	represents the conservation of volume flow rate at a point where three pipes join;
	— 1 —	represents the requirement that the sum of pressure drops around a circuit involving a single flow must sum algebraically to zero when the tour of the circuit is completed.

As might be expected, the existence of 0- and 1-junctions within complex systems is not always obvious, but in succeeding chapters formal techniques for modeling systems using these basic elements are presented.

To make clear the utility of the junctions, four elementary example systems are displayed in Figure 3.12. Note that only two bond graphs are involved. The series and parallel aspects of the junctions are more obvious in the electrical than in the mechanical cases. The reader should study these examples to make sure he understands how the sign conventions are transferred from the physical sketches to the bond graph. Note that the 1-ports have the sign conventions as they were presented in the tables at the beginning of this chapter, but the junction signs are not all inward pointing. When the sign convention arrows are changed from the inward pointing convention used to introduce the 3-ports, then Eqs. (3.17) and (3.20) must be modified with a minus sign for each port with an outward pointing sign. Equations (3.16) and (3.19) remain invariant, however, to changes in sign convention. *A 0-junction has only a single effort and a 1-junction has only a single flow independent of the sign*

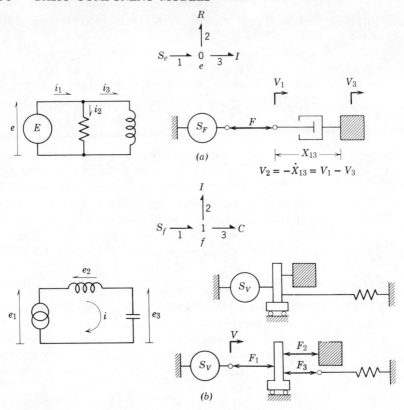

$$V_2 = -\dot{X}_{13} = V_1 - V_3$$

(a)

(b)

Figure 3.12. Example systems involving basic 3-ports. (a) Systems using 0-junctions; (b) systems using 1-junctions.

convention. As an example, consider

$$\xrightarrow[f_1]{e_1} 1 \xrightarrow[f_3]{e_3} ,$$

the equations of which are

$$f_1 = f_2 = f_3,$$
$$e_1 - e_2 - e_3 = 0. \tag{3.21}$$

The reader should verify that the systems and bond graphs are consistant by writing equations such as Eq. (3.21).

The slight generalization from 3-port junctions to 4- or n-port junctions is worth emphasizing. In bond-graph symbolism, two similar 3-ports may be combined into a 4-port thus:

$$-0-0-\ =\ -0-,$$

$$-1-1-\ =\ -1-.$$

An n-port 0- or 1-junction has a common effort or flow on all bonds, and the algebraic sum of the complementary power variables on the bonds vanishes. Occasionally, 2-port junctions arise, and, in some cases, these are precisely equivalent to a single bond. The following bond graph identities are always valid:

$$\rightharpoonup 0 \rightharpoonup\ =\ \rightharpoonup\ ;\qquad \rightharpoonup 1 \rightharpoonup\ =\ \rightharpoonup\ .$$

On the other hand, with some sign patterns, the 2-port 0- and 1-junctions serve to reverse the sign definition of an effort or flow. For example,

$$\frac{e_1}{f_1} \searrow 0 \swarrow \frac{e_2}{f_2} \qquad \text{implies} \qquad e_1 = e_2,\ f_1 = -f_2,$$

and

$$\frac{e_1}{f_1} \searrow 1 \swarrow \frac{e_2}{f_2} \qquad \text{implies} \qquad f_1 = f_2,\ e_1 = -e_2.$$

Such 2-ports are sometimes necessary when two multiports are to be joined by a bond, but the two multiports have been defined with signs that are not compatible with a single bond. In connecting a spring, $-C$, to a mass, $-I$, one could define a common velocity but use a 2-port 1-junction to express the fact that the spring force is the negative of the force on the

TABLE 3.5. Summary of basic 3-ports

Flow junction or 0-junction	$\dfrac{e_1}{f_1} \searrow 0 \swarrow \dfrac{e_3}{f_3}$ $\uparrow\ _{f_2}$	$e_1 = e_2 = e_3$ $f_1 + f_2 + f_3 = 0$
Effort junction or 1-junction	$\dfrac{e_2}{f_1} \searrow 1 \swarrow \dfrac{e_3}{f_3}$ $\uparrow\ _{e_2\,|\,f_2}$	$f_1 = f_2 = f_3$ $e_1 + e_2 + e_3 = 0$

mass. The resulting bond graph would then be $C \leftarrow 1 \rightarrow I$, in which the passive 1-ports have the convenient inward sign convention.

The constitutive relations for 0-junctions and 1-junctions are summarized in Table 3.5.

3.4 CAUSALITY CONSIDERATIONS FOR THE BASIC MULTIPORTS

The concept of causality was already discussed in general terms in Chapter 2, and we may now make some more specific uses of the idea with respect to the basic multiports. Some of the causal properties developed here will be applied in later chapters. For now, we simply note that some of the basic multiports are heavily constrained with respect to possible causalities, some are relatively indifferent to causality, and some exhibit their constitutive laws in quite different forms for different causalities.

3.4.1 Causality for Basic 1-Ports

The effort and flow sources are the most easily discussed from a causal point of view since, by definition, a source impresses either an effort or flow time history upon whatever system is connected to it. Thus, if we use the symbols S_e— and S_f— for the abstract effort and flow sources, the only permissible causalities for these elements are

$$S_e \dashv \quad \text{and} \quad S_f \vdash .$$

The causal forms for effort and flow sources are summarized in the first two rows of Table 3.6.

In contrast to the sources, the 1-port resistor is normally indifferent to the causality imposed upon it. The two possibilities may be represented in equation form as follows:

$$e = \Phi_R(f), \qquad f = \Phi_R^{-1}(e),$$

where we use the convention that the variable on the left of the equality sign represents the output of the resistor or the dependent variable, and that appearing in the function of the right side is the input or independent variable for the element. This convention is used commonly, but not universally, in writing equations and corresponds to the notation used in computer programing, as, for example, in FORTRAN.

TABLE 3.6.   Causal forms for basic 1-ports

Element	Acausal Form	Causal Form	Causal Relation
Effort source	$S_e \rightharpoonup$	$S_e \rightharpoondown$	$e(t) = E(t)$
Flow source	$S_f \rightharpoonup$	$S_f \rightharpoondown$	$f(t) = F(t)$
Resistor	$R \leftharpoonup$	$R \rightharpoondown$	$e = \Phi_R(f)$
		$R \leftharpoondown$	$f = \Phi_R^{-1}(e)$
Capacitor	$C \leftharpoonup$	$C \rightharpoondown$	$e = \Phi_C^{-1}\left(\int^t f\, dt \right)$
		$C \leftharpoondown$	$f = \dfrac{d}{dt}[\Phi_C(e)]$
Inertia	$I \leftharpoonup$	$I \leftharpoondown$	$f = \Phi_I^{-1}\left(\int^t e\, dt \right)$
		$I \rightharpoondown$	$e = \dfrac{d}{dt}[\Phi_I(f)]$

The correspondences between the causally interpreted equations and the causal strokes on the bond of the R— element are shown in the third row of Table 3.6. As long as both the functions Φ_R and Φ_R^{-1} exist and are known, there is no reason for preferring one causality over the other. It is possible, however, that the static relation between e and f shown in Figure 3.1 might be multiple-valued in one direction or the other; that is, either Φ_R or Φ_R^{-1} might be multiple-valued. In such a case, the single-valued causality would be clearly preferable. In the linear case, with a finite slope of the e–f characteristic, the 1-port resistor is indifferent to the causality imposed upon it.

The constitutive laws of the C— and I— elements are expressed as static relations between e and $q = \int^t f\, dt$ and f and $p = \int^t e\, dt$, respectively. In expressing causal relations between es and fs, we will find that the choice of causality has an important effect. Taking the capacitor, we may rewrite the relations from Table 3.2 as follows:

$$e = \Phi_C^{-1}\left(\int^t f\, dt \right), \qquad f = \frac{d}{dt}[\Phi_C(e)], \qquad (3.22)$$

in which causality is implied by the form of the equation. Note that when f is the input to the C—, e is given by a static function of the time integral of f, but when e is the input, f is the time derivative of a static function of e. The correspondences between these causal equations and the causal stroke notation for the capacitance are shown in the fourth row of Table

3.6. The implications of the two types of causality, which are called *integral causality* and *derivative causality*, respectively, will be discussed in some detail in later chapters.

Since inertia is the dual* of the capacitor, similar effects occur with the two choices of causality. Rewriting the inertia element relations from Table 3.3, we have

$$f = \Phi_I^{-1}\left(\int^t e\, dt\right), \qquad e = \frac{d}{dt}[\Phi_I(f)]. \tag{3.23}$$

In this case, integral causality exists when e is the input to the inertia, and derivative causality exists when f is the input. These observations are summarized in the fifth row of Table 3.6. Equations (3.22) and (3.23) are written in a form suitable for nonlinear $C-$ and $I-$ elements, but the distinction between integral and derivative causality remains for the special case of linear elements.

3.4.2 Causality for Basic 2-Ports and 3-Ports

Proceeding now to the basic 2-ports, one might think initially that there would be a total of four possibilities for the assignment of causality of a transformer, namely, any combination of the two possible causalities for each of the two ports. However, there are only two possible causality assignments, as the defining relations, Eqs. (3.9) and (3.11), show. As soon as one of the es or fs has been assigned as an input to the $-TF-$, the other e or f is constrained to be an output by Eq. (3.9). Thus, in fact, the only two possible choices for causality for the transformer are $\vdash TF\vdash$ and $\dashv TF\dashv$. The possible causalities are tabulated in the first row of Table 3.7. In Table 3.7, a simplified naming of the efforts and flows has been achieved by simply numbering the bonds. This technique will be explored in more detail in subsequent chapters. Again, causal equation equivalents to the causal stroke notation are given for all elements in Table 3.7.

For the gyrator, Eqs. (3.11) show that as soon as the causality for one bond has been determined, that for the other is also. Thus, the only permissible causal choices for the $-GY-$ are $\dashv GY\vdash$ and $\vdash GY\dashv$. The choices for the causality for the gyrator are summarized in the second row of Table 3.7.

The causal properties of 3-port 0- and 1-junctions are somewhat similar to those of the basic 2-ports. Although each bond of the 3-ports,

---

* Dual elements have identical constitutive laws except that the roles of effort and flow are interchanged.

TABLE 3.7. Causal forms for basic 2-ports and 3-ports

Element	Acausal Graph	Causal Graph	Causal Relations
Transformer	$\xrightarrow{1} TF \xrightarrow{2}$	$\vdash\xrightarrow{1} TF \vdash\xrightarrow{2}$	$e_1 = me_2$ $f_2 = mf_1$
		$\xrightarrow{1} TF \xrightarrow{2}\dashv$	$f_1 = f_2/m$ $e_2 = e_1/m$
Gyrator	$\xrightarrow{1} GY \xrightarrow{2}$	$\vdash\xrightarrow{1} GY \xrightarrow{2}\dashv$	$e_1 = rf_2$ $e_2 = rf_1$
		$\xrightarrow{1} GY \vdash\xrightarrow{2}$	$f_1 = e_2/r$ $f_2 = e_1/r$
0-Junction	$\xrightarrow{1} 0 \xleftarrow{2}$ $3\uparrow$	$\xrightarrow{1} 0 \xleftarrow{2}\dashv$ $3\uparrow$	$e_2 = e_1$ $e_3 = e_1$ $f_1 = -(f_2 + f_3)$
1-Junction	$\xrightarrow{1} 1 \xleftarrow{2}$ $3\uparrow$	$\vdash\xrightarrow{1} 1 \vdash\xleftarrow{2}$ $3\uparrow$	$f_2 = f_1$ $f_3 = f_1$ $e_1 = -(e_2 + e_3)$

considered alone, could have either of the two possible causalities assigned, not all combinations of bond causalities are permitted by the constitutive relations of the element. For example, the constitutive relations for the 0-junction given in Table 3.5 indicate that all efforts on all the bonds are equal and the flows must sum to zero. Thus, if on any bond the effort is an input to a 0-junction, then all other efforts are determined, and on all other bonds they must be outputs of the 0-junction. Conversely, if the flows on all bonds except one are inputs to the 0-junction, the flow on the remaining bond is determined and must be an output of the junction. A typical permissible causality for a 0-junction is shown in the fifth row of Table 3.7. Here the causal stroke on the end of bond 1 nearest the 0 indicates that e_1 is an input to the junction and that all other bonds must have causal strokes at the end away from the 0. To interpret the diagram another way, the flows on bonds 2 and 3 are inputs to the 0-junction. These considerations are also expressed by the causal equations shown in Table 3.7. For a 3-port 0-junction, then, there are only three different permissible causalities in which each of the three bonds in succession plays the role assigned to bond 1 in the example shown in the table. For an n-port 0-junction this description of the constraints on causality is still valid, and there are exactly n different permissible causal assignments.

For a 1-junction the same considerations apply as for a 0-junction except that the roles of the efforts and flows are interchanged. Table 3.5 indicates that flows on all the bonds are equal and the efforts sum to zero. Thus, if the flow on any single bond is an input to the 1-junction, the flows on all other bonds are determined and must be considered outputs of the junction. On the other hand, when the efforts on all bonds except one are inputs to the 1-junction, the effort on the remaining bond is determined and must be an output of the junction. A typical permissible causality is shown in the sixth row of Table 3.7. In this example, bond 1 plays the special role of determining the common flow at the junction, and the remaining bonds supply effort inputs that suffice to determine the effort on bond 1. Clearly, there are three permissible causalities for a 3-port 1-junction, and there are n permissible different causal assignments for an n-port 1-junction.

Although the causal considerations have been stated for all the basic multiports defined so far (summarized in Tables 3.6 and 3.7), it can hardly be clear what all the implications of causality are. The study of causality is very important, and bond graphs are uniquely suited to this study.

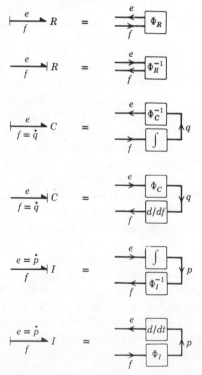

Figure 3.13. Block diagrams for 1-ports.

However, only when some real system models have been assembled is it clear why causal information is so important. In the next chapter, system models are built up using the basic multiports just discussed. Using the rules of causality, it is then possible to predict features of these systems even before the exact characterization of the multiports has been decided. For instance, it will be possible to predict the order of a system model before any equations are written and before a firm decision has been made about whether the model should be linear or nonlinear. In addition, causal considerations will prove invaluable in writing state equations or setting up computational block diagrams.

3.5   CAUSALITY AND BLOCK DIAGRAMS

Block diagrams indicate input and output quantities for each block and, thus, are inherently causal. When causal strokes are added to a bond graph, one may represent the information by a block diagram. For example, the block diagram versions of the causal forms for the R, C, and I 1-ports shown in Table 3.6 are given in Figure 3.13. Similarly, block diagrams for 2-ports and 3-ports corresponding to entries in Table 3.7 are shown in Figures 3.14 and 3.15. It should be possible to correlate the signal flow paths in the block diagrams with the equations in the tables and with the bond-graph representation. Note that when one rigorously

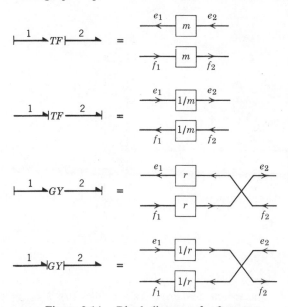

Figure 3.14.   Block diagrams for 2-ports.

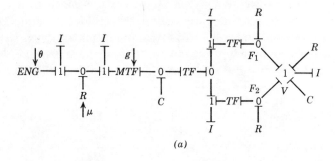

Figure 3.15. Block diagrams for 3-ports.

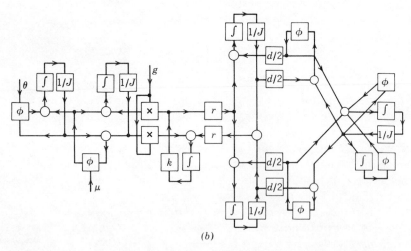

(a)

(b)

Figure 3.16. Interconnected drive train model. (*a*) Bond graph; (*b*) block diagram.

maintains the spatial arrangements with efforts above and to the left of bonds and flows below and to the right, the block diagrams have fixed patterns.

It also may be seen that block diagrams are more complex graphically than bond graphs because a single bond implies two signal flows. Initially block diagrams are easier to understand than bond graphs because they contain redundant information. For systems with some complexity, however, block diagrams rapidly become so complex that the conciseness of bond graphs is an advantage. For example, Figure 3.16 shows a block diagram equivalent to a bond-graph model of the automotive drive train system of Figure 2.7. Note that the sign convention half arrows have yet to be put on the bond graph and the corresponding + and − signs do not appear in the block diagram near the circles representing signal summation.

After procedures for constructing bond-graph models and adding causal strokes to them have been discussed in the following chapters, one option is to construct a block diagram from the bond graph. Block diagrams are particularly useful in setting up analog computer simulations.

3.6 PSEUDO BOND GRAPHS AND THERMAL SYSTEMS

In these introductory chapters, the number of physical domains to be discussed is purposely limited. Later, in Chapter 9, when more sophisticated modeling concepts have already been discussed, the range of physical systems is broadened considerably. Because thermal systems are so important, we briefly introduce some bond-graph representations of thermal elements. Traditionally, thermal systems have been presented as analogous to electric circuits, usually with temperature analogous to voltage and heat flow analogous to current. With this analogy there are then thermal resistors, capacitors, and parallel and series connections (our 0- and 1-junctions) and sources analogous to voltage and current sources. There are no thermal inertias, however.

Since this analogy in which temperature is an effort and heat flow is a flow has proved useful, we present it here. There is one major hitch, however. The product of temperature and heat flow is not a power! Heat flow itself has the dimensions of power. We choose to call any bond graph in which e and f are not power variables a *pseudo bond graph*. Such a pseudo bond graph cannot be coupled to a normal bond graph using power variables except by means of some ad hoc elements that do not obey the rules of normal bond-graph elements. Bond-graph techniques may be usefully applied to any pseudo bond graph as long as the basic elements in the pseudo bond graph correctly relate the e, f, p, and q variables. In Chapter 9, it will be shown that a true bond graph results if temperature and entropy flow are used for effort and flow variables, but the thermodynamic arguments are more sophisticated than those necessary to establish the usefulness of the pseudo bond graph.

In Figure 3.17, two common situations that arise in the study of thermal systems are depicted. In both cases, we assume the temperature gradients

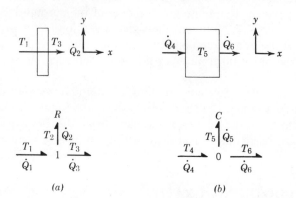

(a)   *(b)*

Figure 3.17. Basic elements for models of conduction heat-transfer systems. (*a*) Thermal resistor and 1-junction; (*b*) thermal capacitor and 0-junction.

and heat flow are present only in the x direction. The case of Figure 3.17a represents a pure resistance. If T_1 and T_3 are the temperatures (in any convenient scale) on the two sides of a slab of material of area, A, we assume that the flow of heat through the slab, $\dot{Q}_2$, is a function of $T_2 = T_1 - T_3$. In the linear case, we say

$$R\dot{Q}_2 = T_1 - T_3 = T_2, \tag{3.24}$$

where $\dot{Q}_2$ may be measured in Btu/sec, cal/sec, or in any other power measure. (Because this is a pseudo bond graph, the choice of units is not critical as long as all element parameters are defined in a consistent manner.) For a material with thermal conductivity, k, thickness, l, and area, A,

$$R = \frac{l}{kA}.$$

Note that the thermal resistor implies a relation between T_2 and $\dot{Q}_2$ and the 1-junction implies $\dot{Q}_1 = \dot{Q}_2 = \dot{Q}_3$ and $T_1 - T_2 - T_3 = 0$. We have simply shown how the 1-port, R, and the 1-junction can be used together to constrain the common flow to be a function of the difference between two efforts.

Figure 3.17b represents a lump of material that changes temperature as a function of the net heat energy stored in it. That is, T_5 is a function of $Q_5 = \int^t \dot{Q}_5 \, dt = \int^t (\dot{Q}_4 - \dot{Q}_6) \, dt$. Since T_5 is an effort and Q_5, the integral of a flow, is a displacement, the element is a capacitor. Again, the bond graph of Figure 3.17b combines a $C-$ with a 0-junction to indicate that $T_4 = T_5 = T_6$ and $\dot{Q}_5 = \dot{Q}_4 - \dot{Q}_6$. In the linear case a thermal capacitance, C, can be defined such that

$$T_5 = T_{50} + \frac{1}{C} \int_{t_0}^{t} \dot{Q}_5 \, dt, \tag{3.25}$$

where T_{50} is the temperature at $t = t_0$. The capacitance, C, can be found by assuming that the element does negligible work by expanding or contracting so that changes in its internal energy are only the result of $\dot{Q}_5$. Then if c is a *specific heat*,

$$c = \frac{\partial u}{\partial T}, \tag{3.26}$$

where u is the internal energy per unit mass, then

$$C = mc,$$

where m is the mass of the substance. Strictly speaking, c is the specific heat *at constant volume*, but for most solids and liquids the work done by expansion is small compared to Q_5 so that c does not vary much even if the material is allowed to expand. (When gases do work as they are heated, we will find it much better to use the true bond-graph representation of Chapter 9.)

Figure 3.18 shows a typical use of the elements shown in Figure 3.17 and introduces a thermal effort source. A unit area of the wall of a pipe carrying hot fluid is modeled. A lumped parameter model consisting of three resistors and two capacitors is used, and the inside and outside temperatures are assumed to be determined by effort (temperature) sources.

If T_1 changes because the temperature of the fluid flowing through the pipe changes, the bond-graph model can give a prediction of the dynamic

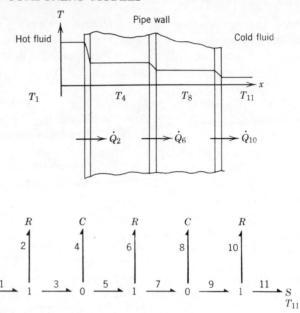

Figure 3.18.   Heat transfer in pipe wall.

changes in the temperature in the pipe wall. If the pipe is really made of a uniform material, then the stepwise distribution of temperature with distance, x, will be fairly crude approximation to the temperature distribution which would be a continuous function of both time and x. A better approximation could be obtained by splitting the pipe into a large number of resistor and capacitor layers, but the model would be more complicated to use.

REFERENCE

1. S. H. Crandall, D. C. Karnopp, E. F. Kurtz, and D. C. Pridmore-Brown, *Dynamics of Mechanical and Electromechanical Systems,* N.Y.: McGraw-Hill, 1968.

PROBLEMS

3-1   A nonlinear dashpot has as its constitutive relation the "absquare law,"

$$F = AV |V|,$$

where F and V are the force and velocity across the dashpot and A is a constant. Plot this relation in a sketch and indicate the bond-graph sign convention implied if $A > 0$ and which causality the equation implies in the form given. Try to invert the constitutive law to yield the velocity as a function of the force.

3-2 A fluid of mass density ρ is pumped into an open topped tank of area A. If P is the pressure at the tank bottom and Q the volume flow rate, the tank is approximately a $-C$ for slow changes in the volume of fluid stored. Is the $-C$ a linear element in this case and, if so, what is the capacitance? *Hint:* It is useful to compute the height of fluid, h, as a function of the total volume of fluid as an intermediate step.

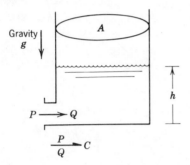

3-3 Reconsider Problem 3-2, but let the tank have sloping walls as shown. What does this do to the constitutive law for the device?

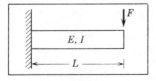

3-4 Consider a uniform cantilever beam of length L, elastic modulus E, and area moment of inertia I. If a force F is applied at the tip of the beam it will deflect. If the beam is supposed to be massless, decide what type of a 1-port it is and compute its constitutive law.

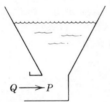

3-5 Consider a given mass of water as a thermal capacitor. Sketch the constitutive law for this element for a range of temperatures including the freezing point. Indicate the effect of the "latent heat" of freezing or melting.

3-6   Linear electrical inductors can be characterized by a law relating the voltage, e, to the rate of change of current,

$$L \frac{di}{dt} = e.$$

Convert this to a law relating the flow, i, to the momentum, λ, which in this case is the time integral of e and is called "flux linkage." Plot a linear flow-momentum constitutive law, and show on the plot where the inductance, L, appears. Now sketch a nonlinear flow-momentum law. Convert the nonlinear law back to a relation between e and di/dt if this is possible.

3-7   A rigid pipe filled with incompressible fluid of mass density, ρ, has length, L, and cross-sectional area, A.

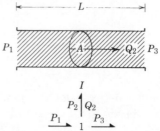

If P_1 and P_3 are pressures at the ends of the pipe and the volume flow rate is Q_2, convince yourself that the bond graph shown correctly represents the pipe in the absence of friction. Show that the correct constitutive law relating pressure momentum and volume flow is

$$p_{P_2} = \int^t P_2 \, dt = \left(\rho \frac{L}{A} \right) Q_2$$

by writing Newton's law for the slug of fluid in the pipe. (It should come as a surprise that small-area tubes have a lot of inertia when P, Q variables are used!)

3-8   An accumulator consists of a heavy piston in a cylinder. If the pressure is determined primarily by the weight of the piston, sketch

the constitutive law for this 1-port device.

3-9   A heat transfer coefficient, α, is given in $[(\text{Btu/hr})/\text{ft}^2\,{}^\circ\text{F}]$. If an area, A, of this surface is involved in a problem, show how the effect of the surface would be represented in a bond graph, and write the constitutive law for the element in terms of A and α.

3-10 A flywheel is a uniform disk of radius, R, and thickness, t, and is made of a material of mass density ρ. Write the constitutive law for the 1-port representing the flywheel in its flow-momentum form. Evaluate the inertance parameter for a steel disk 1 in. in thickness and 10 in. in diameter.

3-11 Assume any needed dimensions for the hydraulic ram of Figure 3.8e and write the constitutive laws for this 2-port.

3-12 Repeat Problem 3-11 for the devices of Figures 3.8b and c.

3-13 In Figure 3.9c assume that the rotor has moment of inertia, J, and spins at a high angular rate, Ω. If the rotor is centered on an axle of length, L, relate F_1, V_1, F_2, V_2 and thus demonstrate that the device is indeed approximately a $-GY-$.

3-14 Draw block diagrams for the following bond graphs, assuming all 1-ports are linear:

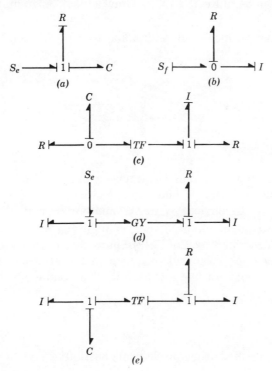

3-15

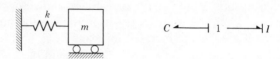

Draw a block diagram for the oscillator using the bond graph shown.

3-16

A block of material with total heat capacitance C is covered with insulating material with total thermal resistance R and surrounded by an atmosphere of temperature T_0. The bond graph is supposed to aid in estimating how fast the block will heat up if it starts at some temperature $T_i(t_0) < T_0$. Draw a block diagram from the bond graph.

3-17 Consider an ideal rack and pinion with no friction losses.

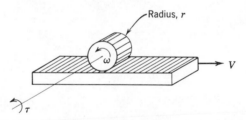

If the pinion has radius r and torque and speed τ and ω and if the rack has velocity V and force F what type of element would represent the device? Write the appropriate constitutive laws.

3-18 An electrodynamic loud speaker is driven by a voice coil transducer described by Eq. (3.12). Show that if only the mass of the speaker cone is considered, then at the electrical terminals the device would act like a capacitor. The bond-graph identity

$$\xrightarrow[i]{e} \dot{G}\dot{Y} \xrightarrow[V]{F} I = \xrightarrow[i]{e} C$$

is to be verified using the element constitutive equations directly.

4

SYSTEM MODELS

You are now ready to model the world, having mastered the basic multiport elements. All that remains is to sally forth—armed with your bond-graph arsenal of C, I, R, S_e, S_f, 0, 1, TF, and GY—ready to represent as a bond graph any physical system you may meet. This chapter is devoted to helping you do just that. However, it is not true that every system you may encounter will be reducible to a simple bond graph. For electrical circuits, certain classes of mechanical systems, hydraulic circuits, and some transducer systems you will be entirely successful. Additional study and experience will enable you to extend your range, but at the end there will always be some problems that do not fall into your carefully spread snare. But be assured that the size of your snare will be amazingly large, with some effort on your part.

In this chapter we show how to represent any electrical circuit in bond-graph form by using a direct, simple modeling procedure. Then we use a similar approach to problems involving mechanical translation. With slight extensions to the procedure, the next class of problems we treat contains fixed-axis rotation. By generalizing our approach slightly we are able to model hydraulic circuits effectively since they are similar in several respects to systems already treated.

All of the previously mentioned systems involve only one type of power within a single system. For that reason they are said to be "single energy-domain" systems. There are many useful devices involving two (or occasionally more) types of power, and these have transducer elements (e.g., motors and pumps) coupling the different energy domains. We introduce some simple transducer system models as one more topic. Bond graphs are ideally suited to the study of multiple energy domain systems, so transducers are important devices worthy of your attention.

4.1   ELECTRICAL SYSTEMS

We begin by observing that any electrical *circuit* can be modeled by a bond graph containing elements of the set $\{0, 1, C, I, R, S_e, S_f\}$. Notice that the elements *TF* and *GY* are not included. That is because these elements are properly used in representing electrical *networks*, a more general class than circuits. First we shall model circuits; then we shall extend the procedure to include network elements, arriving eventually at a complete result.

4.1.1   Electrical Circuits

Inspection Method

Referring to the component Figures 3.1, 3.2, and 3.4, we see that identifying and representing the electrical 1-port elements is simple and direct; one merely identifies the two terminals paired by the element as a port. Hence, inspection will serve as a suitable method for the 1-port elements. The problem of identifying the connection pattern (sometimes called the circuit topology) in terms of parallel and series (0 and 1) joinings is not so readily accomplished by inspection. However, with practice, many circuits can be reduced to bond graphs entirely by inspection. Let us examine two examples using the inspection method. Figure 4.1*a* shows a rather simple circuit with all its terminals labeled. Terminals at the same potential have the same label. In part *b* the ports are emphasized. Notice that the port defined by the pair *a–c* (marked by ×) is a key structural port. To arrive at the bond graph in part *c* we merely argue as follows: the elements *C* and R_1 have the same voltage (V_{ac}), so they are in parallel; the elements *L* and R_2 have the same current (i_{abc}), so they are in series. The graph for the circuit is completed by joining the bonds representing the × port to get the result shown in Figure 4.1*d*.

Because the circuit has reference directions specified for both V_{ac} and i_{abc}, when the two directed variables are paired on a bond they imply a reference power direction. If V_{ac} and i_{abc} are both positive as shown (or both negative), then energy flows from the 0-junction part of the system to the 1-junction part. Hence, positive power on the connecting bond is shown directed from 0 to 1 by the half arrow.

It is possible that the directed variables in a circuit or other schematic imply (1) all of the power directions in the associated bond graph, (2) some of the powers (as in the example just discussed), or (3) none of the powers. Most typical is the case in which some of the powers are implied. Then the rest of the power directions may be chosen conveniently on the bond graph, a very simple matter of putting on half arrows.

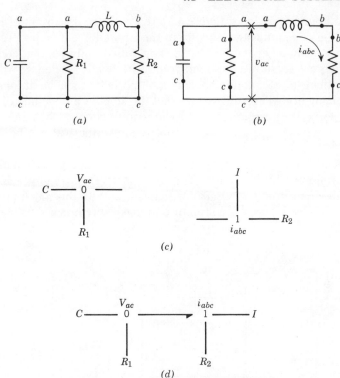

Figure 4.1.   An electric circuit. Example 1.

Many circuits that appear complex at first glance can be reduced to bond graphs quite readily. As an example, consider the circuit shown in Figure 4.2a. Several key node-pairs have been labeled (ab, cd, and ef). Part b shows the bond-graph fragment for the series connection of E and R_1. Part c shows the 2-port fragment between ab and cd; notice that L_1 and C_2 are in parallel, but the entire section is actually in series with the other parts of the circuit. In part d two steps are taken together; L_2 and C_3 are joined in parallel, and the result is joined to R_2 in series. Then in part e the entire fragment is related to the system by observing that it is a 2-port in parallel at terminal pairs cd and ef. The open-circuit condition is added in the form of element S_f ($i = 0$) directly. Finally, the entire model is assembled by joining the appropriate ports in part f. Also, element C_1 is inserted in parallel. Review of the complete graph compared to the circuit shows that we could have argued the case quite directly from the original circuit diagram. For example, E and R_1 are in series; together they are in parallel with C_1 and the rest of the circuit. L_1 and C_2 are in parallel; together they are in series with . . . , and so on.

The bond-graph model of Figure 4.2*f*, representing the circuit of Figure 4.2*a*, can be communicated directly to the ENPORT computer program. To do this we number the bonds, starting with 1 and counting consecutively. Figure 4.3*a* shows the bond graph with bonds labeled. It will now be convenient to refer to the specific multiports by their *bond labels*, so that no ambiguity will result. Notice that this implies a renaming of some of the 1-ports. With experience you will tend to choose labels initially in bond-graph-compatible fashion, so that renumbering will not be necessary.

The bond graph is described to ENPORT by listing each multiport element followed by its bonds. This code is given in Figure 4.3*b*. The order of bonds within a group is not significant. With this much information ENPORT will check the structure to make sure it is a sensible bond

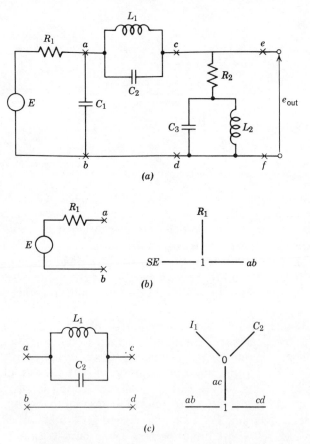

Figure 4.2.   An electric circuit. Example 2.

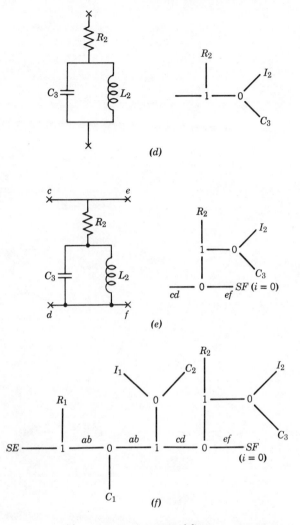

Figure 4.2. (*cont'd*)

graph (e.g., Are there any free bonds? Are all the multiports recognizable?), and, if it is, power directions will be assigned. The power directions for the example graph are listed in Figure 4.3c and shown on the bond graph in Figure 4.3d. Basically, a very routine and orderly pattern is used to assign power directions, as follows:

1. direct power *from* each source;
2. direct power *to* each C, I, and R;

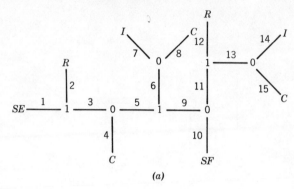

(a)

SYSTEM BOND GRAPH

NODE	BONDS		
SE	1		
R	2		
1	1	2	3
0	3	4	5
C	4		
1	5	6	9
0	6	7	8
I	7		
C	8		
0	9	11	10
SF	10		
1	11	12	13
R	12		
0	13	14	15
I	14		
C	15		

GRAPH ANALYSIS.

THE GRAPH HAS   16 NODES AND   15 BONDS.

(b)

POWER FLOWS

BOND	FROM		TO
1	SE	↦	1
2	1	↦	R
3	1	↦	0
4	0	↦	C
5	0	↦	1
6	1	↦	0
7	0	↦	I
8	0	↦	C
9	1	↦	0
10	SF	↦	0
11	0	↦	1
12	1	↦	R
13	1	↦	0
14	0	↦	I
15	0	↦	C

(c)

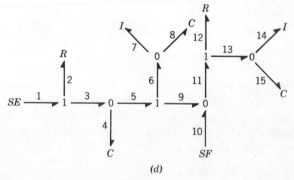

(d)

Figure 4.3.   ENPORT dialog and power directions for Example 2.

78

POWER FLOWS

BOND	FROM		TO
1	SE	⇀	1
2	1	⇀	R
3	1	⇀	0
4	0	⇀	C
5	0	⇀	1
6	0	⇀	1
7	0	⇀	1
8	0	⇀	C
9	1	⇀	0
10	0	⇀	SF
11	0	⇀	1
12	1	⇀	R
13	1	⇀	0
14	0	⇀	1
15	0	⇀	C

THE GRAPH HAS   16 NODES AND   15 BONDS

POWER ON BOND 6 FORCED
FLOW FROM NODE 0   TO NODE 1

POWER ON BOND 10 FORCED
FLOW FROM NODE 0   TO NODE SF

(e)

(f)

Figure 4.3.   (*cont'd*)

3. direct power *through* each *GY* and *TF*;

4. direct power *from* the first junction listed *to* the second.

On the basis of these conventions we conclude that the listed order of all multiport elements *except* 0 and 1 is unimportant. The order of the junction elements can be used to influence the choice of power directions. However, *any* bond(s) can be directed as desired by giving a specific request to ENPORT. In Figure 4.3e the ASSIGN request is used to direct powers on bonds 10 and 6 in particular ways. Part f indicates that this has been done. Thus, the choice of power directions is completely under the control of the bond grapher. The need for thoughtful choice of reference directions will become apparent later, as we interpret the dynamic response patterns of specific physical variables, such as voltage, displacement, and pressure.

Although inspection is a powerful method, there are times when a more routine procedure may be helpful, and there are times when inspection will not work. We now give a construction method that always works. The basic approach is to construct a bond graph that resembles the circuit structurally and then to simplify the bond graph based on selected circuit properties. The following is the circuit construction method.

Circuit Construction Method

1. For each node in the circuit with a distinct potential write a 0-junction.

2. Insert each 1-port circuit element by adjoining it to a 1-junction and inserting the 1-junction between the appropriate pair of 0-junctions (C, I, R, S_e, S_f elements).

3. Assign power directions to all bonds, using a "through-power" convention for the 0-1-0 sections, that is,

$$0 \rightarrow 1 \rightarrow 0.$$

4. If the circuit has an explicit ground potential, delete that 0-junction and its bonds from the graph; if no explicit ground potential is shown, choose any 0-junction and delete it.

5. Simplify the resulting bond graph by replacing 2-port 1- and 0-junctions that have "through powers" by single bonds; for example,

$1 \rightarrow 0 \rightarrow 1$ is replaced by $1 \rightarrow 1$.

To understand the use of this five-part procedure let us apply it to a very simple circuit; one for which the result should be obvious. In Figure 4.4*a* a circuit is shown with three distinct potentials, including an explicit ground node (*g*). Part *b* shows the nodes as 0-junctions, and in part *c* the three 1-port elements (current source, capacitor, and resistor) have been inserted using 1-junctions. Notice that powers have been directed through each 1-junction. The next step is to eliminate the ground junction and its bonds, as in part *d*. Finally, the graph may be simplified to that of part *e* by condensing the 2-port 0- and 1-junctions to simple bonds. The result is just what we expect, a model of a simple series connection.

With regard to the role of signs in the reduction procedure, we may make the following observations. All power directions implied by the circuit should be transferred before any simplification of the graph structure is carried out. In Figure 4.4*a* the assumptions are that positive values at *a* and *b* indicate higher potentials than ground (at *g*). Given the direction of branch currents, the implied powers are indicated in Figure 4.4*c*. All powers are specified because the directions of branch currents in *C* and *R* have been indicated, as well as positive voltage on the capacitance, *C*, and resistance, *R*. The several 2-port junctions in part *d* all have powers directed through them and are equivalent to simple bonds. Therefore they may be simplified, leading to the compact graph of Figure 4.4*e*, in which only three bonds remain. Sign conventions that lead to such simplifications are quite common, but they are not necessary; care must be taken to ensure that simplification is justified or allowed by the sign information of the original circuit.

The resulting bond graph indicates that there is a single (loop) current, I, and that the capacitance voltage, V_{ab}, is given by $V_a - V_b$, as indicated by the directed 1-junction. We should be pleased with these results.

As a final example of circuit modeling by bond graphs we shall study

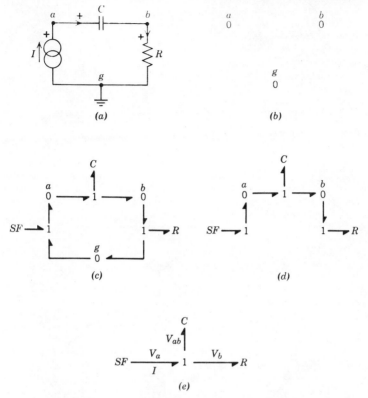

Figure 4.4. An electric circuit. Example 3.

the Wheatstone bridge shown in Figure 4.5*a*. A direct attack on this subtle circuit in terms of series and parallel connections probably would be fruitless, but the construction method works like a charm. First we establish the 0-junctions to represent node potentials, as in part *b*. Next we insert the 1-port elements, R_1, R_2, R_3, and R_4, plus the source, S_e, and the output condition (S_f, $i = 0$) for *e*, using 1-junctions. This is done in part *c*. The structure is now complete.

The most straightforward way to direct powers is to choose reference orientations on the circuit and then to transfer their power implications. An abstracted representation of the bridge circuit is shown in Figure 4.5*d*, with arrows indicating positive currents and plusses indicating positive voltages on the branches (i.e., across the branch elements). Our assumption is that all node potentials are defined positive in the same sense. The directed bond graph of part *e* contains information equivalent to that of the oriented abstract circuit; the 1-ports have been omitted for clarity.

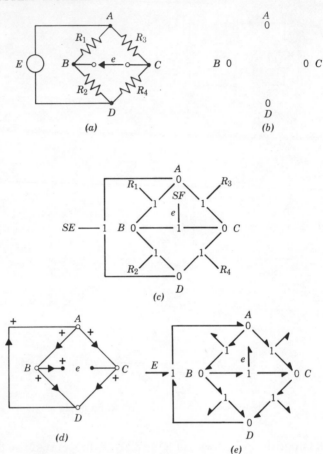

Figure 4.5. A Wheatstone bridge circuit. Example 4.

Note that power is defined positive into all the 1-ports except the source, E. Also observe that the junction-structure powers *always* "flow through" 1-junctions according to the corresponding branch-current directions. This is a major feature of directed bond-graph models derived from circuits.

The complete, directed bond graph is shown in Figure 4.6a. The next step is to select a ground node and eliminate it and its bonds. This has been done in part b for node D, in which a 0-junction and three bonds were removed from the graph. Finally, junctions which have 2-ports and through powers can be simplified to directed bonds. This has been done for R_2 and R_4, and the resulting graph has been rearranged to reveal a beautiful structural symmetry in Figure 4.6c. Because of its form it is

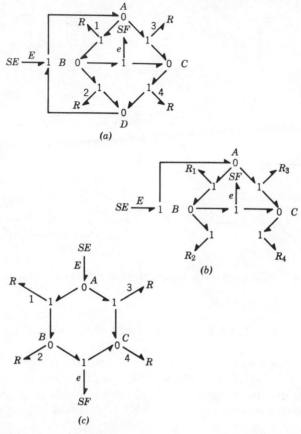

Figure 4.6. Simplification of the Wheatstone bridge model.

sometimes called the "benzene ring" structure. Quite nice, don't you think?

4.1.2 Electrical Networks

In order to extend our modeling approach to the broader class of electrical networks, we need to consider the *TF* and *GY* elements and the type of interaction represented by "controlled sources." We shall discuss two examples, one containing a transformer for isolation purposes, the other containing a controlled-source element (from a transistor model).

Figure 4.7*a* shows a transformer with its four terminals labeled. In part *b* an equivalent bond-graph fragment representing the transformer is shown. Sometimes the transformer has a common reference connection

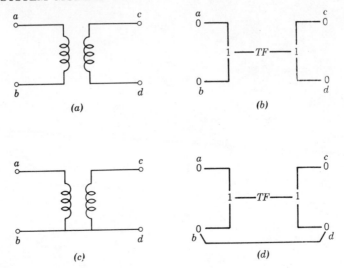

Figure 4.7.   The ideal electrical transformer and its bond-graph model.

shown, as in part c. In such a case, the bond-graph model shown in part d may be used.

The bond-graph model in Figure 4.7b is a structural model. In discussing signal orientations and power directions we must first remark on typical network conventions. The explicit way to show circuit orientations for a transformer (or any 2-port) is to direct both currents and both voltages, as in Figure 4.8a. The corresponding relations are given in part b, and an equivalent bond graph is shown in part c. It is often convenient to write the relations each with a positive sign, and the type of orientation which leads to this situation is the "through-power" convention. Figure 4.8d illustrates a through-power orientation on the bond graph, and part e gives the corresponding oriented circuit. The definition of the modulus m will be discussed subsequently in relation to an actual system study.

Now consider the electrical network shown in Figure 4.9a. It is rather complicated and contains an isolating transformer. We first establish the 0-junctions, as in part b. Then the elements are inserted, as shown in part c, and powers are directed. Two grounds have been chosen, one in each circuit part of the network; the corresponding nodes, e and h, have been eliminated in part d. Simplification leads to the bond-graph result shown in part e, portions of which can be checked by inspection. In making the reductions in part e we have used the fact that two adjacent 1-junctions may be reduced to a single junction. It also might be convenient to choose

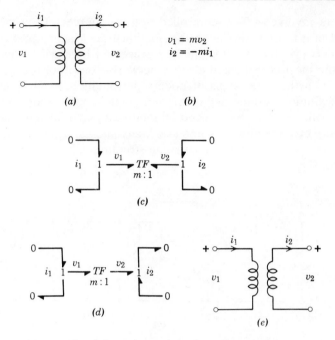

$$v_1 = mv_2$$
$$i_2 = -mi_1$$

Figure 4.8.   Orientation conventions for transformers.

a circuit convention to eliminate the 1-junction on the S_f element, if possible.

Another example illustrating the construction method of obtaining bond-graph models from electrical networks is shown in Figure 4.10a. Most of the network can be modeled by inspection, but the unusual 2-port element called a "controlled source," marked by terminal pairs uv and xy, requires comment. The basic idea is that there is a current source whose value depends on a voltage elsewhere in the network. The coupling is one-directional, in that there is no back effect upon the input port, so that an active bond can be used to show that the current, I, depends on the voltage, e, in a manner to be specified in the definition of S_f.

First the required 0-junctions and 1-port elements are inserted, as shown in Figure 4.10b. For convenience the node f has been repeated several times. Next the ground node, f, is eliminated, resulting in a much simpler graph. Also, considerable simplification has been accomplished, as shown in part c, with acknowledgment of the power direction requirements. Observe that the voltage controlling the current source is V_{cf} (or e in part a). Typically, junction transistors involve voltage-controlled current sources, whereas vacuum tubes, field-effect transistors, and amplifiers involve voltage-controlled voltage sources. They can all be handled in a fashion similar to the one illustrated in Figure 4.10.

The network construction method presented here will allow you to model the most complicated electric networks, once you have acquired facility with the basic steps of node representation, element insertion, power definition, ground definition, and graph simplification. With practice you will begin to observe certain recurrent patterns, and your ability to do more by inspection will increase. Many interesting discoveries await you as you explore systems containing structures like "ladders," "pi's," and "tees."

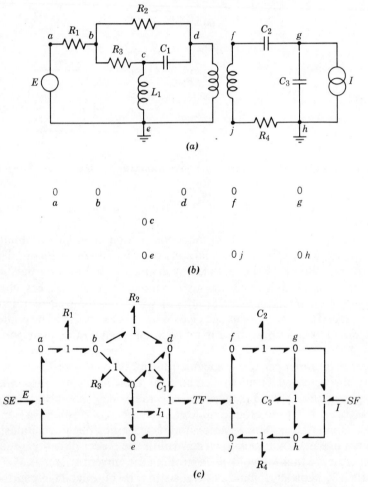

Figure 4.9. A network example with an isolating transformer.

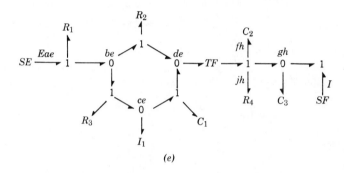

(d)

(e)

Figure 4.9. *(cont'd)*

Speaking of "ladders," let us make use of our ability to model the topology of circuits by bond-graph junction structures by examining the bond-graph form of a ladder network. An example of a resistive ladder circuit is given in Figure 4.11*a*. We are not really concerned with the nature of the 1-ports (source and resistances), but rather with their interconnection pattern (or circuit topology). The bond graph of part *b* may be found by inspection, using arguments like "..., R_2 in parallel, R_3 in series, R_4 in parallel ...." Or the formal construction procedure may be used. To display the system structure without regard to what is (or might be) connected to it we use a bond graph like that of Figure 4.11*c*, which is a direct representation of the ladder structure. It is presented in 2-port form, connecting a source port to a load port. So ladders are nothing more than chains of alternating 3-port 0s and 1s.

While you are still recovering from the surprise of looking at the structure, or connection pattern, expressed as a bond graph, we shall use such a graph to explore the idea of dual topologies in circuits. A pi network is shown in Figure 4.12*a* in resistive form. The structure is represented as "parallel, series, parallel" in part *b* by the bond graph.

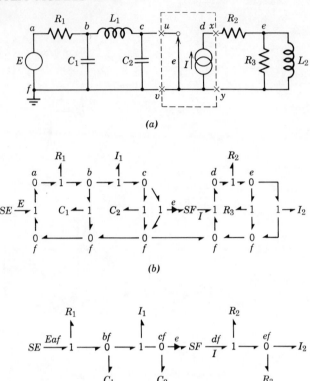

Figure 4.10.   A network example with a controlled source.

Resistances 1, 2, and 3 would go on the corresponding bonds to complete the model. In Figure 4.12c a T network is shown, again in resistive form. The bond graph in part d is a "series, parallel, series" type. We observe that part d can be obtained from part b by switching the role of 0 and 1, and part b can be obtained from part d similarly. The formal idea behind the switching is that if the roles of voltage and current are interchanged, a *dual* network results.* In terms of bond-graph structure, this implies a switching of 0 and 1 elements. By this technique the topological dual of a complex circuit may be obtained from its bond graph in a very simple fashion.

---

* See for example [1] p. 267.

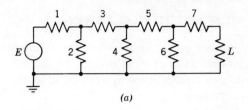

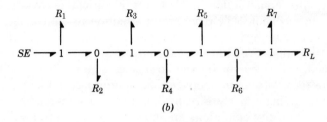

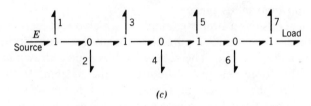

Figure 4.11. A ladder network example.

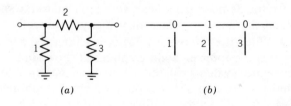

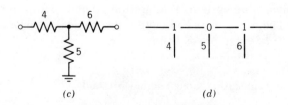

Figure 4.12. Pi and T networks.

89

4.2   MECHANICS OF TRANSLATION

Two of the common types of systems in engineering mechanics involve the translation and rotation of rigid bodies. In this part we shall make bond-graph models of translational systems, and in the next section we study fixed-axis rotation.

The systematic modeling procedure for mechanical translation is similar to the construction method for electrical networks in many respects. In mechanics we pay careful attention to the (scalar) velocity components including the inertial reference, denoting each distinct velocity by a 1-junction. Care must be taken to ensure that velocities of bodies are defined properly with respect to inertial space. Elements involving forces are then "strung in" between the appropriate velocity pairs using 0-junctions, giving concreete realization to the action–reaction principle of Newton. Velocity conditions are imposed by sources of velocity, and the velocity reference junction may be deleted along with its bonds. Simplification then takes place.

First let us consider a simple example in which direct inspection can be used to check the result. Figure 4.12a shows a mass-spring-damper oscillator in a gravity field. The schematic has been redrawn in part b to emphasize the force-velocity structure of the system. Velocity V_1 is defined with respect to inertial space, and reference velocity V_{ref} has been set to zero. Then the ground shown can be considered at rest in inertial space. Each component is modeled separately in part c. Velocity V_{ref} is given by a velocity source; the spring generates a force related to the integral of the velocity difference between V_{ref} and V_1 (the relative displacement); the damper generates a force related to the same velocity difference; and the mass moves with velocity V_1 directly, as does the gravity force. When the reference velocity is eliminated and the graph is simplified, we get the compact bond-graph result shown in Figure 4.13e.

Many mechanical systems will require no more than an extension of the direct method just used in order to obtain a bond-graph model. In fact, with experience it will become possible for you to produce the final result directly, skipping all the intermediate steps in simple cases. However, there will always be systems that are too complicated to be reduced by inspection. Therefore, we shall state a mechanical translation construction method in five steps, as follows:

Mechanical Translation Construction Method

1. For each distinct velocity establish a 1-junction. Some 1-junctions will represent absolute velocities and some relative velocities.

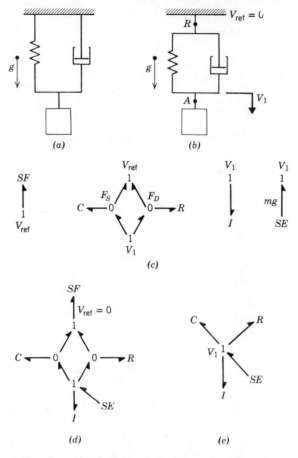

Figure 4.13. A mechanical translation example. Inspection method.

2. Insert the 1-port force-generating elements between appropriate pairs of 1-junctions, using 0-junctions; also, add inertias to their respective 1-junctions (be sure they are defined properly with respect to an inertial frame).

3. Assign all power directions.

4. Eliminate any zero velocity 1-junctions and their bonds.

5. Simplify the resulting graph by condensing 2-port 0- and 1-junctions into bonds.

Now let's apply this procedure to the example considered previously (Figure 4.13b). Figure 4.14a presents the schematic again. In part b the

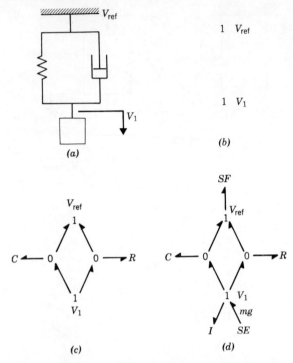

Figure 4.14.   Mechanical translation example. Construction method.

1-junctions have been established. Both 1-junctions represent absolute velocities. The spring and damper effects have been inserted in part c, while the inertia, gravity, and reference conditions have been added in part d. This is the same graph as that shown in Figure 4.13d, and it simplified to the same result as that given in Figure 4.13e.

There is an interesting graph which is equivalent to that of Figure 4.14d, but is simpler in some respects. Observe that the effective velocity on the spring is $(V_1 - V_{ref})$, and the effective velocity on the damper is $(V_1 - V_{ref})$, as well. Then we might use a single 0-junction to "take the velocity difference" as shown in Figure 4.15a. To "distribute" the common velocity difference to both C and R we use a 1-junction, as shown in part b. This last 1-junction represents a relative velocity. The bond graph still has four junctions, but now it has one fewer bonds and one fewer loops. Many bond graphs can be simplified by the use of this simple equivalence procedure when the power directions (i.e., signs) permit it.

For the sake of variation let's interpret the bond graph of Figure 4.15b back into a physical example. There is an inertial velocity, V_1, with an

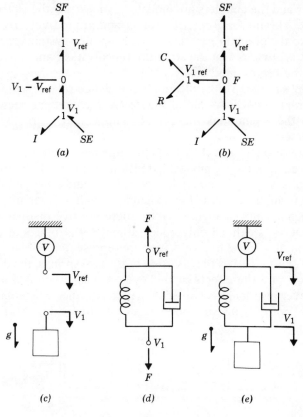

Figure 4.15.   Equivalent model for the oscillator example.

inertia and gravity force present. There is a reference velocity, V_{ref}, defined with respect to the inertial frame. Put it in as a velocity source, as in Figure 4.15c. A spring and damper share the same velocity (difference), and their forces are summed; the net force, F, acts at V_1 and V_{ref}, as shown in part d. What we get when the pieces are assembled (Figure 4.15e) should not be a shock; it is a slight generalization of the system we originally considered.

Next we shall apply the construction method to model a string of railroad cars impacting on a "snubber." The schematic of Figure 4.16a shows the system just as contact occurs. The applied force is due to an engine, and it may be zero. We shall include it in the model for generality.

As the first step we establish a 1-junction for each distinct inertial velocity, as shown in part b of Figure 4.16. Notice that the ground velocity is shown, even though we shall later take it to be zero. In part c

the snubber and the couplers are modeled by inserting the elements using 0-junctions. The inertias, external force, and ground velocity condition are added to the graph in part *d*, and the simplified system graph is given in part *e*. It is obvious that the structure repeats itself and any number of cars could be represented easily.

If we apply the loop-simplification equivalence rule developed earlier, the bond graph reduces to that of Figure 4.16*f*. Now the structure has no loops, and the 1-junctions for C_{12}, R_{12} and C_{23}, R_{23} directly represent relative velocities.

In addition to mechanical systems having one-directional translational motion, it requires only a modest extension of our methods to deal with translational systems containing levers, pulleys, and other simple motion-force transforming devices. For example, a pulley-lever mechanism is shown schematically in Figure 4.17*a*. There are five distinct velocities; observe that V_2 is defined only once, although it also acts at point P in the direction of V_1. The five velocities are represented in Figure 4.17*b* by 1-junctions. Part *c* shows the insertion of the two springs, the damper, the force source, and the effects of the lever. Because points 3 and 4 have different lever arm lengths two separate transformer elements are used: one relates V_3 to V_2, and the other relates V_4 to V_2. The next two steps

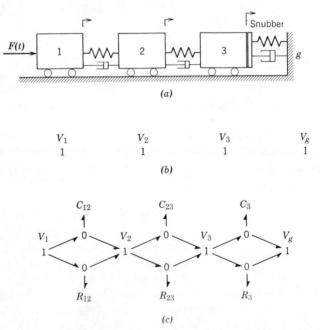

Figure 4.16.   Model of impacting train.

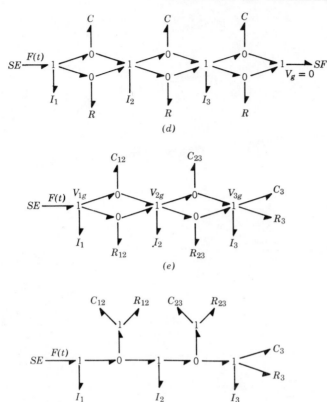

(d)

(e)

(f)

Figure 4.16. *(cont'd)*

involve eliminating the reference velocity junction and its bonds, and simplifying. When these steps have been taken, the resulting system graph is given in Figure 4.17d.

One of the valuable features of a mathematical model in graphical terms is that it can be added to and modified easily. For example, in Figure 4.18a a mechanical system is depicted that is closely related to the one just considered (see Figure 4.17a). The major changes are that a pulley has been removed and a mass has been added at the left end of the lever. A simplified system graph is given in Figure 4.18b. Compare the system bond graphs of Figures 4.17d and 4.18b. The only structural difference is the addition of the I element to the new graph. The zero velocity source appears as a special case of the arbitrary velocity source. For convenience the bonds have been numbered.

Since we are dealing with springs it is necessary to take into account

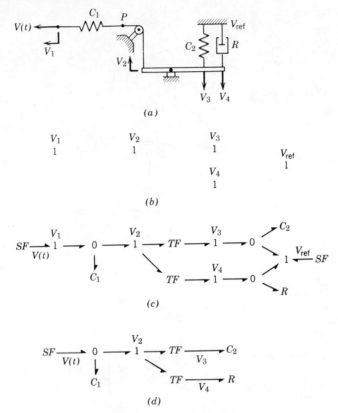

Figure 4.17. A pulley-lever mechanism.

their free lengths. The required information about geometry is given in Figure 4.18c. The free length of each spring is 12 in.; the distances between successive connection points on the lever arm are 12, 18, and 6 in. This means that if the system is put into (or achieves) this configuration with zero velocities everywhere, it will remain in this configuration since we neglect gravity.

The parameters of interest for this mechanism are:

C_1, stiffness is 15.0 lb/ft;
C_2, stiffness is 7.5 lb/ft;
I_3, mass is 0.33 slugs ($= 0.33$ lb-sec^2/ft);
R_8, resistance is 0.2 lb-sec/ft;
TF_{62}, modulus is 1.50 [$V_2 = (1.5)V_6$; $F_6 = (1.5)F_2$];
TF_{78}, modulus is 2.00 [$V_8 = (2.0)V_7$; $F_7 = (2.0)F_8$].

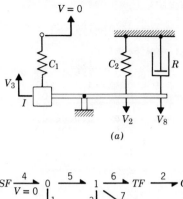

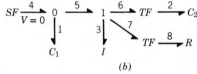

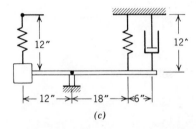

Figure 4.18. An oscillating mechanism related to Figure 4.17a.

The specification of the bond graph and the element parameters to ENPORT is shown in Figure 4.19. For the springs the stiffnesses are given; for the inertia the inverse mass value is used; for the damper the resistance is specified; for the transformers their moduli are given; the other elements require no values at this time. (Of course, junctions never do.) The program responds by indicating the power directions it has assigned to the graph. This completes the structural specification of the model.

In order to study the dynamic behavior of this mechanism we must specify a value for the velocity source. This is given as zero (see "constant sources") in Figure 4.19, indicating that the upper end of spring one will be held fixed. Finally, we must give an initial state for the entire system. This is indicated as spring one in extension by 1 in. (0.083 ft),

```
ENPORT IS IN CONTROL.

LEVER MECHANISM --- FIG.4.11A ... SYSTEM DYNAMICS

THE GRAPH WILL NOW BE READ.
   1  C          1 0 0 0 0    1.500E 01
   2  C          2 0 0 0 0    7.500E 00
   3  I          3 0 0 0 0    3.000E 00
   4  R          8 0 0 0 0    2.000E-01
   5  SF         4 0 0 0 0    0.000E 00
   6  0          4 5 1 0 0    0.000E 00
   7  1          5 3 7 6 0    0.000E 00
   8  TF         2 6 0 0 0    6.700E-01
   9  TF         7 8 0 0 0    2.000E 00


GREND

THE GRAPH HAS    9 NODES AND    8 BONDS.
```

	POWER	
BOND	FROM	TO
1	0	C
2	TF	C
3	1	I
4	SF	0
5	0	1
6	1	TF
7	1	TF
8	TF	R

```
THE CONSTANT SOURCES ARE

0.0000E 00

THE INITIAL STATE IS

8.3000E-02 -1.2450E-01   0.0000E 00



STEADY-STATE X-VECTOR IS
0.0000E 00   0.0000E 00   0.0000E 00
```

Figure 4.19.   ENPORT dialog for Figure 4.18.

spring two in compression by 1.5 in. or 0.1245 ft (hence, negative extension), and the inertia not moving (thus zero momentum). Such a situation can arise if the left end of the lever is pushed down 1 in. from the configuration of Figure 4.18c and held. If we suddenly release it, the mechanism "vibrates."

The program responds by describing the steady-state condition for the system. The three state variables, already used as initial conditions, are the extension of spring one (C_1) from free length, the extension of spring two (C_2) from free length, and the momentum of the inertia (I_3). Other output data (not shown) informs us of that, and it confirms what we anticipate from the model. Recall from the diagram in Figure 4.18c that when both springs are at free length (zero extension) and the inertia has no velocity (hence zero momentum), then the system is in a configuration in which it can and will remain.

The response of the mechanism is shown in Figure 4.20a, in which the extensions of the two springs are plotted versus time. The extensions are measured in feet, and time is measured in seconds. Clearly, the variable starting at 0.083 refers to spring one, and the variable starting at -0.1245 refers to spring two. The extensions oscillate back and forth in harmony, eventually dying down to zero. They die down because the damper is always extracting energy from the system and dissipating it; eventually all the system energy has been extracted. While the springs are stretching and compressing, the inertia is also moving. The momentum of the inertia (in units of lb-sec) is plotted versus time (in seconds) in Figure 4.20b. The inertia starts off with relatively big swings in momentum (hence relatively big swings in velocity) which also die out as time goes on.

There is much to be learned from careful study of traces such as those of Figure 4.20. What is the approximate period, or time between successive peaks, of the mechanism's oscillations? How are the spring extensions related to the inertia motion? How fast does the energy leave the system? What effects on dynamic system response will altering the inertia value have? The damping factor? The ability to answer such questions with reasonable quantitative estimates is an important element in your capability to do good dynamic system design and evaluation.

4.3 FIXED-AXIS ROTATION

Mechanical systems involving fixed-axis rotation occur in the study of machinery of many types and are very important practically. Fortunately, their modeling is not too difficult. The same construction procedure used in translation needs only to be modified slightly to apply to these systems. We shall omit a discussion of the inspection approach since you will naturally develop a "feel" for it as you do examples.

Mechanical Fixed-axis Construction Method

1. For each distinct angular velocity establish a 1-junction. Some angular velocities will be defined with respect to inertial space, and some will be relative angular velocities.

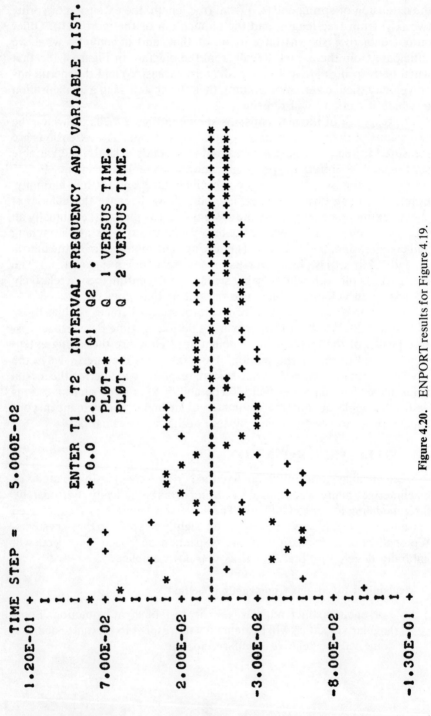

Figure 4.20. ENPORT results for Figure 4.19.

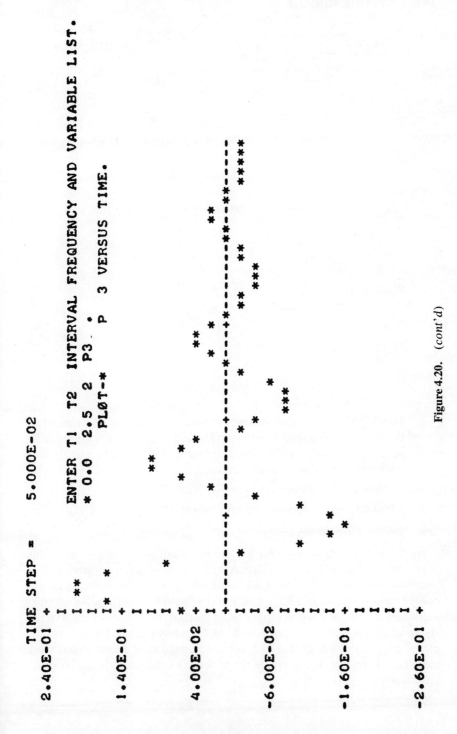

Figure 4.20. (cont'd)

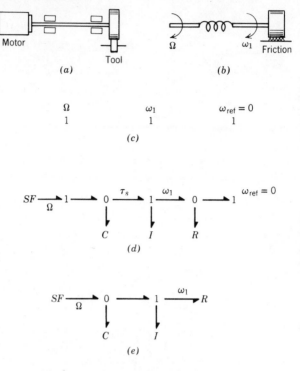

Figure 4.21.   A grinding-wheel model.

2. Insert the torque-generating components into the graph using 0-junctions; also, add inertias and velocity constraints directly. Be sure than angular velocities associated with inertias are defined with respect to an inertial frame.

3. Assign power directions.

4. Eliminate the zero velocity ground reference and its bonds.

5. Simplify the bond graph.

As the first example consider a grinding wheel at the end of a flexible shaft, as shown in Figure 4.21a. A schematic is given in part b that assumes that the motor acts as a speed source, the shaft only has compliance, and the grinding action is dissipative in nature. All angular velocities are defined with respect to an inertial reference, so the explicit showing of the zero reference is done for convenience in part c. The elements are added in part d just as their translational counterparts were handled previously. The graph is simplified, and the result is the compact structure of Figure 4.21e.

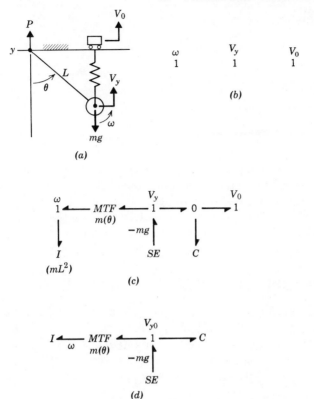

Figure 4.22.   A nonlinear oscillator.

We do not deal extensively in the early parts of this book with nonlinear mechanical systems.* Nonetheless, let us try to model one just to see that our methods are quite applicable. Figure 4.22a shows a mass fixed at the end of a link that can rotate about P in the plane of the paper. Gravity acts to draw the mass down, and a sliding spring acts to draw it up. Without friction the system can oscillate nonlinearly for an indefinite time. In part b the three distinct velocities—ω, V_y, and V_0—are shown. Part c states that the rotary inertia (mL^2) moves with angular velocity ω; gravity (mg) acts with velocity V_y, and the spring state depends on both V_y and V_0. In addition, there is a geometric relation between V_y and ω that depends on

* This discussion may be skipped without impeding the main development.

θ. To see this we may write

$$y = L \cos \theta, \tag{4.1}$$

$$\frac{dy}{dt} = V_y = L \sin \theta \left(\frac{d\theta}{dt}\right), \tag{4.2}$$

and

$$V_y = (L \sin \theta)\omega. \tag{4.3}$$

That is, V_y depends on ω in the usual transformer manner, but the modulus is now a function of θ, rather than a constant. Hence, we write *MTF*, for Modulated Transformer, to relate ω and V_y. Simplification reduces the graph to that shown in Figure 4.22*d*, which may be recognized as a nonlinear oscillator with a forcing term. The notation for the *MTF* modulus, $m(\theta)$, reminds us that it is not a constant.

4.4 HYDRAULIC SYSTEMS

Fluid mechanical systems are among the most difficult to model, analyze, and predict with regard to dynamic behavior. However, there is a subset of fluid mechanical systems—hydraulic systems—that is very important in engineering practice, and it is hydraulic systems which we shall study in this part. They are relatively simple to model using bond graphs.

Hydraulic systems are characterized by high-pressure, low-velocity flows, and the models we will make, at least initially, are based on the assumption that all fluid power of interest is the product of a pressure times a volume flow rate. Many hydraulic elements, such as rams, pistons, hydraulic pumps and motors, valves, and lines, have useful models based on that assumption.

When volume flow is conserved and the fluid is nearly incompressible, hydraulic systems can be treated as circuits. We can "bookkeep" the quantity of fluid at all times and approach the system with a straightforward construction procedure. The basic procedure is given below.

Hydraulic Circuit Construction Method

1. For each distinct pressure establish a 0-junction.
2. Insert the component models between appropriate 0-junction pairs using 1-junctions; add pressure and flow sources.
3. Assign power directions.

4. Define all pressures relative to reference (usually atmospheric) pressure, and eliminate the reference 0-junction and its bonds.

5. Simplify the bond graph.

Figure 4.23a shows a hydraulic pump-motor unit with a pressure-relief valve bypass. We shall assume the pump to be a flow source, the relief valve and motor both to be resistances, and the reservoir to be at atmospheric pressure. For now we shall not deal with the fact that both the pump and the motor transduce hydraulic and mechanical powers. We also shall ignore the filter. In Figure 4.23b the two distinct pressures are identified. In part c the pump, valve, and motor are inserted. This is a correct bond-graph model. Next we define the pressure at B with respect to atmospheric and eliminate the P_A junction. Simplification yields the system bond graph of Figure 4.23d. We are now dealing with "gage" pressures.

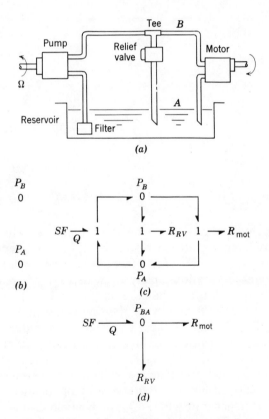

Figure 4.23. A hydraulic pump-motor unit.

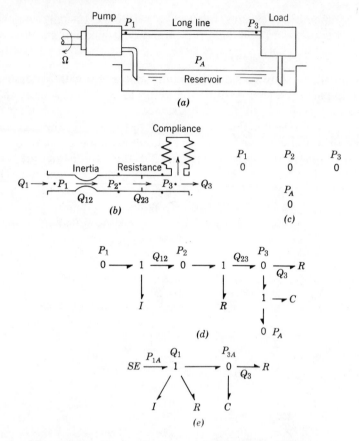

Figure 4.24. Hydraulic system with a long, flexible line.

Frequently, hydraulic elements must be represented which do not show their pressure points quite so distinctly as in the previous example. Consider a length of flexible line connecting a pump to a load. If the line is long, if the pressure varies rapidly or widely, or if great accuracy in prediction is required, it may be necessary to include the effects of fluid inertia in the line, the flexibility of the line walls, and the internal line resistance to flow. In fact, all these effects are distributed throughout the line. Fortunately, it is often possible to get good predictive models by making a few simple assumptions. A long, flexible line is shown in Figure 4.24a. In part b a simple dynamic model indicating inertia, resistance, and compliance is shown; notice the pressure points defined in the process. The inertia acts to generate a flow (Q_{12}) in dynamic response to a pressure difference (P_1, P_2). The resistance also generates a flow (Q_{23}) in response to a pressure difference (P_2, P_3). Wall compliance generates a pressure (P_3) (relative to atmosphere) in dynamic response to a net flow. These pressures are displayed in Figure 4.24c. Elements are inserted as shown in part d. The inertia and resistance effects are inserted directly; the compliance goes between local pressure and atmospheric. Finally, the pump is added as a pressure source, the load as a resistance, and the graph simplified. All pressures are measured relative to atmospheric.

Inspection of Figure 4.24e with regard to the line model itself shows that a simple 1–0 structure with I, R, and C attached is the basic 2-port line representation. By subdividing the line, making such a model for each unit, and then adjoining them port-to-port in cascade, a model of greater accuracy (albeit more complexity) can be constructed. This approach applies equally well to many other types of devices that are primarily distributed devices, such as rods, shafts, plates, and electrical lines. This technique is discussed again in later chapters.

Two related hydraulic devices are shown in Figure 4.25. The first of these, a hydraulic jack, is shown in part a. A highly accurate model that could predict input-output behavior of the device for sudden large loads (e.g., dropping a heavy weight on lift 1) and also be useful for giving exact motion predictions for extremely high pressures would involve considerable study and probably result in a rather complex bond graph. However, the most basic characteristics of the device (indeed, the engineering design purpose) can be modeled very directly by the bond graph shown in Figure 4.25b. The two motions are identified by 1-junctions (V_1 and V_2), and the motions are related by a transformer with modulus $(A_2: A_1)$. The modulus is derived by noting that the fluid is assumed to be incompressible and the device walls rigid, so that a volume change at port 1 must be accompanied by an instantaneous volume change at port 2. Thus

$$V_1 A_1 = V_2 A_2, \qquad \text{or} \qquad V_1 = \frac{A_2}{A_1} V_2. \qquad (4.4)$$

If full power is transmitted from port 1 to port 2, then

$$F_2 V_2 = F_1 V_1, \qquad \text{or} \qquad F_2 = \frac{V_1}{V_2} F_1 = \frac{A_2}{A_1} F_1. \qquad (4.5)$$

You might observe from Eq. (4.5) that F_2/A_2 equals F_1/A_1, implying that the pressure in the working fluid is the same everywhere. This assumption is reasonable if the fluid acceleration forces are small and if gravity forces are unimportant.

A device related to the jack is shown schematically in Figure 4.25c.

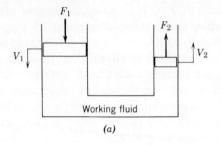

(a)

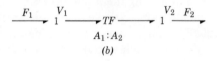

(b)

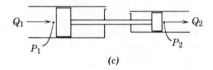

(c)

$$\xrightarrow[Q_1]{\quad} 0 \xrightarrow{\quad P_1 \quad} TF \xrightarrow{\quad P_2 \quad} 0 \xrightarrow[Q_2]{\quad}$$

$A_2 : A_1$

(d)

Figure 4.25.   Two hydraulic transformer devices.

This time the working fluid is on the "outside" of the two pistons, which are joined by a shaft. Behind each piston is free space, indicated by the vented chambers. By making a set of simplifying assumptions, the reasonableness of which must be judged by experience with actual performances of such devices, we can arrive at the bond-graph model shown in part d. The two characteristic equations are

$$P_1 A_1 = P_2 A_2, \tag{4.6}$$

which implies a single force throughout the mechanical part, and

$$\frac{Q_1}{A_1} = \frac{Q_2}{A_2}, \tag{4.7}$$

which implies a single velocity for the part. Taken together, these assumptions are equivalent to modeling a part which has no mass and is rigid. This is a reasonable place to begin.

Probably the most important set of effects to model next would be the friction and fluid leakage effects which cause a loss of power (and performance). These could be represented by introducing dissipation elements (Rs) at key places. We will not pursue this example further here.

Hydraulic systems involve phenomena and devices somewhat more complicated and varied than the electrical circuit and mechanical translation elements that we have dealt with previously. Consequently, the reduction of such systems to bond-graph models sometimes requires the exercise of considerable judgment. As usual, experience is a great teacher in the process. The next section, which is about transducers, will discuss a few other hydraulic devices which play a role in developing purposeful engineering systems.

4.5 SIMPLE TRANSDUCER MODELS

Transducers are devices that couple distinct energy domains. They may have two or more ports, and they may be active or passive devices. Some of the power of bond-graph modeling (no pun intended) begins to emerge when we study electromechanical, hydromechanical, and other mixed domain systems and reduce their initially nonhomogeneous representations to unified bond-graph form.

Several transducers and their simplest useful models are shown in Figure 4.26. Part a shows a single-acting hydraulic cylinder, which relates a hydraulic port to a mechanical translation port. The TF of Figure 4.25d may be made from a pair of these in cascade. Part b shows a bond-graph model of the cylinder, together with the constitutive equations. The key design parameter is the area, "A," and it appears as a transformer modulus. Figure 4.26c is a rack-and-pinion transducer which converts power between translation and rotation. Part d shows a simple 2-port model and its equations. In this case the radius, "R," is the key parameter. A tachometer is shown in Figure 4.26e with two ports. The bond-graph model in part f indicates that, under normal operating conditions, the voltage is proportional to the speed. *And* the torque is proportional to the current. The coupling parameter is "T." Gyrator coupling is very common in electromechanical devices such as generators, motors, and electrodynamic loud speakers. The initial studies for an engineering system design often involve models of this level.

Two basic pump models are shown in Figure 4.27. In part a a positive-displacement type of pump is sketched, and a bond graph is given

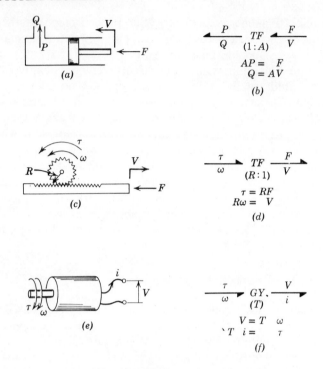

Figure 4.26.   Three 2-port transducers.

in part *b* along with some equations. Notice that the pump is a 3-port (shaft power in, fluid power in, fluid power out) and the bond graph is a 3-port composed of a 2-port *TF* and a 1-junction. A centrifugal type of pump, in which the pressure rise depends on speed, is shown in part *c* with a bond-graph model given in part *d*. Once again a 3-port device may be represented by a 2-port and a junction, although this model should only be considered as an incremental one about an operating point since it is linear. Actual pump characteristics are quite complex and contain important losses which should not be ignored.

Now let us flex our mental muscles by considering a positioning system for a machine-tool table bed that involves electrical, hydraulic, and mechanical power. Figure 4.28*a* shows a motor-driven pump that provides the primary power in the form of a closed hydraulic circuit. The hydraulic motor has a valve control ahead of it to control the power it delivers to the pinion shaft, which in turn drives the table bed. The fluid return line has a filter in it. The bond-graph model in part *b* reflects the overall structure without going into details of component models.

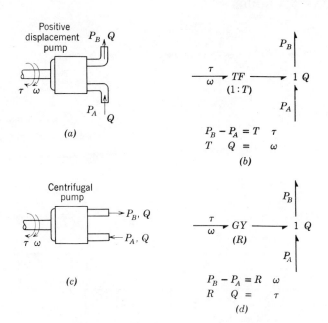

Figure 4.27. Two pump transducers.

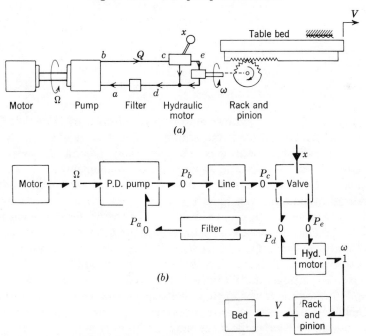

Figure 4.28. A table-positioning system.

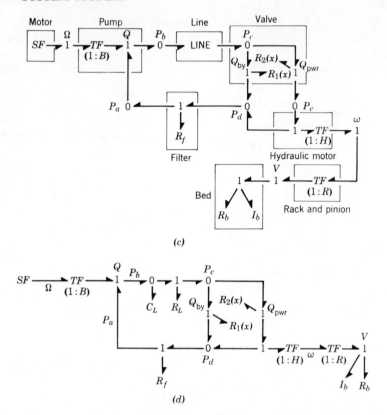

Figure 4.28. (*cont'd*)

In Figure 4.28*c* detailed models are developed for most of the components. The electric motor is taken to be a speed source, S_f. The pump is the positive-displacement type, and operates among the ports with variables Ω, P_b, and P_a. The basic pump parameter is B, a transformer modulus with dimensions of Q/Ω, and its flow is Q, working between pressures P_b and P_a. We have not committed ourselves to a detailed model of the line at this stage, but simply introduced the place holder LINE. The control valve is a 3-port (P_c, P_d, and P_e) which splits the flow into Q_{by} (bypass) and Q_{pwr} (power) according to the control lever position (x). The two fluid streams recombine at pressure point P_d, and the total flow passes through the filter (R_f). The hydraulic motor is a "reverse pump," turning the flow Q_{pwr} into shaft speed with transformer modulus, H. Finally shaft speed, ω, is transformed to a linear velocity, V, by the rack-and-pinion arrangement, moving the inertia of the table (I_b) against friction (R_b).

The 2-port LINE has been left undefined to emphasize the idea that we wish to examine the effects that various line models will have on overall system performance predictions. In order to write a complete set of equations or do computing it will be necessary to specify a model for LINE.

The entire system graph has been simplified considerably, but not completely, in Figure 4.28d. All the initial important coupling variables are still displayed explicitly (Ω, P_a, P_b, etc.), but many junctions have been condensed. The line has been modeled as having compliance (C_L) and flow resistance (R_L), but no significant inertia. Now, more detailed studies can commence.

It is not uncommon for a systems engineer to approach a complex system such as the one we have been discussing and find that he is not familiar with one or more of the devices. Nonetheless, the methods illustrated will work very nicely to describe the overall system structure in compatible (bond graph) terms. The specification of quantitative models for the various parts may require study or consultation with

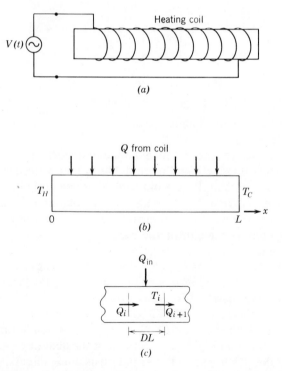

Figure 4.29.   Temperature distribution in a uniformly heated bar.

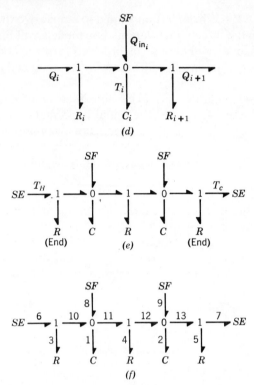

Figure 4.29.   (*cont'd*)

specialists, but the results can all be coordinated into one unified system model at the end.

In concluding this chapter let us examine one more example, this time involving thermal conduction. Figure 4.29*a* shows a uniform bar being heated by an electric coil. Part *b* indicates that one end of the bar is at a hot temperature (T_H) and the other end is at a cold temperature (T_C). We shall assume that the electric coil introduces heat in a uniformly distributed manner along the length of the bar.

To study the temperature and heat flow dynamics of the bar when it is subjected to such conditions requires that we develop a model describing the thermal conduction properties of the bar. A simple element of length DL is isolated in part *c*. It is assumed to have a single uniform temperature (T_i), thermal resistance to heat flow lumped at the two ends, and three significant heat flows (Q_i, Q_{i+1}, and Q_{in}). A bond-graph model of the element is shown in Figure 4.29*d*. Each of the heat flows is associated with a port, the Rs represent thermal resistance effects, and the C represents thermal capacitance of the element (DL in length).

```
M S U   6 5 0 0   ENPORT          VERSION 4.1          UPDATE  50772          12/21/73
-----------------------------------------------------------------------------------------


CONTENTS OF TAPE2  (INPUT)

CARD   1 = [HEADER                                                                    ]
CARD   2 = [FIG. 4.30 UNIFORMLY-HEATED BAR                                            ]
CARD   3 = [GRAPH                                                                     ]
CARD   4 = [C 1 , C 2 , R 3 , R 4 , R 5 , SE 6 , SE 7 , SF 8 , SF 9 , 1 6 3 10 ,      ]
CARD   5 = [0 10 8 1 11 , 1 11 4 12 , 0 12 9 2 13 , 1 13 5 7 .                        ]
CARD   6 = [ASSIGN 7 1 SE .                                                           ]
CARD   7 = [GREND                                                                     ]
CARD   8 = [PARAMETERS                                                                ]
CARD   9 = [C1= 2.0  C2= 2.0  R3= 0.5  R4= 1.0  R5= 0.5                               ]
CARD  10 = [PREND                                                                     ]
CARD  11 = [SOURCES                                                                   ]
CARD  12 = [E6= CONSTANT 100.0                                                        ]
CARD  13 = [E7= CONSTANT  20.0                                                        ]
CARD  14 = [F8= CONSTANT 50.0                                                         ]
CARD  15 = [F9= CONSTANT  50.0                                                        ]
CARD  16 = [SREND                                                                     ]
CARD  17 = [VARIABLES                                                                 ]
CARD  18 = [F3 F4 F5                                                                  ]
CARD  19 = [VREND                                                                     ]
CARD  20 = [SOLUTION                                                                  ]
CARD  21 = [TIN= 0.0  TFIN= 2.5  DT= 0.05                                             ]
CARD  22 = [ICS= 20.0 20.0 .                                                          ]
CARD  23 = [BPLOT 0.0 2.5 1 Q1 Q2 E6 E7 .                                             ]
CARD  24 = [BPLOT 0.0 2.5 1 F3 F4 F5 F8 .                                             ]
CARD  25 = [SLEND                                                                     ]
CARD  26 = [JOBEND                                                                    ]

END-OF-FILE
```

Figure 4.30. ENPORT-4. Input for uniformly heated bar example.

Figure 4.29e gives a bond-graph model of the entire bar and its environment based on two "lumps." A labeled, directed bond graph is shown in part f.

Figure 4.30 gives a description of the entire problem to ENPORT-4 in batch mode. Cards 4 and 5 define the bond graph; card 9 defines the capacitance and resistance parameter values; cards 12 and 13 specify the left-end $(x = 0)$ temperature (T_H) as 100.0 and the right-end $(x = L)$ temperature (T_C) as 20.0, respectively; the heat inputs from the coil (Q_{in}) are given as 50.0 by cards 14 and 15. The other cards are control cards used to aid in the program processing of the physical data. Initial conditions (thermal energies) are given by card 22.

The following table of variable correspondences might be helpful in interpreting output results.

ENPORT variable	Physical interpretation
$Q1, Q2$	Internal energy of lump 1, lump 2
$E6, E7$	End temperatures, T_H and T_C
$F8, F9$	Coil heat flow inputs
$F3, F4, F5$	Bar heat flows at $x = 0, x = L/2, x = L$, respectively

The temperatures of the lumps 1 and 2 are related to the $Q1$ and $Q2$ energy variables by the capacitance relations, namely,

$$T1 = (C1)Q1 = 2.0\,Q1, \tag{4.8a}$$

$$T2 = (C2)Q2 = 2.0\,Q2. \tag{4.8b}$$

(Note that $C1$ and $C2$ are inverse capacitance values.)

In Figure 4.31 a number of intermediate results are shown. The state variables are identified as $Q1$ and $Q2$ (independent energy variables). The A and B matrices are given, implying the state equations

$$\dot{Q}1 = -6.\,Q1 + 2.\,Q2 + 2.\,E6 + 1.\,F8, \tag{4.9a}$$

$$\dot{Q}2 = \ \ 2.\,Q1 - 6.\,Q2 + 2.\,E7 + 1.\,F9, \tag{4.9b}$$

which the reader may verify independently.

The thermal system is diffusive in nature, characterized by two real, negative roots; the associated time constants, being the inverses of the respective roots, are also given in Figure 4.31. Output equations for the

THERE ARE   9 FIELD BONDS AND   4 JUNCTION BONDS.

THE BONDS INTERNAL TO THE JUNCTION STRUCTURE ARE --
13 12 11 10

THE 2 INDEPENDENT ENERGY VARIABLES ARE --

Q 1 Q 2

THE 3 DISSIPATION VARIABLES ARE --

F 3 F 4 F 5

THE 4 SOURCE VARIABLES ARE --

E 6 E 7 F 8 F 9

THE A MATRIX

```
 -6.000E+00   2.000E+00
  2.000E+00  -6.000E+00
```

THE B MATRIX

```
 2.000E+00   0.          1.000E+00   0.
 0.          2.000E+00   0.          1.000E+00
```

THE EIGENVALUES OF THE A MATRIX...

NUMBER	REAL PART	IMAGINARY PART	TIME CONSTANT
1	-8.0000E+00	0.	-1.2500E-01
2	-4.0000E+00	0.	-2.5000E-01

THE 3 OUTPUT VARIABLES ARE --

F 3   F 4   F 5

THE C MATRIX.
```
 -4.0000E+00  0.
  2.0000E+00 -2.0000E+00
  0.          4.0000E+00
```

THE D MATRIX.
```
 2.0000E+00  0.          0.          0.
 0.          0.          0.          0.
 0.         -2.0000E+00  0.          0.
```

THE STEADY STATE X-VECTOR IS...
```
 .5250E+02       .3250E+02
```

THE STEADY STATE Y-VECTOR IS...
```
-.1000E+02       .4000E+02      .9000E+02
```

Figure 4.31.   Intermediate results for heated bar example.

117

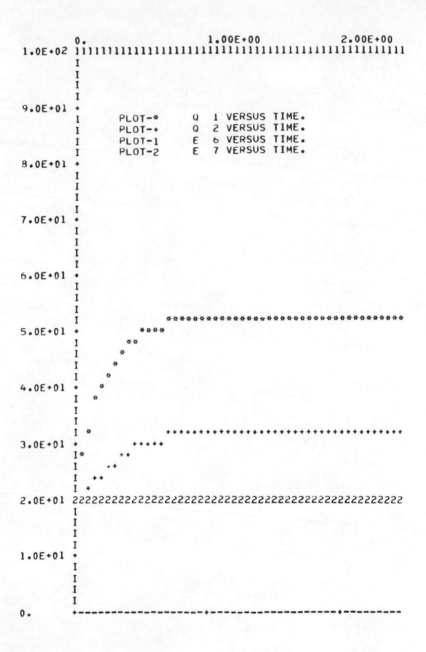

Figure 4.32. Temperature transients for heated bar example.

118

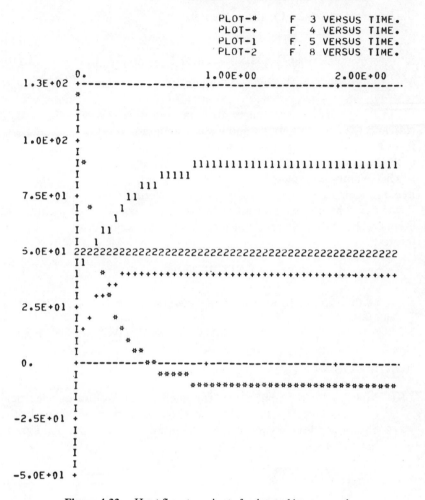

Figure 4.33. Heat flow transients for heated bar example.

heat flows ($F3, F4, F5$) are implied by the C and D matrices, where

$$F3 = -4.\,Q1 \qquad\qquad + 2.\,E6, \qquad\qquad (4.10a)$$

$$F4 = \;\;2.\,Q1 - 2.\,Q2, \qquad\qquad\qquad (4.10b)$$

$$F5 = \qquad\qquad 4.\,Q2 \qquad -2.\,E7. \qquad\quad (4.10c)$$

This result may also be checked independently.

The system settles into a steady state for which all derivatives are zero (i.e., $\dot{Q}1 = 0$ and $\dot{Q}2 = 0$); the values are

$$Q1 = 52.5 \qquad (\text{and } T1 = 105.0), \qquad\qquad (4.11a)$$

$$Q2 = 32.5 \qquad (\text{and } T2 = 65.0). \qquad\qquad (4.11b)$$

The steady-state heat flows are given at the bottom of Figure 4.31 (steady-state Y vector). The exponential form is used for the numbers.

Finally the energy and temperature transients are shown in Figure 4.32 ($Q1$ and $Q2$ are plotted; $T1$ and $T2$ are implied), with the end temperatures (T_H and T_C) plotted as references. Figure 4.33 shows the heat flows, with the coil heat input ($F8$) plotted for reference.

This example completes our current study of the use of bond graphs to make models of various physical and engineering devices and systems. In the next chapter, we present methods by the use of which the intrepid bond grapher can derive suitable state and output equations for models of interest.

REFERENCES

1. A. G. Bose and K. N. Stevens, *Introductory Network Theory*, N.Y.: Harper and Row, 1965.
2. R. Rosenberg, *A User's Guide to ENPORT-4*, N.Y.; J. Wiley & Sons, 1974.

PROBLEMS

4-1   Make a bond-graph model for each of the following electrical circuits. Use the inspection method whenever possible. If the circuit has voltage and current orientations, make the bond graph have equivalent power directions.

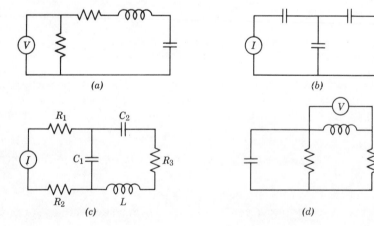

(a)   (b)

(c)   (d)

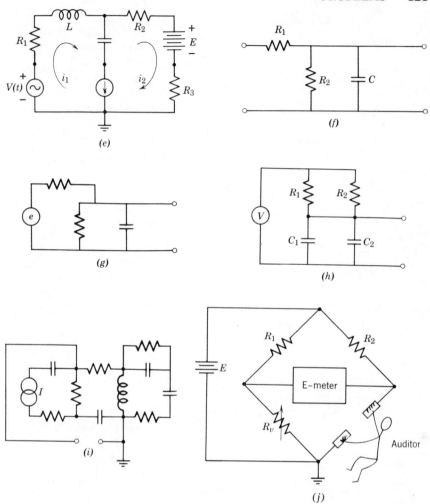

(e)

(f)

(g)

(h)

(i)

(j)

Note. The E-meter may be modeled as an open-circuit pair, and the auditor may be modeled as a resistance, R_A.

4-2 Make a bond-graph model of each of the electrical networks shown below.

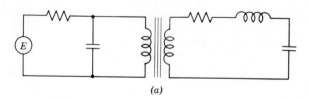

(a)

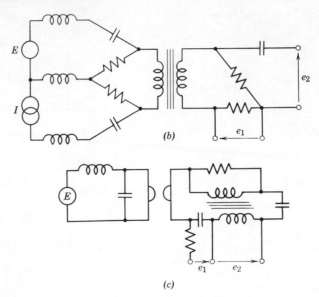

(b)

(c)

4-3 Make a bond-graph model of each of the translational mechanical systems shown below.

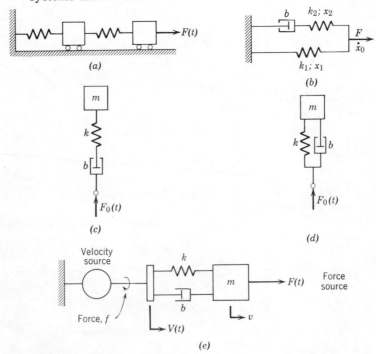

(a)

(b)

(c)

(d)

(e)

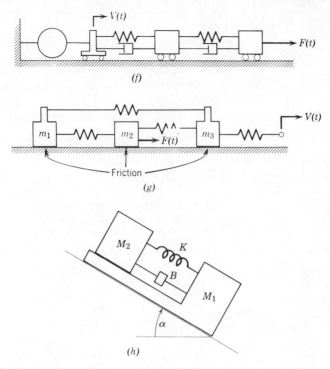

(f)

(g)

(h)

4-4 Find a bond-graph model for the following mechanical networks.

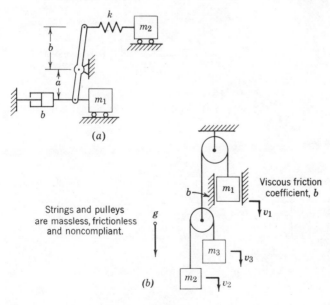

(a)

Strings and pulleys
are massless, frictionless
and noncompliant.

Viscous friction
coefficient, b

(b)

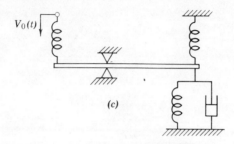

$V_0(t)$

(c)

4-5 For the following mechanical systems involving fixed-axis rotation, find a bond-graph model in each case.

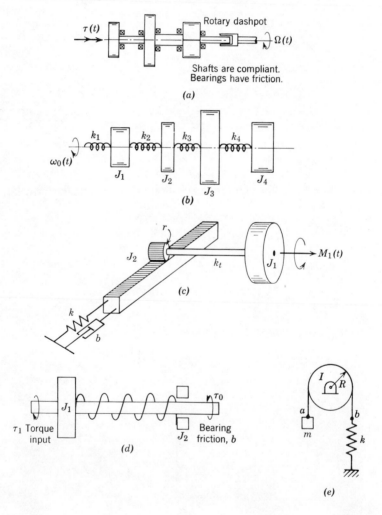

$\tau(t)$

Rotary dashpot

$\Omega(t)$

Shafts are compliant.
Bearings have friction.

(a)

$\omega_0(t)$ k_1 k_2 k_3 k_4

J_1 J_2 J_3 J_4

(b)

r

J_2

k_t

J_1

$M_1(t)$

(c)

k

b

J_1

τ_1 Torque input

τ_0

J_2 Bearing friction, b

(d)

I R

a b

m

k

(e)

In the rack and pinion system shown several frictionless guides and bearings are omitted for clarity.

4-6 Two hydraulic systems that are closely related, but not identical, are shown below. Make a bond-graph model for each.

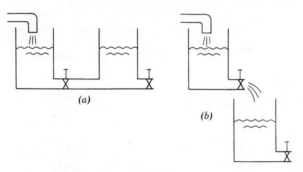

4-7 A simplified model of a water storage system is shown below. Assume the three tanks behave as (nonlinear) capacitances. Assume the three conduits have resistance only.

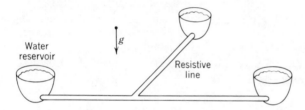

(a) Make a bond-graph model.
(b) Introduce inertance effects for each of the conduits into your model for (a).

4-8 In the hydraulic system shown below the pipes containing flows Q_4 and Q_5 have both inertia and friction effects present. The outflow line (Q_6) has only friction effects. The control pump maintains a desired flow, Q_C, independent of pressure. Make a bond-graph model of the system.

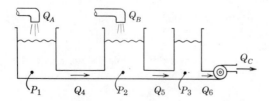

4-9 In the positive-displacement pump shown below, the piston moves back and forth and the check valves act like unsymmetrical resistances, allowing relatively free forward flow, while impeding the back flow greatly. Inertia effects are important in the outlet line, while the nozzle is a resistive restriction.

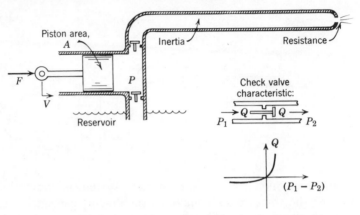

(a) Make a bond-graph model that can be used to relate the mechanical port variables (F, V) to the fluid port variables (P, Q).

(b) Noting that an electrical diode as shown below is analogous to a fluid check valve, use the bond graph of part (a) to find an equivalent electrical network to the pump system.

4-10 Consider a permanent magnet generator of the type often used to power bicycle lights.

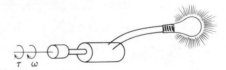

(a) Suppose the generator to be an ideal gyrator in its transducer action. Make a bond-graph model of the system.

(b) Suppose the lamp short-circuits. Is your model from (a) still useful? If not, modify it so that it can handle this situation.

4-11 Two related hydraulic devices are shown in Figure 4.25, a hydraulic jack and ram.

(a) If the inertia of the sliding parts and friction at the seals are important, modify the models given in the figure for the devices to include these effects.

(b) If compliance of the working fluid for the jack and of the shaft for the ram is important, modify your models from (a) above to include these effects.

4-12 A schematic diagram of a basic d-c machine is shown below. The principal electrical effects are armature resistance and inductance

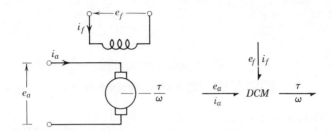

and field resistance and inductance. The coupling between armature and shaft powers is modulated by the strength of the magnetic field, set by i_f. A simple assumption for coupling is

$$\tau = T(i_f)i_a = (Ki_f)i_a,$$

and

$$e_a = T(i_f)\omega = (Ki_f)\omega,$$

where $T(i_f)$ is the modulating value, assumed linear with i_f.

(a) Make a bond-graph model of the basic d-c machine, using a modulated gyrator (MGY) as the heart of the coupling.

(b) Introduce mechanical rotational inertia and dissipation effects into your model from (a) above.

(c) Operate your model from (b) in the generator mode. What changes must be made?

(d) Calculate the power efficiency for the d-c machine operating as a motor and as a generator.

4-13 The network shown below includes a voltage-controlled current

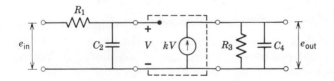

source. Make a bond-graph model of the network, using an active 2-port as the coupling element.

4-14 A load of mass M is suddenly dropped onto piston 1 from height, L.

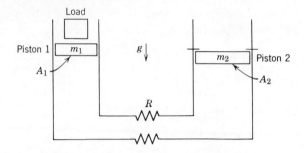

(a) Make a bond-graph model of the system, assuming piston 2 is locked in place by pins, Also assume the load stays joined to piston 1.

(b) If piston 2 is not locked in place, but pistons 1 and 2 were in equilibrium before the load is dropped, modify the model from (a) above.

4-15   Model the spring-loaded accumulator device shown below. Include

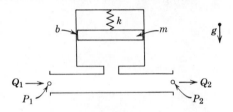

inertia and dissipation effects.

4-16

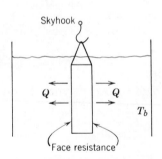

A slab is shown above being quenched in a cooling bath. Assume the bath temperature is maintained constant at T_b. You may neglect heat transfer from the ends and edges.

(a) Make a bond-graph model of the cooling process.

(b) Suppose the water bath is in a tank of volume V_b and the temperature is not maintained constant. Modify the model to cover this case.

5

STATE-SPACE EQUATIONS

5.1 STANDARD FORM FOR SYSTEM EQUATIONS

One of the most remarkable features of bond graphs is that a study of equation formulation may be carried out prior to writing any equations! To understand how this can be so, we first consider some particular forms of equations that are used to represent a system, and select one form—the state-space type—as our goal.

Briefly, we may state that there are two limit forms for systems of differential equations, plus a wide range of possibilities within these limits. An nth-order system may be represented by

1. a single nth-order equation in terms of one unknown variable;

2. n first-order coupled equations in terms of n unknown variables; or

3. various combinations of unknowns and equations of appropriate orders (not necessarily equal).

Many important mathematical problems, methods, and results are organized in terms of the first form. Until relatively recently, almost all engineering mathematics was cast in that form. The second form has certain advantages that have recommended it to mathematicians* for development of theory, but, even more important from our point of view, it is a convenient form for use by engineering system analysts, control engineers, and people conducting digital and analog computer studies. An interesting example of the third form is to be found in the sets of

---

* See, for example, G. D. Birkhoff, *Dynamical Systems*, Providence, R.I.: American Mathematical Society, 1966.

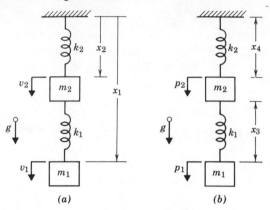

Figure 5.1. Mechanical double-oscillator example.

second-order equations generated by the Lagrangian approach to system analysis.

To illustrate each of the forms, and to give an appreciation of some of the differences, let us consider the mechanical double-oscillator example shown in Figure 5.1. There are several choices of unknowns available. Let us choose x_2 and express the system behavior in terms of that single displacement. The system equation in terms of displacement, x_2, turns out to be

$$\ddddot{x}_2 + \left[\frac{k_2}{m_2} + k_1\left(\frac{1}{m_1} + \frac{1}{m_2}\right)\right]\ddot{x}_2 + \left[\frac{k_1 k_2}{m_1 m_2}\right]x_2 = k_1\left(\frac{1}{m_1} + \frac{1}{m_2}\right)g. \tag{5.1}$$

If a Lagrangian approach is used, with x_3 and x_4 (the spring extensions) shown in Figure 5.1b as the generalized coordinates, the following pair of coupled second-order equations may be developed:

$$\ddot{x}_3 + k_1\left(\frac{1}{m_1} + \frac{1}{m_2}\right)x_3 - \frac{k_2}{m_2} x_4 = 0,$$

$$\ddot{x}_4 - \frac{k_1}{m_2} x_3 + \frac{k_2}{m_2} x_4 = g. \tag{5.2}$$

Finally, if one concentrates closely upon the energy in the system, associating a momentum or displacement variable with each distinct energy element, a set of four coupled first-order equations in terms of p_1, p_2, x_3, and x_4 may be found, as shown in Eq. (5.3). Clearly, this set could also be converted to a geometric-variable form by replacing the momentum variables by the velocities to which they are related.

$$\dot{p}_1 = -k_1 x_3 + m_1 g$$

$$\dot{p}_2 = k_1 x_3 - k_2 x_4 + m_2 g$$

$$\dot{x}_3 = \frac{p_1}{m_1} - \frac{p_2}{m_2} \tag{5.3}$$

$$\dot{x}_4 = \frac{p_2}{m_2}$$

It is possible, in principle, to transform from one of the given forms to any other, whether the system is linear or nonlinear. However, for nonlinear systems the desired transformations may be very difficult to achieve. Furthermore, if one starts with the form of Eq. (5.1), for example, choosing additional unknowns is a rather haphazard process unless one has considerable insight into the system being studied. On the other hand, if a physically meaningful set of variables is already available [e.g., as displayed by Eqs. (5.3)], elimination of unwanted variables can be carried out with considerable insight.

In studying engineering systems using bond graphs there is an ideal opportunity to start the formulation in terms of significant physical variables and to generate simultaneous sets of first-order equations from the bond graphs. This is the approach we shall follow. When the system being studied is nonlinear, the form we seek is given by

$$\dot{x}_1(t) = \phi_1(x_1, x_2, \ldots, x_n; u_1, u_2, \ldots, u_r),$$

$$\dot{x}_2(t) = \phi_2(x_1, x_2, \ldots, x_n; u_1, u_2, \ldots, u_r),$$

$$\vdots \tag{5.4}$$

$$\dot{x}_n(t) = \phi_n(x_1, x_2, \ldots, x_n; u_1 u_2, \ldots, u_r),$$

where the x_i are the *state* variables, the $\dot{x}_i$ are the time derivatives of the x_i, the u_i are *inputs* to the system, and the ϕ_i are a set of static (or algebraic) functions.*

---

* This simply means that, given the values of the arguments on the right-hand side of Eq. (5.4), a set of values for the derivatives may be found by algebraic means.

If the system is linear, Eqs. (5.4) take on a simpler form, as shown below.

$$\dot{x}_1(t) = a_{11}x_1 + a_{12}x_2 + \cdots + a_{1n}x_n + b_{11}u_1 + b_{12}u_2 + \cdots + b_{1r}u_r,$$

$$\dot{x}_2(t) = a_{21}x_1 + a_{22}x_2 + \cdots + a_{2n}x_n + b_{21}u_1 + b_{22}u_2 + \cdots + b_{2r}u_r,$$

$$\vdots \tag{5.5}$$

$$\dot{x}_n(t) = a_{n1}x_1 + a_{n2}x_2 + \cdots + a_{nn}x_n + b_{n1}u_1 + b_{n2}u_2 + \cdots + b_{nr}u_r,$$

where the a_{ij} and b_{ij} are constants in most cases. For a linear "time-varying" system the a_{ij} and b_{ij} may vary with time, but they must not depend on the x variables.

Our task in the rest of this chapter is to learn how to select significant system variables from among the many possible ones available and to organize them in relations of the form of Eqs. (5.4) or (5.5), as appropriate.

5.2 AUGMENTING THE BOND GRAPH

Before writing any equations it is desirable to prepare the bond graph with additional information that will make the writing of equations take on a very orderly pattern. The major steps are threefold:

1. name all the bonds in the graph (i.e., by numbering them consecutively);

2. assign to each bond a reference power direction; and

3. assign to each bond a causal sense for the pair of variables e and f.

Naming the bonds means that it is possible to refer to every variable in the system directly and unambiguously (e.g., e_4, f_7, p_3, q_{11}).

Assigning power directions, if not done in the modeling process itself, may be done in one of two basic ways. Either choose a reference direction in the original system for every power variable (e.g., voltage and current, etc.) and transfer to the bond graph the reference power directions thereby implied, or put the directions on the bond graph directly and interpret the implications in the original system as the need arises. Experience will show that often the latter approach is to be preferred for its ease and simplicity, as well as its high correspondence with typical engineering practice in selecting reference directions.

As an illustration of steps one and two we shall label and assign reference powers to the bond-graph model of the circuit in Figure 5.2.

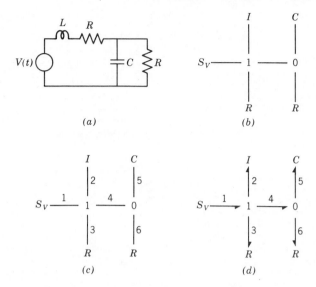

Figure 5.2.   Partially augmented bond graph.

Part *a* of the figure shows an unlabeled circuit, and part *b* shows a corresponding bond-graph model obtained by inspection. In Figure 5.2*c* the bonds have been labeled by numbering them from 1 to 6 in arbitrary order. Now we may conveniently refer to resistance element 3, for example, or inductor current i_2. We may also refer to charge q_5, which is the charge associated with capacitance element 5.

In Figure 5.2*d* a set of power reference directions has been chosen directly on the bond graph. They seem reasonable in the sense that, if all powers are positive at some instant of time, then energy is flowing from the voltage source into each of the storage elements (*I*2 and *C*5), out of the system through resistance elements (*R*3 and *R*6), and generally "down the system from left to right."

To give further insight into the correlation between the physical variables and their bond-graph representation, Figure 5.3 develops a set of oriented voltages and currents corresponding to the directed bond graph of Figure 5.2*d*. In part *a* of Figure 5.3 the bond graph is undirected except for bond 1, which shows power taken positive from the source into the system. Part *b* indicates a corresponding choice of voltage and current orientations for the source branch. In part *c* bond 2 and 3 have been directed; part *d* gives corresponding branch orientations for *I*2 and *R*3. It is important to note that the branch current directions must be the same as that for *V*1 because the 1-junction connection among the several

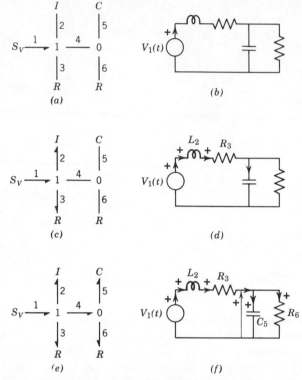

Figure 5.3. Correspondence between a directed bond graph and an oriented circuit.

elements implies that they have a common current. In Figure 5.3e the remaining bonds are directed. Given the current direction associated with the 1-junction, we choose a voltage orientation as shown in part ƒ by the arrow to correspond to bond 4 power. The 0-junction then indicates the branch voltage orientation for branches 5 and 6. Finally, the directions on bonds 5 and 6 lead directly to the current orientations on branches $C5$ and $R6$.

It is rarely necessary to transfer every single bond direction to the circuit. Usually a few key orientations are sufficient. Furthermore, common patterns of bond direction and circuit branch orientation occur repeatedly, making the transferance very routine with practice. Similar remarks and interpretations may be made for bond-graph directions and mechanical variable (i.e., force and velocity, torque and angular velocity) orientations, as well as all other power variable types. To a certain extent the visualization and selection of variable orientations is a personal matter, and we shall not pursue the matter further in a formal fashion.*

At this point, it should be noted that system equations can be written for a bond graph that has been labeled and directed. For a graph with N bonds there are $2N$ bond variables (N efforts and N flows). Each n-port implies n constraints among its associated bond variables. Each bond is adjacent to two multiports. From basic graph theory we can find that the number of bond variables and constraints are always equal (provided there are no open bonds). It is possible to write down all the constraints for each multiport and arrive at a pile of system equations, essentially unsorted and unorganized, but correct. Manipulation of the resulting equations to obtain insight and to achieve a useful form is very difficult in a complicated system in the absence of some algorithm. One reason is the excessive number of variables used in the initial formulation.

For example, in Figure 5.3e the bond graph has six bonds, hence twelve bond variables ($v_1 \cdots v_6$, $i_1 \cdots i_6$). The number of constraints imposed by the elements is as follows: $SE1$ imposes one, $I2$ imposes one (dynamic), $R3$ imposes one, 1 1, 2, 3, 4 imposes four (three identities and one summation), 0 4, 5, 6 imposes three (two identities and one summation), $C5$ imposes one (dynamic), and $R6$ imposes one. Try writing twelve relations among the twelve bond variables and then reducing them to a pair of coupled first-order differential equations. There must be a better way; there is, and we will discuss it next!

The third step in the augmentation process is the *assignment of causality*. Basic considerations of causality for the various elements were introduced earlier (Section 3.4). It is now appropriate to apply such information to the entire system in an orderly fashion.

In a causal sense there are two distinct types of bond-graph elements. Source elements (S_e and S_f) and junction elements (0, 1, TF, and GY) must meet certain causal conditions or their basic definitions are no longer valid. As an obvious example, if a source of force does not have a causality showing that it defines a force on the system 'to which it is connected, then it has no meaning. We generalize this observation and assert that *every source element must have its appropriate causal form assigned to it*.

For each of the 2-port junction elements, TF and GY, there are two possible causal forms that preserve the basic definition of the element. If neither of these forms can be assigned, the concept of input and output

---

* For complex systems, the assignment of sign conventions is only partly arbitrary, with certain forms being impossible for a given physical system. If a complete set of effort and flow conventions is chosen for the physical system and then the information is transferred to the bond graph, no ambiguities will arise.

associated with the element is not valid. Consequently we assert that *every TF and GY must have one of its two allowable causal forms assigned* to it. The choice of form will generally be indicated by the adjoining system on the basis of other considerations to be discussed.

In a similar vein we argue with respect to the ideal multiport junctions 0 and 1 that *each 0 and 1 must have one of its appropriate causal forms assigned* to it or the basic definition of the particular element will not be valid. The selection of particular causal form will be motivated by other system considerations in general.

If a system cannot meet the causal conditions outlined above, the basic physical model on which the bond graph is based must be restudied. The indications are that an impossible situation has been created that is not capable of sensible mathematical resolution. Two such examples are shown in Figure 5.4. In part *a*, source element 2 is defined invalidly. Inspection of an electrical circuit interpretation indicates the nature of the difficulty; obviously, a modeling error has been made in putting two supposedly independent current sources in series. In part *c* of Fig. 5.4, a *TF* is found to have invalid causality. If this bond graph were derived from a fluid circuit, as shown in part *d*, the interpretation would be that two independent pressure sources have been joined by an ideal transformer of pressure (the *TF*), leading to a physically incompatible situation. The next step is up to the system modeler, who must correct the model in an appropriate way.

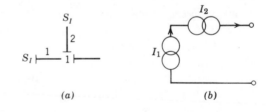

(a)   (b)

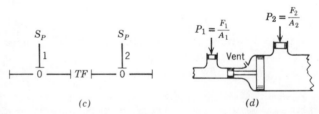

(c)   (d)

Figure 5.4.   Two examples of invalid causality and physical interpretations.

Continuing our discussion of assigning causality, we come to the C and I elements. The energy variables on these elements (p on I, q on C) are the basis of the system state variables. That is, a knowledge of the values of a necessary and sufficient set (i.e., just enough and not too many), together with the inputs, will enable us to predict the system response in time. In an intuitive way we observe that the energy variables can be used to determine the system energy and its distribution, and the bond-graph structural constraints serve to define the powers that cause the energy to flow subject to input conditions, resulting in the particular system dynamics.

As we shall soon see, it is not always possible to make all C and I elements have integration causality. If a storage element has differentiation causality forced upon it, the meaning is that its energy variable is not algebraically independent of other energy variables and source constraints. Hence, it is not an independent state variable and can be eliminated from the final state equations.

At the system level causality associated with R elements is largely a matter of indifference. The major exception is in the case of a nonlinear constitutive law which is not biunique (e.g., Coulomb friction). Then the causality associated with a unique input-output relation for the element should be used. Otherwise, R elements accept whatever causality they are assigned (and are grateful for it).

The basic causality assignment procedure is summarized below.

Assignment of Causality Procedure

1. Choose any source (SE, SF), and assign its required causality. Immediately extend the causal implications through the graph as far as possible, using the constraint elements ($0, 1, GY, TF$).

2. Repeat step (1) until all sources have been used.

3. Choose any storage element (C or I), and assign its preferred (integration) causality. Immediately extend the causal implications through the graph as far as possible, using the constraint elements ($0, 1, GY, TF$).

4. Repeat step (3) until all storage elements have been assigned a causality. In many practical cases all bonds will be causally oriented after this stage. In some cases, however, certain bonds will not yet have been assigned. We then complete the causal assignment as follows:

5. Choose any unassigned R element and assign a causality to it (basically arbitrary). Immediately extend the causal implications

through the graph as far as possible, using the constraint elements
(0, 1, *GY*, *TF*).

6. Repeat step (5) until all *R* elements have been used.

7. Choose any remaining unassigned bond (joined to two constraint
    elements) and assign a causality to it arbitrarily. Immediately
    extend the causal implications through the graph as far as possible,
    using the constraint elements (0, 1, *TF*, *GY*).

8. Repeat step (7) until all remaining bonds have been assigned.

The procedure is straightforward and orderly. Some practice on
examples will convince you of the ease and rapidity with which causality
can be assigned. It is important to recognize that the constraint elements
represent the physical structural ties in the system (e.g., Kirchoff's
voltage and current laws; Newton's law and geometric compatibility), and
assigning causality to them means that they will be used in a particular
input-output fashion correctly. Further discussion of the use and interpretation of causality in bond graphs is given in Chapter 7.

There are several situations that can arise when applying causality
according to the procedure given. They are:

1. All storage elements have integration causality, and the graph is
    completed after step (4). This simple, common case is discussed in
    Section 5.3.

2. Causality is completed by using *R* elements or bonds, as indicated
    in steps (5–8). This situation is discussed in Section 5.4.

3. Some storage elements must have differentiation causality. This
    case is discussed in Section 5.5.

Now let us consider several examples of using the procedure.

Assignment of causality is carried out step by step in the example in
Figure 5.5. Part *a* shows a labeled bond graph without causality. A graph
without causality is sometimes said to be *acausal*. In part *b* bond 1 is
directed according to the meaning of the source element (a source of
flow). Since the 0-junction has only one flow variable determined, the
other bonds cannot yet be assigned a causality. There are no more sources
so we turn to step (3). In part *c* bond 2 is causally directed based on
integration causality for the *C* element. Immediately, bond 3 may be
directed because of the 0-junction, which can have only one effort input.
However, bonds 4 and 5 are as yet undirected. Bond 5 is directed as
shown in Figure 5.5*d* to give integration causality to the *I* element.
Immediately bond 4 may (nay, must!) be causally directed as shown, due
to the 1-junction, which can have only one flow input. In the grand finale

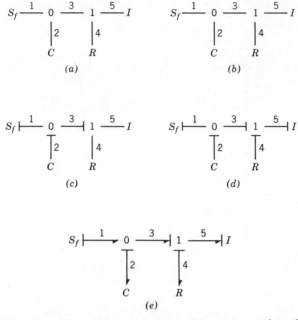

Figure 5.5. Causality assignment and complete augmentation of a bond graph. Example 1.

of part *e* we have added a set of power directions to the graph; the result is a completely augmented bond graph. That is, the bonds are labeled, power directions have been chosen, and causality has been assigned. Such a graph will yield its state equations to us with very little resistance, as we shall show in the next section. However, let us first augment another graph or two to gain some experience.

Figure 5.6 shows a bond graph derived from a fluid example involving pipes and reservoirs. There are three pressure sources (*SE*1, *SE*2, and *SE*3) feeding through three pipes (roughly, the 1-junction complexes) into a tank (0 and *C*10). In part *a* the graph is labeled. In part *b*, bonds 1, 2, and 3 associated with the sources have been assigned causality. In each case no extension of causality is possible. Also, bond 10 has been directed according to integration causality for the *C*10 element, and causality has been extended to bonds 11, 12, and 13 by the 0-junction, using the effort identity condition. In part *c* bond 4 has been directed; consequently, so has bond 7 due to the 1-junction, and so on for bonds 5 and 8 and bonds 6 and 9. Finally, power directions have been added to produce the fully augmented graph shown in part *d*. In the next section, we shall show how this graph yields four first-order equations with three inputs to the system.

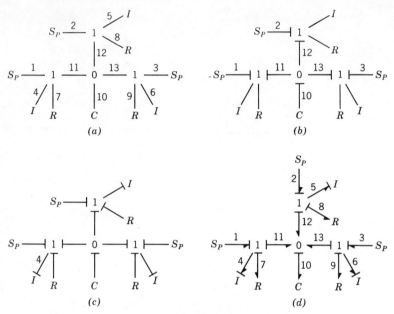

Figure 5.6.   Augmentation of a bond graph. Example 2.

As a final example in this section, consider the acausal bond graph in Figure 5.7*a*. The model is derived from a study of a pressure-controlled valve and includes both mechanical and fluid mechanical power. The elements *SE*1 and *SE*2 represent sources of pressure and force, respectively, and the *TF* element couples the two power domains. In part *b* bonds 1 and 2 are causally assigned, one at a time. The causal information cannot be extended using constraint elements at this point. Bonds 3 and 4 are assigned next and causality extended as shown in part *c*. Bond 4 causality does not extend to other bonds, but bond 3 has implications for bonds 8 and 9, and also for bonds 10 and 11. Inspection of Figure 5.7*d* reveals that assigning causality to bond 5 (associated with element *I*5) determines the causality on bonds 6 and 7. With powers assigned as shown in part *e* the bond graph is completely augmented.

Two further points may be made with respect to augmentation. The first is that, in assigning causality, the results do not depend on the order of bonds chosen except in special circumstances. These will be discussed in Section 5.4. The second point is that assignment of causality and assignment of power directions are two entirely independent operations. Either one may be performed first. Typically, power directions will be first, but on occasion one may not bother to assign powers in studying aspects of system structure.

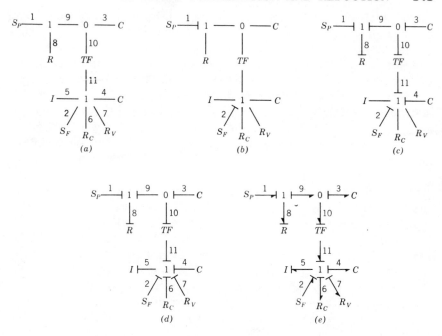

Figure 5.7. Augmentation of a bond graph. Example 3.

5.3 BASIC FORMULATION AND REDUCTION

Once a fully augmented bond-graph model is available, the equations for the system can be developed in a very orderly fashion. Frequently, when the system is relatively small or uncomplicated in structure, state-space equations can be written down directly. However, as system size and complexity grow, the need for an organized procedure for equation generation becomes apparent.

Very general and powerful procedures are available for producing sets of system equations. In this section we shall concentrate on a basic pattern which is applicable in a large majority of cases encountered in engineering practice. There are three simple steps to be followed:

1. select input, energy, and co-energy variables;

2. formulate the initial set of system equations; and

3. reduce the initial equations to state-space form.

Selection of *inputs* is straightforward. For each source element write on the graph the input variable to the system. These variables will appear in the final state-space equations if they have any effect on system behavior. The list of input variables will be called U.

Selection of *state variables* is accomplished by choosing the energy variable for each storage element in the graph. When integration causality can be assigned on all C and I elements we know that each energy variable is statically independent of all of the others.* We choose as our state variables the p variable on every I element and the q variable on every C element. The list of state variables will be called X. On the bond graph we write $\dot{p}$ and $\dot{q}$ on the appropriate bonds, representing the effort and flow corresponding to each p and q.

In addition to writing input and energy variables it will prove useful to write one other set on the graph. This is the co-energy set, consisting of the f on each I and the e on each C. These variables will appear in the initial formulation and then be eliminated in the reduction process.

In Figure 5.8a the augmented bond graph given in Figure 5.5e is repeated. In part b of Figure 5.8 the input flow variable, $F_1(t)$, is introduced. The energy variables for the example are q_2 and p_5. Their derivatives are written on bonds 2 and 5, respectively, as shown in part c. Finally, the co-energy variables, e_2 and f_5, are also written on bonds 2 and 5, respectively, resulting in the graph of part d. From this graph, the equations will be written.

The lists are identified as follows:

$$X = \begin{bmatrix} q_2 \\ p_5 \end{bmatrix} \quad \text{and} \quad U = [F_1].$$

The formulation result we seek is, for a linear system,

$$\dot{q}_2 = a_{11}q_2 + a_{12}p_5 + b_1F_1,$$
$$\dot{p}_5 = a_{21}q_2 + a_{22}p_5 + b_2F_1,$$

$$(5.6)$$

or, if the system is nonlinear,

$$\dot{q}_2 = \phi_1(q_2, p_5; F_1),$$
$$\dot{p}_5 = \phi_2(q_2, p_5; F_1),$$

$$(5.7)$$

as indicated by Eqs. (5.5) or (5.4) with $n = 2$ and $r = 1$.

Our plan is first to write the constitutive relations for the storage elements in appropriate form (i.e., for $C2$ and $I5$). Then we shall write

---

* The case when some storage elements have differentiation causality is discussed in Section 5.5.

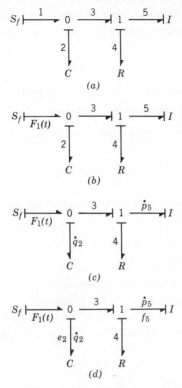

Figure 5.8. Identification of key variables (input, energy, and co-energy). Example 1.

equations for the derivatives of the energy variables, which are bond variables, and try to express them in terms of inputs and co-energy variables. These relations represent the structural constraints of the system, as well as dissipation. This will yield an initial set of equations, which we then shall reduce to state-space form.

Consider element $C2$ of Figure 5.8d. Its constitutive equation in linear form is

$$e_2 = \frac{1}{C_2} q_2. \tag{5.8a}$$

For element $I5$ the linear constitutive equation is

$$f_5 = \frac{1}{I_5} p_5. \tag{5.8b}$$

Now the co-energy variables are expressed in terms of the energy variables and subsequently may be replaced by them.

Turning now to the structure, we seek an expression for $\dot{q}_2$, which is the flow on bond 2. Using the 0-junction constraint and *following the causality assigned* we get

$$\dot{q}_2 = F_1(t) - f_3. \tag{5.9a}$$

(Look at the causal strokes. They indicate that $\dot{q}_2$ is an output of the 0-junction and F_1 and f_3 are inputs. The signs in Eq. (5.9a) follow from the sign half arrows.) But the causality further indicates that f_3 is expressible in terms of co-energy variable f_5, using the 1-junction. So

$$\dot{q}_2 = F_1(t) - f_5. \tag{5.9b}$$

To eliminate f_5 in favor of a state variable we can use Eq. (5.8b). The result is

$$\dot{q}_2 = F_1(t) - \frac{1}{I_5} p_5, \tag{5.9c}$$

which involves only X and U variables.

Let us develop an equation for the other state variable, p_5, in terms of its structural relations. Starting with bond 5 in Figure 5.8d, we identify $\dot{p}_5$ as the effort. So, following causality,

$$\dot{p}_5 = e_3 - e_4, \tag{5.9d}$$

using the effort summation relation of the 1-junction and the sign convention shown. But the causal strokes also show that e_3 may be replaced by e_2, using the 0-junction. This yields

$$\dot{p}_5 = e_2 - e_4. \tag{5.9e}$$

Referring to the causality on bond 4 we see that e_4 may be replaced by its f_4 equivalent as indicated by the R causality. Where does f_4 come from? The 1-junction causality indicated f_5 as the determiner of f_4. This sequence is

$$\dot{p}_5 = e_2 - R_4 f_4 = e_2 - R_4 f_5, \tag{5.9f}$$

Since e_2 and f_5 are both co-energy variables (this is no accident!), they may be replaced in Eq. (5.9f) by their equivalents in Eq. (5.8). Thus,

$$\dot{p}_5 = \frac{1}{C_2} q_2 - R_4 \frac{1}{I_5} p_5. \tag{5.9g}$$

Now p_5 is expressed in terms of X and U variables only.

As a final step, it is useful to order the terms of Eqs. (5.9c) and (5.9g) according to the order of the elements of X and U. This is good practice and leads to a natural representation by matrices in the linear case. Then the desired state equations become

$$\dot{q}_2 = \qquad \frac{1}{I_5} p_5 + F_1(t),$$

$$\dot{p}_5 = \frac{1}{C_2} q_2 - \frac{R_4}{I_5} p_5. \qquad (5.10)$$

To emphasize the pattern, all zero terms have been implied by the spacing. Compare Eqs. (5.10) to Eqs. (5.5) with $n = 2$ and $r = 1$ to see that the basic information we have derived is the sets of a_{ij} and b_{ij} coefficients.

At this point, you might well ask yourself what would happen to the formulation if some constitutive laws were nonlinear. Retrace the formulation steps to see that the same pattern holds but the manipulation is not as simple as for the linear system. For example, let us assume that elements $C2$ and $I5$ are nonlinear, with relations

$$e_2 = \phi_2^{-1}(q_2), \qquad (5.11a)$$

$$f_5 = \phi_5^{-1}(p_5). \qquad (5.11b)$$

Since the structural equations are unchanged we can turn directly to Eqs. (5.9b) and (5.9f) and substitute Eqs. (5.11). The result is

$$\dot{q}_2 = F_1(t) - \phi_5^{-1}(p_5),$$

$$\dot{p}_5 = \phi_2^{-1}(q_2) - R_4 \phi_5^{-1}(p_5). \qquad (5.12)$$

Equations (5.12) may be compared to Eqs. (5.4) to see that we have attained our objective.

A second example is shown in Figure 5.9a, which is the augmented bond graph of Figure 5.3e. The graph with inputs, energy variables, and co-energy variables identified is shown in Figure 5.9b. The input list, U, contains $V_1(t)$; the energy variable list, X, contains λ_2 and q_5; and the co-energy variables are i_2 and v_5.

For convenience, let us assume that the storage and dissipation elements are linear with constant coefficients. Then we may write the

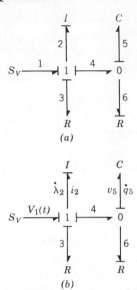

Figure 5.9.   Identification of key variables. Example 2.

constitutive laws for $I2$ and $C5$ as

$$i_2 = \frac{1}{L_2} \lambda_2$$

and   (5.13)

$$v_5 = \frac{1}{C_5} q_5,$$

respectively, where L_2 is the inductance and C_5 the capacitance in the electric circuit in Figure 5.3f.

The derivatives of the energy variables are given by structural Eqs. (5.14c) and (5.14f), which are written by following the implications of the causal strokes and the sign convention until the state variable derivatives are given in terms of co-energy and input variables.

$$\dot{\lambda}_2 = V_1(t) - v_3 - v_4,$$ (5.14a)

or

$$\dot{\lambda}_2 = V_1(t) - R_3 i_3 - v_5,$$ (5.14b)

or

$$\dot{\lambda}_2 = V_1(t) - R_3 i_2 - v_5,$$ (5.14c)

and

$$\dot{q}_5 = i_4 - i_6, \tag{5.14d}$$

or

$$\dot{q}_5 = i_2 - \frac{1}{R_6} v_6, \tag{5.14e}$$

or

$$\dot{q}_5 = i_2 - \frac{1}{R_6} v_5. \tag{5.14f}$$

Following the typical reduction pattern, we use Eqs. (5.13) to eliminate i_2 and v_5 in Eqs. (5.14c) and (5.14f) to obtain unordered state-space equations in terms of the energy variables, given by Eqs. (5.15); namely,

$$\dot{\lambda}_2 = V_1(t) - \frac{R_3}{L_2} \lambda_2 \quad - \frac{1}{C_5} q_5$$

and   (5.15)

$$\dot{q}_5 = \qquad \frac{1}{L_2} \lambda_2 - \frac{1}{R_6 C_5} q_5.$$

When the terms in Eqs. (5.15) are ordered according to X and U, with zero terms implied, we get

$$\dot{\lambda}_2 = -\frac{R_3}{L_2} \lambda_2 - \frac{1}{C_5} q_5 + V_1(t)$$

and   (5.16)

$$\dot{q}_5 = \frac{1}{L_2} \lambda_2 - \frac{1}{R_6 C_5} q_5.$$

One more example should be sufficient to demonstrate the power and consistency of the formulation and reduction pattern in terms of energy variables. Consider the augmented bond graph for a pressure-controlled valve shown in Figure 5.7e. Bonds 1, 3, 8, 9, and 10 represent fluid power and have P, Q variables. Bonds 2, 4, 5, 6, 7, and 11 represent mechanical power and have F, V variables.* The TF has an area modulus, A, that couples the two power domains. Figure 5.10 shows the augmented bond graph of Figure 5.7e with input, energy, and co-energy variables identified. We anticipate a third-order system in terms of the X variables: V_3, a fluid volume; x_4, a spring extension; and p_5, a momentum. The inputs U are $P_1(t)$, a control pressure, and $F_2(t)$, a loading force.

---

* Note that V_3 is a volume and v_5 is a velocity. If you don't like using the letter V for voltage, volume, and velocity, try using f for all flows and e for all efforts.

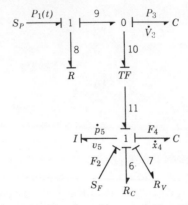

Figure 5.10. Identification of key variables. Example 3.

The X and U lists are

$$X = \begin{bmatrix} V_3 \\ x_4 \\ p_5 \end{bmatrix} \quad \text{and} \quad U = \begin{bmatrix} P_1 \\ F_2 \end{bmatrix}.$$

To begin we write the constitutive equations in the order of X for $C3$,

$$P_3 = \phi_3(V_3); \tag{5.17a}$$

for $C4$,

$$F_4 = \phi_4(x_4); \tag{5.17b}$$

and for $I5$,

$$v_5 = \frac{1}{m_5} p_5. \tag{5.17c}$$

In the above expressions both $C3$ and $C5$ are assumed to be nonlinear, but the mass is taken as linear.

Next we write the X derivative equations for the structure, of which there are three, by following the causal strokes and the sign convention. We always follow the causal marks until $\dot{X}$ is expressed in terms of input or co-energy variables.

$$\dot{V}_3 = Q_9 - Q_{10}, \tag{5.18a}$$

or

$$\dot{V}_3 = Q_8 - A v_{11}, \tag{5.18b}$$

or

$$\dot{V}_3 = \frac{1}{R_8}(P_1(t) - P_3) - Av_5. \tag{5.18c}$$

$$\dot{x}_4 = v_5, \tag{5.18d}$$

$$\dot{p}_5 = F_2(t) + F_{11} - F_6 - F_7 - F_4, \tag{5.18e}$$

$$\dot{p}_5 = F_2(t) + AP_{10} - R_6v_6 - R_7v_7 - F_4, \tag{5.18f}$$

$$\dot{p}_5 = F_2(t) + AP_3 - (R_6 + R_7)v_5 - F_4. \tag{5.18g}$$

Finally, using Eqs. (5.17), we obtain Eqs. (5.19):

$$\dot{V}_3 = \frac{1}{R_8}P_1(t) - \frac{1}{R_8}\phi_3(V_3) - \frac{A}{m_5}p_5, \tag{5.19a}$$

$$\dot{x}_4 = \frac{1}{m_5}p_5, \tag{5.19b}$$

$$\dot{p}_5 = F_2(t) + A\phi_3(V_3) - \phi_4(x_4) - \frac{R_6 + R_7}{m_5}p_5. \tag{5.19c}$$

Inspection of these state-space equations shows that Eq. (5.19a) is a volume flow equation (in terms of its physical dimension), Eq. (5.19b) is a velocity equation, and Eq. (5.19c) is a force equation. As a check on the correctness of the final equations, it is useful to work out the dimensions of all the terms to make sure that they are consistent.

Summarizing the contents of this section, we recall that by starting with an augmented bond graph with the inputs, energy variables, and co-energy variables identified, it is possible to write two sets of equations initially, the constitutive relations and the structural relations, and then to reduce them to state-space form in terms of the energy variables.

The balance of this chapter is directed to exploring further aspects of formulation and reduction, including such questions as

1. What if causality is not completed by source and storage elements alone?

2. What if differentiation causality shows up on certain storage elements?

3. What if we need expressions for variables other than energy variables?

5.4 EXTENDED FORMULATION
METHODS—PART 1

As an example of a system in which causality is not completed by source and storage element causal assignment, consider the electric circuit and its bond graph shown in Figure 5.11a and b, respectively. The two diagrams have been labeled in direct correspondence. In part c causality has been assigned to bonds 1, 3, and 6. Bond 1 is determined by $SE1$, and bond 3 and 6 by $C3$ and $I6$, respectively. No further causal extensions are possible based on source and storage elements. In order to obtain a complete causal assignment we must choose another bond and make it causal. In part d bond 2 has been causally directed as shown. When this choice is extended, using the 1- and 0-junctions, bonds 4 and 5 are directed, and the graph is completed. To show that the causal choice is

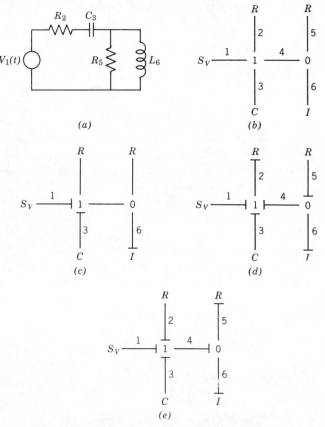

Figure 5.11. An example of completing causality.

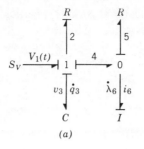

(a)

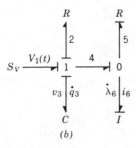

(b)

Figure 5.12. Two variations of an augmented bond graph.

not unique, in Figure 5.11e bond 2 was directed the other way causally, yielding the graph indicated. Each of these cases, parts d and e, leads to a particular formulation of equations. The two sets are similar, but not identical, and reduce to the same state-space set, as we shall show.

Referring to Figure 5.12a, which is the graph of Figure 5.11d with powers assigned and key variables labeled, including i_2, we see that $X = \begin{bmatrix} q_3 \\ \lambda_6 \end{bmatrix}$ and $U = [V_1]$. The set of constitutive relations is

$$v_3 = \frac{1}{C_3} q_3,$$

and   (5.20)

$$i_6 = \frac{1}{L_6} \lambda_6,$$

for $C3$ and $I6$, respectively.

The energy variable derivatives are given by

$$\dot{q}_3 = i_2 \qquad\qquad (5.21a)$$

and

$$\dot{\lambda}_6 = v_5 = R_5 i_5 = R_5(i_4 - i_6), \qquad (5.21b)$$

or

$$\dot{\lambda}_6 = R_5(i_2 - i_6). \qquad\qquad (5.21c)$$

The reason we stop at this point is that we anticipate that i_2 will be related to itself by an algebraic equation. This insight was gained in the process of assigning causality. Recall that the causal assignment of i_2 on bond 2 finally determined all other variables.

Let us see, by writing equations for i_2, that this is so. By reading the causal strokes in Fig. 5.12a, the following sequence of equation is found;

$$i_2 = \frac{1}{R_2} (v_2) \tag{5.22a}$$

$$= \frac{1}{R_2} [V_1(t) - v_3 - v_4], \tag{5.22b}$$

$$i_2 = \frac{1}{R_2} [V_1(t) - v_3 - v_5]. \tag{5.22c}$$

But v_3 is a co-energy variable, and v_5 is given by

$$v_5 = R_5 i_5 \tag{5.23a}$$

$$= R_5(i_4 - i_6), \tag{5.23b}$$

$$v_5 = R_5(i_2 - i_6), \tag{5.23c}$$

and i_6 is a co-energy variable. Examining Eqs. (5.22c) and (5.23c) we see that i_2 may be found in terms of input $V_1(t)$ and the co-energy variables, v_3 and i_6, and i_2 itself. The equation is

$$i_2 = \frac{1}{R_2} [V_1(t) - v_3 - R_5(i_2 - i_6)]. \tag{5.24a}$$

Solving this algebraic equation for i_2 yields

$$i_2 = \frac{1}{(R_2 + R_5)} [V_1(t) - v_3 + R_5 i_6]. \tag{5.24b}$$

To complete the reduction we put the results given by Eq. (5.24b) into Eqs. (5.21a) and (5.21c) and also use Eqs. (5.20) to eliminate the co-energy variables, as we have done before. The grand and glorious state-space results are given by Eqs. (5.25); namely,

$$\dot{q}_3 = -\frac{1}{(R_2 + R_5)C_3} q_3 + \frac{R_5}{(R_2 + R_5)L_6} \lambda_6 + \frac{1}{(R_2 + R_5)} V_1(t)$$

and

$$\dot{\lambda}_6 = -\frac{R_5}{(R_2 + R_5)C_3} q_3 - \frac{R_2 R_5}{(R_2 + R_5)L_6} \lambda_6 + \frac{R_5}{(R_2 + R_5)} V_1(t). \tag{5.25}$$

Now let us consider the other causal variation, as shown by the augmented graph of Figure 5.12b. The constitutive equations are identical to those derived before, namely, Eqs. (5.20). The equations for structure will be slightly different, and again we anticipate simultaneous dependence involving an internal variable. For the state variable derivatives we write, following the causality of Figure 5.12b,

$$\dot{q}_3 = i_4, \tag{5.26a}$$

or

$$\dot{q}_3 = i_5 + i_6, \tag{5.26b}$$

or

$$\dot{q}_3 = \frac{1}{R_5} v_5 + i_6 = \frac{1}{R_5}(V_1 - v_2 - v_3) + i_6, \tag{5.26c}$$

and

$$\dot{\lambda}_6 = v_4, \tag{5.26d}$$

or

$$\dot{\lambda}_6 = V_1(t) - v_2 - v_3. \tag{5.26e}$$

At this point, we recognize that an expression must be found for v_2 in terms of key system variables ($V_1(t)$, v_3, and i_6). This is because when we chose the causality on R_2 such that v_2 was an input to the remainder of the system, all causality was thereby determined.

$$v_2 = R_2 i_2 = R_2 i_4 = R_2(i_5 + i_6), \tag{5.27}$$

but

$$i_5 = \frac{1}{R_5} v_5 \tag{5.28a}$$

or

$$i_5 = \frac{1}{R_5} v_4 \tag{5.28b}$$

or

$$i_5 = \frac{1}{R_5}(V_1 - v_2 - v_3). \tag{5.28c}$$

Putting Eq. (5.28c) into Eq. (5.27) and solving for v_2, we get

$$v_2 = \frac{R_2}{R_2 + R_5}[V_1(t) - v_3 + R_5 i_6]. \tag{5.29}$$

To eliminate the intermediate variable, v_2, from Eqs. (5.26), we use the results of Eqs. (5.29). This gives the following pair of equations:

$$\dot{q}_3 = \frac{1}{R_2 + R_5} V_1(t) - \frac{1}{R_2 + R_5} v_3 + \frac{R_5}{R_2 + R_5} i_6$$

and   (5.30)

$$\dot{\lambda}_6 = \frac{R_5}{R_2 + R_5} V_1(t) - \frac{R_5}{R_2 + R_5} v_3 - \frac{R_2 R_5}{R_2 + R_5} i_6.$$

When v_3 and i_6 are eliminated from Eqs. (5.30) by use of Eqs. (5.20), and the terms are reordered on the right-hand side of the resulting equations, Eqs. (5.25) are obtained, thus demonstrating that the same state-space equations in terms of energy variables may be derived in more than one way.

What we have done in these two variations of the same example is to use a single intermediate variable to assist in the formulation (i.e., i_2 in the first case and v_2 in the second). It is possible to use more than one intermediate variable and obtain more than one algebraic equation to be solved. For example, both i_2 and v_5 could have been used in case one. Then a pair of algebraic equations involving both variables could have been solved in the reduction to state-space form. With experience you can learn to introduce as many or few intermediate variables as you find convenient. *The minimum number of intermediate variables which must be algebraically related is equal to the number of bonds which must be causally assigned after all source and storage element causality has been completed and the resulting causal implications have been extended throughout the graph.*

At this stage in our development we defer further consideration and interpretation of the physical meaning of the causal situation with which we have been dealing. In Section 7.2 a general discussion of the subject of R fields is to be found, and some of the reasons for the pattern described here will be put in a more profound perspective.

We shall study one other variation, again by example, to further our understanding of the process of completing causality and deriving state-space equations. Consider the bond graph of Figure 5.13a, in which bonds are labeled. In part b causality is assigned to bonds 1, 4, and 7 and extended as far as possible. In order to complete causality for the graph we arbitrarily choose to direct bond 3 as shown in part c. The consequence of this choice is that causality is completed on the graph, as shown in part d. Part e shows key variables labeled and powers directed so that the graph is augmented.* Note that e_3 and f_3 also have been written on the graph, for convenience in formulation.

---

*Since this graph contains a loop, sign convention is important and should really be determined from the physical system variable orientations.

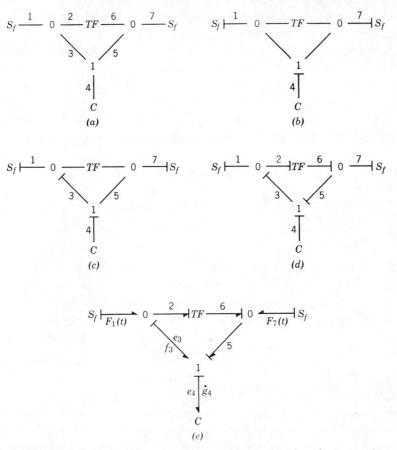

Figure 5.13. Completing causality by arbitrary bond assignment. An example.

Formulation begins with the constitutive equations, of which there is one, for $C4$, namely,

$$e_4 = \frac{1}{C4} q_4. \tag{5.31}$$

Formulation continues with the state variable derivative equations for the structure, of which there is one, namely,

$$\dot{q}_4 = f_3. \tag{5.32}$$

The reason we stop here is that we anticipate an algebraic equation relating f_3 to itself, as indicated by the causality-assignment process, and f_3 has been written as an intermediate variable.

Now let us seek an expression for f_3 in terms of key variables (i.e., $F_1(t)$, $F_7(t)$, e_4, f_3).

$$f_3 = F_1(t) - f_2 \tag{5.33a}$$

$$= F_1(t) - mf_6 \tag{5.33b}$$

$$= F_1(t) - m[-F_7(t) + f_5], \tag{5.33c}$$

$$f_3 = F_1(t) - m[-F_7(t) + f_3]. \tag{5.33d}$$

Note that m is the modulus of the transformer element. In this case, Eq. (5.33d) shows that f_3 depends on itself, and we solve for it explicitly to find

$$f_3 = \frac{1}{1+m} F_1(t) + \frac{m}{1+m} F_7(t). \tag{5.34}$$

The state equation is found by using Eq. (5.34) in Eq. (5.32), yielding

$$\dot{q}_4 = \frac{1}{1+m} F_1(t) + \frac{m}{1+m} F_7(t). \tag{5.35}$$

The interested reader is encouraged to experiment with other causal choices in the graph of Figure 5.13b and other intermediate bond variables (e.g., e_5, f_5, or e_2). The final result should be Eq. (5.35), of course.

In summarizing this section, we observe that if causality is not completed by source and storage element assignments, it may be completed by selecting appropriate bonds and causally directing them to complete the graph. In this process we may need to use one or more intermediate variables in the initial formulation of equations. The reduction to final state-space form follows an orderly pattern of elimination, as shown by several examples. Chapter 7 considers a physical interpretation of completing causality in the broader framework of R fields and junction structures, and Reference [1] describes a formal vector-field method for generating state-space equations.

5.5   EXTENDED FORMULATION METHODS—PART 2

In this section, we study stystems in which one or more storage elements have differentiation causality. This situation arises for systems

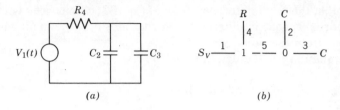

(a)   (b)

(c)   (d)

Figure 5.14. Differentiation causality in a bond graph. An example.

in which storage elements are not dynamically independent. For example, consider the electrical circuit of Figure 5.14a. The labeled bond graph is shown in part b, and causality has been assigned and extended in part c. Observe that integration causality on bond 2 has led to differentiation causality on bond 3. In part d key variables have been identified, and powers have been directed, so that the graph is augmented. The energy variables are q_2 and q_3, the co-energy variables are v_2 and v_3, and input is $V_1(t)$.

Because of the differentiation causality on bond 3 we anticipate only one state variable (i.e., q_2), and we expect to eliminate q_3 from the dynamic equations.* The general rule is that the energy variables for the storage elements in integration causality in a completely augmented bond graph can serve as state variables. *Any storage elements with differentiation causality do not contribute state variables.* (In the foregoing statements it is assumed that causality is assigned according to the procedure outlined in Section 5.2.)

The constitutive equations for the example are

$$v_2 = \frac{1}{C_2} q_2 \tag{5.36a}$$

and

$$q_3 = C_3 v_3. \tag{5.36b}$$

---

*Causality suggests the following: if q_2 is given, then e_2 is specified. But e_2 is related to e_3, and e_3 is related to q_3, both in an algebraic way. Therefore q_2 algebraically determines q_3.

Observe that the relation for element $C3$, Eq. (5.36b), has been written in inverse form because in differentiation causality v_3 is the input and $\dot{q}_3$ is the output from the point of view of $C3$.

An expression for the derivative of q_2 is

$$\dot{q}_2 = i_5 - \dot{q}_3, \qquad (5.37a)$$

or

$$\dot{q}_2 = i_4 - \dot{q}_3, \qquad (5.37b)$$

or

$$\dot{q}_2 = \frac{1}{R_4} [V_1(t) - v_2] - \dot{q}_3. \qquad (5.37c)$$

The input to element $C3$ is v_3, and this is the other structure relation.

$$v_3 = v_2. \qquad (5.37d)$$

Equations (5.36a), (5.36b), (5.37c), and (5.37d) contain four unknowns: q_2, q_3, v_2, and v_3. However, careful inspection of those equations reveals that they include three algebraic equations in terms of four unknowns, and q_3 may be expressed directly in terms of q_2. The result is

$$q_3 = C_3 v_3,$$

or

$$q_3 = C_3 v_2,$$

or

$$q_3 = C_3 \frac{1}{C_2} q_2. \qquad (5.38)$$

Therefore, the derivative of q_3 may be found in terms of q_2 from Eq. (5.38) as follows:

$$\dot{q}_3 = \frac{C_3}{C_2} \dot{q}_2. \qquad (5.39)$$

Now Eq. (5.37c) may be rewritten more nearly in state-space form by using Eqs. (5.36a) and (5.39) to eliminate v_2 and q_3, respectively. The result is

$$\dot{q}_2 = \frac{1}{R_4} V_1(t) - \frac{1}{R_4 C_2} q_2 - \frac{C_3}{C_2} \dot{q}_2. \qquad (5.40)$$

It is typical of systems having differentiation causality that they give rise to equations of the form of Eq. (5.40), in which derivatives appear on both sides of the equation. The final step in reduction is to solve Eq. (5.40) for $\dot{q}_2$ explicitly, obtaining Eq. (5.41), namely,

$$\dot{q}_2 = -\frac{1}{R_4(C_2+C_3)}\, q_2 + \frac{C_2}{R_4(C_2+C_3)}\, V_1(t). \tag{5.41}$$

Just as we had anticipated, Eq. (5.41) shows that a first-order system describes the original circuit or bond graph. A very important additional equation for the system is Eq. (5.38), which expresses q_3, the other energy variable, in terms of q_2.

It should be noted that another acceptable causal form exists for the graph in Figure 5.14b. That form would have integration causality on element $C3$ and differentiation causality on element $C2$. In that case, q_3 would be the state variable, and an additional relation [the inverse of Eq. (5.38), in fact] would be obtained, relating q_2 to q_3 algebraically. With some experience in dealing with such situations the bond-graph analyst is able to select the state variables desired from among the energy variables and to organize the equations in an appropriate fashion. In every case involving differentiation causality, knowledge of the state variables and inputs implies knowledge of the remaining energy variables through algebraic relations, as in Eq. 5.38.

One more example involving differentiation causality may serve to illustrate sufficiently the pattern of formulation and reduction. Consider the mass-spring-lever mechanism depicted in Figure 5.15a. A labeled bond-graph model is shown in part b, in which the lever-arm ratio is specified as $(a:b)$. The element parameters are the masses, m_1, m_2, and the spring constant, k_3, for elements $I1$, $I2$, and $C3$, respectively; $F(t)$ is a force input.

When causality is assigned to the graph the result is that element $I1$ has differentiation causality imposed, as Figure 5.15c indicates. Physically, only one of the momenta, p_1 and p_2, is independent. For example, if p_2 is given, then v_2 is algebraically fixed. But v_1 is algebraically determined by v_2, due to the lever arrangement. Now, v_1 algebraically specifies p_1, so we conclude that p_2 algebraically determines p_1, just as causality indicates. The equations will subsequently bear this out.

After key variables are identified and power directions are specified, a completely augmented graph is available from which equations may be written. We anticipate a second-order system with the state variables, p_2 and x_3, plus an additional algebraic relation specifying p_1 in terms of p_2. The key lists are $X = \begin{bmatrix} p_2 \\ x_3 \end{bmatrix}$ and $U = [F_4(t)]$.

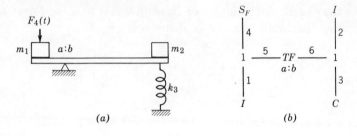

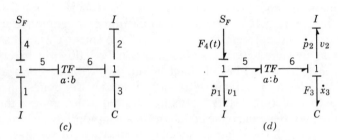

Figure 5.15.   Augmented bond graph for a lever mechanism.

We begin the formulation with the constitutive relations.

$$p_1 = m_1 v_1, \qquad (5.42a)$$

$$v_2 = \frac{1}{m_2} p_2, \qquad (5.42b)$$

$$F_3 = k_3 x_3. \qquad (5.42c)$$

Equation (5.42a) is written in the form shown because of the causality on element $I1$.

The next step is to write an expression for the input to each storage element in terms of the key variables. That is, we need equations for v_1, $\dot{p}_2$, and $\dot{x}_3$. Following the causality shown on the graph, we have

$$v_1 = v_5,$$

or

$$v_1 = \frac{a}{b} v_6,$$

or

$$v_1 = \frac{a}{b} v_2. \tag{5.43a}$$

$$\dot{p}_2 = F_6 - F_3,$$

or

$$\dot{p}_2 = \frac{a}{b} F_5 - F_3,$$

or

$$\dot{p}_2 = \frac{a}{b} [F_4(t) - \dot{p}_1] - F_3. \tag{5.43b}$$

$$\dot{x}_3 = v_2. \tag{5.43c}$$

Since we expect to be able to relate p_1 to p_2 directly, we examine the equations and find that Eqs. (5.42a), (5.43a), and (5.42b) may be used to derive the result:

$$p_1 = m_1 \frac{a}{b} \frac{1}{m_2} p_2. \tag{5.44}$$

From Eq. (5.44) an expression for $\dot{p}_1$ may be found in terms of variable $\dot{p}_2$. This expression together with Eq. (5.42c) may be put into Eq. (5.43b), and also Eq. (5.43c) may be simplified, to yield

$$\dot{p}_2 = \frac{a}{b} F_4(t) - \frac{a}{b} m_1 \frac{a}{b} \frac{1}{m_2} \dot{p}_2 - K_3 x_3, \tag{5.45a}$$

$$\dot{x}_3 = \frac{1}{m_2} p_2. \tag{5.45b}$$

As the final step in reduction we solve for $\dot{p}_2$ explicitly in Eq. (5.45b), giving the state-space equations

$$\dot{p}_2 = -\frac{k_3}{[1 + (m_1/m_2)(a/b)^2]} x_3 + \frac{a/b}{[1 + (m_1/m_2)(a/b)^2]} F_4(t) \tag{5.46a}$$

and

$$\dot{x}_3 = \frac{1}{m_2} p_2. \tag{5.46b}$$

We have indeed obtained the equations for a second-order system, in fact those of a forced oscillator, plus an additional equation specifying the energy variable p_1 in terms of the state-variable p_2 directly [Eq. (5.44)]. The reader may wish to try an alternate causality pattern, in which differentiation causality is imposed upon element $I2$. The state equations will be somewhat different, but the behavior of the system will be predicted equally well by either set of state equations.

Further discussion of systems with differentiation causality is to be found in Section 7.1, where a physical point of view is taken, and also in Reference [1], where formal methods of equation generation and reduction are presented.

5.6   OUTPUT   VARIABLE   FORMULATION

So far we have seen how state variables can be chosen from the set of energy variables and how state-space equations governing the system dynamics can be developed. Frequently in engineering work the need arises to find expressions for particular output variables which may not have been chosen as state variables. Once a graph has been augmented and the key variables defined, it is a straightforward matter to derive expressions for any other system variables in terms of the state and input variables.

The list of output variables is typically denoted by Y. For linear systems the output equations are usually written as

$$y_1 = c_{11}x_1 + c_{12}x_2 + \cdots + c_{1n}x_n + d_{11}u_1 + \cdots + d_{1r}u_r,$$

$$y_2 = c_{21}x_1 + \cdots \qquad + c_{2n}x_n + d_{21}u_1 + \cdots + d_{2r}u_r,$$

$$\vdots \qquad\qquad\qquad\qquad\qquad\qquad\qquad\qquad (5.47)$$

$$y_m = c_{m1}x_1 + \cdots \qquad + c_{mn}x_n + d_{m1}u_1 + \cdots + d_{mr}u_r,$$

where there are n state variables, r input variables, and m output variables. In other words, each output variable is a linear combination of state and input variables.*

As an example, consider the problem of finding the volume flow drawn from the pressure supply in the bond-graph model of the pressure-control valve shown in Figure 5.10. We wish to express Q_1 in terms of state

---

* It is possible for outputs to depend upon the time derivatives of input variables, so that an extra set of f_{ij} coefficients may also be necessary. This is not a common case. By a mathematical redefinition of input and state variables, it is always possible to relate output variables as indicated in Eq. (5.47) but in this text we prefer to use state variables closely related to the physics of the system. One should recognize that there are also variables which are not *linearly* related to state variables; examples are power flows and stored energies. If these output variables are of interest, the system of output equations is not linear even if the state equations are linear.

variables V_3, x_4, and p_5, plus the inputs $P_1(t)$ and $F_2(t)$. We have, writing equations implied by the causal strokes,

$$Q_1 = Q_8,$$

or

$$Q_1 = \frac{1}{R_8}(P_1(t) - P_9),$$

or

$$Q_1 = \frac{1}{R_8}(P_1(t) - P_3),$$

or

$$Q_1 = \frac{1}{R_8}\left(P_1(t) - \frac{1}{C_3}V_3\right), \tag{5.48}$$

provided the constitutive laws of both R_8 and C_3 are constant-coefficient linear.

Let us now assume that R_8 is characterized by

$$P_8 = \phi_8(Q_8) \tag{5.49}$$

and C_3 is characterized by

$$P_3 = \phi_3^{-1}(V_3), \tag{5.50}$$

where ϕ_8 is a resistance function and ϕ_3 is a capacitance relation. An expression for Q_1 may be found, using indicated causality as follows:

$$\begin{aligned}
Q_1 &= Q_8, \\
Q_1 &= \phi_8^{-1}(P_8) \qquad (\phi_8^{-1} \text{ is the inverse of } \phi_8), \\
Q_1 &= \phi_8^{-1}(P_1 - P_9), \\
Q_1 &= \phi_8^{-1}(P_1 - P_3), \\
Q_1 &= \phi_8^{-1}[P_1 - \phi_3^{-1}(V_e)].
\end{aligned} \tag{5.51}$$

Equation (5.51) gives the desired result, namely, Q_1 as a function of the state variables (V_3) and inputs (P_1).

Another common problem is that of estimating the effective power delivered to a resistive load. A mechanical example is shown in Figure 5.16a, in which an angular velocity source feeds through a compliant

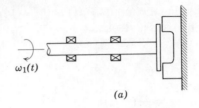

(a)

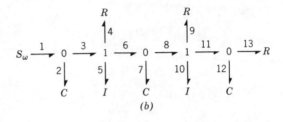

(b)

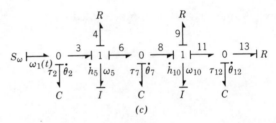

(c)

Figure 5.16.   Augmented bond graph for a shaft-driven load.

shaft supported by two bearings to drive a resistive load. We wish to estimate the power delivered to the load under various conditions.

A bond-graph model is given in Figure 5.16b in which R_{13} is the load. We seek expressions for load variables, τ_{13} and ω_{13}, in terms of the state variables, θ_2, θ_7, θ_{12}, h_5, and h_{10}, and input, ω_1, where θ denotes angular displacement and h denotes angular momentum. The augmented graph is given in Figure 5.16c. The key lists are

$$X = \begin{bmatrix} \theta_2 \\ h_5 \\ \theta_7 \\ h_{10} \\ \theta_{12} \end{bmatrix} \quad \text{and} \quad U = [\omega_1(t)].$$

The linear constitutive relations are

$$\tau_2 = k_2\theta_2, \tag{5.52a}$$

$$\omega_5 = \frac{1}{J_5} h_5, \tag{5.52b}$$

$$\tau_7 = k_7\theta_7, \tag{5.52c}$$

$$\omega_{10} = \frac{1}{J_{10}} h_{10}, \tag{5.52d}$$

and

$$\tau_{12} = k_{12}\theta_{12}. \tag{5.52e}$$

The structure equations are

$$\dot{\theta}_2 = \omega_1(t) - \omega_5, \tag{5.53a}$$

$$\dot{h}_5 = \tau_2 - R_4\omega_5 - \tau_7, \tag{5.53b}$$

$$\dot{\theta}_7 = \omega_5 - \omega_{10}, \tag{5.53c}$$

$$\dot{h}_{10} = \tau_7 - R_9\omega_{10} - \tau_{12}, \tag{5.53d}$$

and

$$\dot{\theta}_{12} = \omega_{10} - \frac{1}{R_{13}} \tau_{12}. \tag{5.53e}$$

Substituting Eqs. (5.52) into Eqs. (5.53), we arrive at the state-space equations (in unordered form),

$$\dot{\theta}_2 = \omega_1(t) - \frac{1}{J_5} h_5, \tag{5.54a}$$

$$\dot{h}_5 = k_2\theta_2 - \frac{R_4}{J_5} h_5 - k_7\theta_7, \tag{5.54b}$$

$$\dot{\theta}_7 = \frac{1}{J_5} h_5 - \frac{1}{J_{10}} h_{10}, \tag{5.54c}$$

$$\dot{h}_{10} = k_7\theta_7 - \frac{R_9}{J_{10}} h_{10} - k_{12}\theta_{12}, \tag{5.54d}$$

and

$$\dot{\theta}_{12} = \frac{1}{J_{10}} h_{10} - \frac{k_{12}}{R_{13}} \theta_{12}. \tag{5.54e}$$

To estimate the power delivered to the load we need values for τ_{13} and

ω_{13}. For τ_{13} we have

$$\tau_{13} = \tau_{12},$$

or

$$\tau_{13} = k_{12}\theta_{12}. \qquad (5.55)$$

For ω_{13} we have

$$\omega_{13} = \frac{1}{R_{13}} \tau_{13},$$

or

$$\omega_{13} = \frac{1}{R_{13}} k_{12}\theta_{12}. \qquad (5.56)$$

Thus, the power on bond 13 is given by

$$P_{13}(t) = \tau_{13}\omega_{13} = \frac{1}{R_{13}} (k_{12}\theta_{12})^2. \qquad (5.57)$$

Although the state equations are linear in this case, the second-order term in the expression for P_{13} indicates that the problem of predicting the power flow is not strictly linear. Thus, the relation between inputs and state variables is linear, but the relation between inputs and the output P_{13} is not.

We note in closing this chapter that, since any bond variable can be expressed in terms of the state and input variables, any function of a bond variable (or set of them) can be expressed in terms of state and input variables. Therefore, the real challenge in systems analysis is to obtain information about the dynamic behavior of the state variables. This is the topic of the next chapter.

REFERENCE

1. R. C. Rosenberg, "State Space Formulation for Bond Graph Models of Multiport Systems," *Trans. ASME, J. Dyn. Sys., Meas. Control*, **93,** Ser. G, (1) (March, 1971), 35–40.

PROBLEMS

5-1 For each of the following bond graphs (p. 167), assign causality, predict the number of state variables, and write a set of state equations. All elements may be assumed to be linear with constant coefficients.

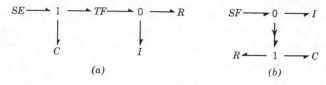

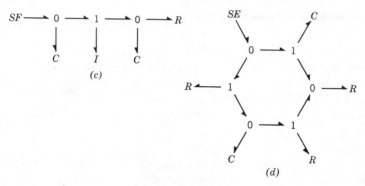

Problem 5-1.

5-2 In the problems below you are asked to find an equivalent expression for the system given. Try it first assuming constant coefficient elements and then for nonlinear characteristics, using the first case as a guide.

5-3 For the problems listed below, in each case make a bond-graph model of the circuit, schematic, or network diagram. Then augment the graph and write state equations. Interpret the equations physically. You may assume the elements are linear.

(a) Problem 4-1(e), electrical circuit;

(b) Problem 4-3(h), mechanical translation;

(c) Problem 4-5(a), mechanical rotation;

(d) Problem 4-6(a) and (b), hydraulic;

(e) Problem 4-5(c), mechanical transduction;

(f) Problem 4-5(e), mechanical transduction.

5-4 For the electrical circuit shown below verify the bond graph, write state equations, and develop an output equation for the voltage across the load resistor (R_L) in terms of state variables and input variables. Assume all elements have constant coefficients.

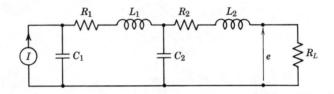

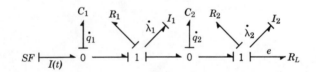

5-5 The mechanical system shown below has two nonlinear springs with constitutive laws as shown (p. 169). Note that δ is a deflection. The friction relations are also nonlinear, with a signum (sign) characteristic as shown. Make a bond graph and write state equations.

5-6 Make a bond-graph model of the pulley system shown in Figure 4.4b. Augment the graph and predict the number of state variables. Write a suitable set of state equations.

Modify the graph to include pulley inertias and pin-joint friction. Write state equations for the resulting graph.

5-7 A ball is suspended at the end of a spring in a container of fluid, as shown (p. 169).

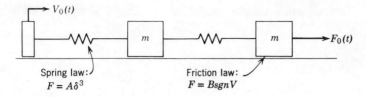

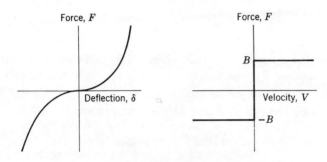

Problem 5-5.

The extension of the spring measured from free length is x_s, and the velocity of the mass is v, measured downward.

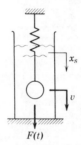

The spring is nonlinear, characterized by a relation

$$F_s = \phi_s(x_s).$$

The damping effect of the fluid is proportional to the square of the velocity, corrected for sign.

$$F_D = b \, |v| \, v.$$

(a) Formulate state equations for the system in terms of x_s and p (mass momentum).

(b) Transform the equations from x_s, p variables to x_s, v variables.

5-8   Consider the simple model of a vehicle shown below, where

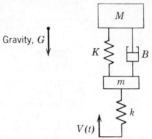

suspended mass is M; main suspension stiffness is K; damper coefficient is B; tire mass is m; tire spring stiffness is k; and velocity input from roadway is $V(t)$.

(a) Make a bond-graph model, augment the graph, and write state equations.

(b) If the tire mass is assumed to be negligible, let $m \rightarrow 0$, but retain all other values. Make a bond graph and derive state equations.

5-9

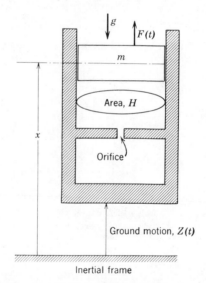

A commercial type of air spring isolator is shown. The mass is supported by air pressure contained by means of a virtually frictionless bellows arrangement (not shown). Damping is provided by adjusting the area of the orifice between the two chambers. The orifice resistance law is generally nonlinear, as is the capacitance relation, since the pressure volume relation for the gas is approximately $PV^n = \text{const}$.

(a) Make a bond graph for the system which allows both ground motion $Z(t)$ and force $F(t)$ as inputs.

(b) Write state equations using general functions for resistance and capacitance relations.

(c) Using the bond graph, show an analogous all-mechanical system.

5-10 For the water storage system of Problem 4.7, including the inertia effects introduced in part *b*, augment the graph, predict the number of state variables, and write state equations. (If you encounter formulation difficulties, indicate clearly the steps to be taken to achieve state equations in standard form.)

5-11

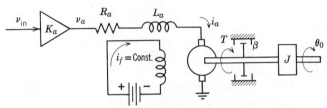

Open-loop positioning system.

On the diagram physical-system variables and parameters are identified, where

θ_0 = output position angle;

ν_{in} = input voltage;

ν_a = output voltage of linear amplifier;

i_a = motor armature current;

i_f = motor field current, assumed constant;

K_a = gain of linear amplifier, assumed to have no significant time constants;

R_a = resistance of armature winding;

L_a = inductance of armature winding;

J = inertial load;

β = viscous-damping constant;

K_T = torque constant of motor;

K_ν = back-emf constant of motor.

The differential equations that govern the dynamics of the system are

$$J\ddot{\theta}_0 + \beta\dot{\theta}_0 = K_T i_a,$$

$$L_a \dot{i}_a + R_a i_a = V_a - K_\nu \dot{\theta}_0.$$

A conventional schematic diagram and corresponding equations are shown above.

(a) Construct a bond graph for the system.

(b) Write state space equations and verify that your equations are equivalent to those listed above.

(c) Compare the two methods for analyzing this system. For example, is the system third or second order? Are K_T and K_ν related in any way?

512   Consider the seismometer sketched below:

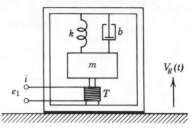

The input is ground motion, $V_g(t)$, and an electrical transducer using a permanent magnet moving in a coil reacts to the relative motion between the case and the seismic mass, m.

(a) Construct a bond-graph model of the device, leaving the electrical port as a free bond.

(b) Assume the device is connected to a voltage amplifier, so that $i = 0$. Find an expression relating $V_g(t)$ and e_1.

(c) Sometimes it is preferable to use current rather than voltage as output, due to noise considerations. Suppose a current amplifier is used, so that $e_1 = 0$. Relate output current, i, to input signal, $V_g(t)$, for the case when coil resistance and inductance are neglected.

(d) Reconsider the problem of (c) with coil resistance included.

5-13   The three systems shown below each possess a feature which makes the writing of state space equations interesting. In each case assign causality sequentially and predict any difficulties. Where possible, write state space equations assuming linear element characteristics.

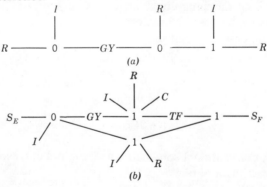

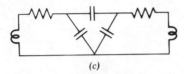

(c)

Problem 5-13

5-14 Consider the mechanical system shown below.

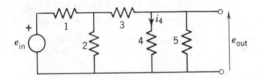

f is force here required to achieve specified $V(t)$

The mass, m_1, plays a somewhat unusual role in the system.

(a) Show that a state space for the system can be found which does not even depend on m_1.

(b) Show that most system output variables depend statically on the state variables and the inputs, F and V, but because of m_1 this is *not* true of f (defined above).

5-15 The resistive circuit shown below offers an opportunity to exploit the use of causality and helping (or auxiliary) variables.

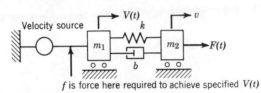

(a) Make a bond-graph model of the circuit, and assign causality.

(b) Find e_{out} in terms of e_{in}.

(c) Find i_4 in terms of e_{in}.

(d) If a load resistance, R_L, is put on at the e_{out} port, modify the graph and your solution to part (b) above.

6

ANALYSIS OF LINEAR SYSTEMS

6.1 INTRODUCTION

Historically, a major reason for the study of linear systems was convenience, both in model formulation and in the solution of the equations. But, in this book, we have taken pains to show that formulation techniques using bond graphs need not be much more difficult for nonlinear elements than for linear ones. This is particularly true if $-I$, $-C$, or $-R$ elements are the only nonlinear elements in the model. In Chapter 7 nonlinear elements that cause a further loss in convenience of formulation will be introduced. Since systematic formulation techniques exist, a loss in convenience would seem a small price to pay for a system model better able to indicate real system performance.

A more serious reason for the study of linear systems was the complete analytical theory of linear equations, which provides information about the nature of all possible responses of a linear system, as well as recipes for finding particular responses under given initial conditions and forcing functions. But the existence of powerful computers now makes it possible to solve nonlinear equations almost as rapidly and economically as linear ones.

One could argue that it would be best to study only nonlinear techniques for state equation formulation and solution and to consider the linear system as a special case in which some special simpler techniques are available. In a sense, this is the viewpoint of this book, but in a practical sense, linear systems are more important than their special mathematical status would indicate.

There are fields such as acoustics, electric-circuit design, structural vibration, and parts of thermofluid system design in which methods based on linear systems have been remarkably successful. Collective experience in these fields allows practitioners to use linear techniques with great confidence. Another area in which linear systems are prominent is the theory and practice of automatic control. There are two reasons for this: (1) The complete theory that exists for linear systems allows the synthesis of controllers with desired characteristics. (2) The linear description of the system to be controlled may be almost tautologically valid. In regulator design, if the controller can be designed to maintain the system near its desired state in the face of disturbances, then the linear system description will be a valid approximation in practice. This is not to say that in automatic control, as in all fields, many of the most important and challenging problems do not lie outside of the domain of linear systems.

Since linear differential equations are treated in virtually all engineering and scientific curricula, we do not attempt to give a mathematically rigorous and complete account here. Our purpose is to discuss the techniques for linear systems that are important for systems theory and design. This will be done in a style that carries over to large-scale systems, but here we will not introduce the formalism of matrices in order to concentrate on fundamental concepts. We will, however, concentrate on sets of n first-order equations, rather than on the equivalent single nth-order equation as is more common in elementary treatments. This equivalence is discussed briefly near the end of this chapter. Bond graphs naturally produce first-order equations, and this is the form most convenient for high-order systems and machine computation.

6.2 SOLUTION TECHNIQUES FOR ORDINARY DIFFERENTIAL EQUATIONS

Before discussing the characteristics of linear state equations it is useful to note that it is not difficult in principle to find one or more solutions to general state equations using a computer if necessary. In most of the examples encountered so far, the physical system model yielded state equations of the form

$$\dot{x}_1 = f_1(x_1, x_2, \ldots, x_n; u_1, u_2, \ldots, u_r),$$
$$\dot{x}_2 = f_2(x_1, x_2, \ldots, x_n; u_1, u_2, \ldots, u_r),$$
$$\cdot$$
$$\cdot \qquad\qquad\qquad\qquad\qquad\qquad\qquad (6.1)$$
$$\cdot$$
$$\dot{x}_n = f_n(x_1, x_2, \ldots, x_n; u_1, u_2, \ldots, u_r),$$

where $x_1, \ldots, x_n$ are *state variables* (typically ps or qs in bond-graph terms) and $u_1, \ldots, u_r$ are *input variables* (typically es or fs from sources).

In addition, there may be *output variables* $y_1, \ldots, y_s$ statically related to the state and input variables:

$$y_1 = g_1(x_1, x_2, \ldots, x_n; u_1, u_2, \ldots, u_r),$$
$$y_2 = g_2(x_1, x_2, \ldots, x_n; u_1, u_2, \ldots, u_r),$$
$$\cdot$$
$$\cdot \qquad\qquad\qquad\qquad\qquad\qquad\qquad (6.2)$$
$$\cdot$$
$$y_s = g_s(x_1, x_2, \ldots, x_n; u_1, u_2, \ldots, u_r).$$

In general, the f and g functions are nonlinear, but we mainly shall consider linear functions. The linear versions of Eqs. (6.1) and (6.2) are

$$\dot{x}_1 = a_{11}x_1 + a_{12}x_2 + \cdots + a_{1n}x_n + b_{11}u_1 + b_{12}u_2 + \cdots + b_{1r}u_r,$$
$$\dot{x}_2 = a_{21}x_1 + a_{22}x_2 + \cdots + a_{2n}x_n + b_{21}u_1 + b_{22}u_2 + \cdots + b_{2r}u_r,$$
$$\cdot$$
$$\cdot \qquad\qquad\qquad\qquad\qquad\qquad\qquad (6.1a)$$
$$\cdot$$
$$\dot{x}_n = a_{n1}x_1 + a_{n2}x_2 + \cdots + a_{nn}x_u + b_{n1}u_1 + b_{u2} + \cdots + b_{nr}u_r;$$

$$y_1 = c_{11}x_1 + c_{12}x_2 + \cdots + c_{1n}x_n + d_{11}u_1 + d_{12}u_2 + \cdots + d_{1r}u_r,$$
$$y_2 = c_{21}x_1 + c_{22}x_2 + \cdots + c_{2n}x_n + d_{21}u_1 + d_{22}u_2 + \cdots + d_{2r}u_r,$$
$$\cdot$$
$$\cdot \qquad\qquad\qquad\qquad\qquad\qquad\qquad (6.2a)$$
$$\cdot$$
$$y_s = c_{s1}x_1 + c_{s2}x_2 + \cdots + c_{sn}x_n + d_{s1}u_1 + d_{s2}u_2 + \cdots + d_{sr}u_r.$$

Now the f and g functions and their linear versions involving the coefficients represented by subscripted as, bs, cs, and ds are called *static functions* because whenever arguments x_1 to x_n and u_1 to u_r are known, then the functions may be evaluated, that is, the expressions for the $\dot{x}$s and the ys are known. In many applications, the f and g functions and the a, b, c, and d coefficients do not change with time. In this case the system is stationary, and in the linear case the system is called a *constant coefficient linear system*. We will be primarily concerned with this case, but it is also possible for systems to arise in which the a, b, c, and d coefficients vary with time (e.g., $a_{ij} = a_{ij}(t)$). Such a system is often called a *time variable linear system*. The general analog of such a system would be a system in which the f and g functions varied with time, for example,

$$x_1 = f_1(x_1, x_2, \ldots, x_n; u_1, u_2, \ldots, u_r; t). \tag{6.3}$$

The question of time dependence in the state equations occasionally causes confusion. Since the input variables, u_i, are known functions of time, it is true that the f and g functions are functions of time even in the stationary case. Also, when the equations are solved for a particular trajectory the state variables become functions of time. Nonstationary or time-variable systems contain an explicit time dependence, as indicated in Eq. (6.3). In such a system, one cannot evaluate the functions given only the values of the xs and the us: one must also know the time. Stated another way, the $\dot{x}$s and ys in a stationary system depend only on the instantaneous values of the xs and us, and if the same values of xs and us recur at any other time, the $\dot{x}$s and ys would be identical. In a nonstationary system, the functions change with time in a way not influenced by the xs and us that may exist for any particular solution.

Equations (6.1), (6.1a), (6.2), and (6.2a) describe state-determined systems. To completely specify a problem to be solved, we need more information. The input functions must be known time functions:

$$u_1 = u_1(t),$$
$$u_2 = u_2(t),$$
$$\cdot$$
$$\cdot \tag{6.4}$$
$$\cdot$$
$$u_r = u_r(t).$$

Also, a particular trajectory in state space must be singled out for attention. Our main concern will be with *initial condition problems* in which the state at some initial time, t_0, is known:

$$x_1(t_0) = x_{10},$$
$$x_2(t_0) = x_{20},$$
$$\cdot$$
$$\cdot \tag{6.5}$$
$$\cdot$$
$$x_n(t_0) = x_{n0},$$

where x_{10} to x_{n0} are the initial states. Note that given the information in Eqs. (6.4) and (6.5) the initial output variables are determined, for example,

$$y_1(t_0) = g_1[x_{10}, x_{20}, \ldots, x_{n0}; u_1(t_0), u_2(t_0), \ldots, u_r(t_0)].$$

One may also determine $\dot{x}_1(t_0)$ to $\dot{x}_n(t_0)$ using Eq. (6.1), for example,

$$\dot{x}_1(t_0) = f_1[x_{10}, x_{20}, \ldots, x_{n0}; u_1(t_0), u_2(t_0), \ldots, u_r(t_0)].$$

In principle, it is easy to compute how the system changes in a short interval of time, Δt, using the concept of a derivative directly. For example,

$$x_1(t_0 + \Delta t) \cong x_1(t_0) + \dot{x}_1(t_0)\Delta t \cong x_{10} + \dot{x}_1(t_0)\Delta t. \tag{6.6}$$

This equation is really just a rearrangement of the definition of a derivative,

$$\dot{x}_1 \equiv \frac{dx_1}{dt} = \lim_{\Delta t \to 0} \left[\frac{x_1(t_0 + \Delta t) - x_1(t_0)}{\Delta t} \right],$$

in which Δt is small but finite. Since $\dot{x}_1(t_0)$ depends on the initial state and the us at t_0 which are known, one can use Eq. (6.6) to find the state at $t_0 + \Delta t$. This is sometimes called *Euler's formula* for integrating equations, and it may be applied to all the state equations at $t = t_0$. It can be reapplied at $t = t_0 + \Delta t$ since the us are known at any time, and after applying Eq. (6.6) at t_0, the state at $t_0 + \Delta t$ is then known. The equation may then be applied recursively to march the solution along in time as far as desired.

The process described above is readily adapted to automatic digital computation, but there is a snag. Equation (6.6) is not exactly correct except in the limit as $\Delta t \to 0$. As Δt is made very small the approximation generally becomes increasingly accurate, but it takes more steps and more evaluations of the functions to cover any given time span of interest. The problems of choosing suitable time increments, Δt, and recursion formulas analogous to Eq. (6.6) to optimize digital integration accuracy falls in the domain of *numerical methods*, and there is a wealth of literature on this subject. Here it suffices to note that methods to integrate state equations exist in profusion and that, despite its obviousness, Euler's formula turns out to be quite inefficient.

The integration of state equations is relatively straightforward, but when the equations are linear, one may say a great deal about the totality of possible system behavior without the necessity of finding many solutions for various initial conditions and forcing functions. As will be demonstrated, linear systems obey the *principle of superposition*, which implies that any scaled sum of solutions to linear state equations is also a solution. The superposition property allows one to consider a particular solution to the state equations to be the sum of a *response due to initial conditions or free response* (in which all input variables vanish) and a *response due to inputs, or forced response*. In addition, since the free response components of total system response has to do only with the system equations, this response may be used to characterize the system in a useful manner. We may discover natural frequencies, time constants, and stability properties that, in contrast to the general case, have nothing to do with initial conditions or forcing terms and help in understanding how the system would behave in a variety of situations.

6.3 FREE RESPONSE

Consider now the linear case as represented by Eqs. (6.1a) and (6.2a). For simplicity, we will study a first-order system with a single state equation at the outset and then move on to an example with multiple state variables.

6.3.1 A First-Order Example

In Figure 6.1, a block of hot steel is shown after it has been lowered into a large oil bath. We can predict the time history of the temperature of the steel using a simple lumped-parameter thermal-system model if we make several assumptions:

1. We assume that the temperature gradients within the steel block are small compared to the gradient between the cool oil and the hot steel. We then define a single average temperature for the steel, T_s.
2. We assume that the oil temperature does not change much as the steel cools so that the temperature, T_o, is nearly constant.
3. We assume that the heat flow rate, Q, from the steel to the oil is proportional to $T_o - T_s$.
4. We assume that the change in T_s is proportional to the total heat exchanged, Q. Clearly, this simplifying set of assumptions is physically reasonable for some situations, but many details of the thermal process have been glossed over. It would be easy to complicate the model in the style of Figure 3.18 to allow more detailed prediction of temperatures, but our purpose here is to construct a first-order system.

Two pseudo bond graphs using temperature as an effort and heat flow rate as a flow are shown in Figures 6.1a and b. The constitutive laws for R

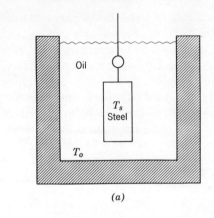

(a)

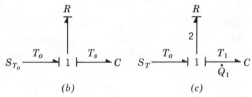

(b)   (c)

Figure 6.1. First-order example system. (a) Schematic diagram; (b) bond graph showing temperatures, T_o for oil and T_s for steel; (c) bond graph with bonds numbered for convenience in writing equations.

and C can be found as follows. Using assumption 3, we write

$$\dot{Q} = \frac{1}{R}(T_s - T_o),\qquad (6.7)$$

where the resistance R has the dimensions °F/(Btu/sec), for example. Often R is estimated from a heat transfer coefficient, h, and the total area of the steel block, A,

$$\frac{1}{R} = hA.\qquad (6.8)$$

In this case, h has the dimensions of

$$(\text{Btu/sec})/(\text{°F ft}^2).$$

Using assumption 4, we write

$$T_s - T_{s0} = \frac{Q}{C},\qquad (6.9)$$

where T_{so} is the initial temperature of the steel when no heat has been transferred. The capacitance, C, with dimensions Btu/°F, may be found from

$$C = mc, \tag{6.10}$$

where the specific heat per unit mass, c, has dimensions Btu/°F lbm if mass is measured in pounds mass. (For a solid, if we neglect the small amount of work due to volume change, the specific heats at constant pressure and constant volume are identical.) Clearly, any compatible set of temperature scales and measures of heat energy may be used for the R and C parameters.

Using the methods of Chapter 5, we may easily write the state equation for the bond graph of Figure 6.1c, which embodies the relations discussed above.

$$\dot{Q}_1 = \dot{Q}_2 = \frac{1}{R}(T_o - T_1) = \frac{1}{R}\left(T_o - \frac{Q_1}{C} - T_{10}\right),$$
$$\dot{Q}_1 = -\frac{Q_1}{RC} + \frac{1}{R}(T_o - T_{10}), \tag{6.11}$$

where $T_1 = T_s$ and $T_{10} = T_{so}$ in Eq. (6.9). Another version of the state equation which is found by substituting Eq. (6.9) into Eq. (6.11) is

$$RC\dot{T}_s = -T_s + T_o, \tag{6.12}$$

where we have returned to T_s in place of T_1. It is sometimes convenient to measure temperatures from some datum so that some constants are absorbed in the definitions. For example, if we define

$$T = T_s - T_o, \tag{6.13}$$

then Eq. (6.12) becomes

$$RC\dot{T} = -T. \tag{6.14}$$

Also, if we modify Eq. (6.9) to read

$$T = T_s - T_o = \frac{Q_1}{C} = T_1 - T_o, \tag{6.15}$$

then Eq. (6.11) becomes

$$\dot{Q}_1 = \frac{1}{R}(T_o - T_1) = \frac{1}{R}\left(T_o - \frac{Q_1}{C} - T_o\right),$$

or

$$\dot{Q}_1 = \frac{-1}{RC} Q_1. \tag{6.16}$$

In Eq. (6.16), when $Q_1 = 0$, $T_s = T_o$ and the initial value of Q_1 is related to the initial value of T_s through Eq. (6.15).

For simplicity, we will use Eq. (6.16) for our discussion, but the presence of extra constants in state equations such as those in Eqs. (6.11) and (6.12) will cause no problems since the constants simply act as constant forcing functions for the system.

The solution to a first-order equation such as Eq. (6.16) or Eq. (6.14) is well known. First we note that one solution to Eq. (6.16) is obviously $Q_1 = 0$. This solution is called the *trivial solution* since it is of little use. In particular, if $Q_1(0)$, the initial value of Q_1, is nonzero, then the trivial solution is of no use in fitting the initial condition. A general assumption that proves useful for linear systems is to assume that the solution is exponential, for example,

$$Q_1(t) = Ae^{st}, \tag{6.17}$$

where the constants, A and s, are to be adjusted, if possible, to make sure that $Q_1(t)$ fits the differential equation and the initial conditions.

Substituting Eq. (6.17) into Eq. (6.16), we find

$$Ase^{st} = \frac{-1}{RC} Ae^{st},$$

or

$$\left(s + \frac{1}{RC}\right)Ae^{st} = 0. \tag{6.18}$$

Now, either

$$Ae^{st} = 0,$$

in which case we return to the trivial solution, or

$$s = \frac{-1}{RC}, \tag{6.19}$$

in which case the solution is

$$Q_1(t) = Ae^{-t/RC} = Ae^{-t/\tau}, \tag{6.20}$$

where the *time constant*, τ, has been identified with RC. To complete the

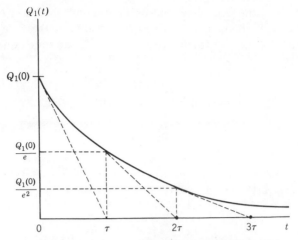

Figure 6.2. Free response of system of Figure 6.1.

solution, the constant, A, can be found from the initial value of Q_1, $Q_1(0)$,

$$Q_1(0) = Ae^{-0} = A, \tag{6.21}$$

where $Q_1(0)$ can be related to the initial temperature through Eq. (6.15). The solution is then

$$Q_1(t) = Q_1(0)e^{-t/\tau}, \qquad t \geq 0. \tag{6.22}$$

This solution is shown in Figure 6.2. For this simple exponential solution, it is readily shown that, at any point on the curve, if one extends a straight line along the local slope of the curve to the steady state value of Q_1 (zero in this case), it will intersect after τ time units have elapsed. Thus, τ sets the basic time scale of the response. If a system response is measured experimentally, one can easily estimate τ as shown in Figure 6.2, and one can check to see how well a linear, first-order constant-parameter system would reproduce the measured response.

6.3.2 Second-Order Systems

Although a knowledge of first-order system response will prove very useful, it must be obvious that more complex systems respond in ways much more complicated than the exponential of Figure 6.2. Surprisingly, the general pattern of analysis, somewhat generalized, does carry over for arbitrarily high-order systems. In fact, by studying a second-order system, the pattern will become evident, and the extension to nth-order systems is very straightforward.

Consider, then, the simple second-order example shown in Figure 6.3. To make the correspondence between the system of the figure and the general equations, Eqs. (6.1a) and (6.2a), we write the state equations in a standard form:

$$\dot{x} = 0x - \frac{1}{m}p + 1V + 0F,$$

$$\dot{p} = kx - \frac{b}{m}p + bV + 1F. \tag{6.23}$$

The force, f, is just one of several possible output variables,

$$f = kx - \frac{b}{m}p + bV + 0F. \tag{6.24}$$

The system has two state variables, $x_1 = x$, $x_2 = p$, two input variables,

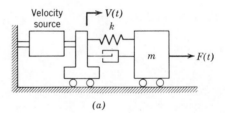

(a)

$$k\!:\!C \longleftarrow \!\!\!|\; 1 \;|\!\!\!\longrightarrow R\!:\!b \qquad m\!:\!I$$
$$\underset{x}{} $$
$$f\! \uparrow \qquad\qquad \dot{p}\! \uparrow$$
$$S_V \longmapsto 0 \longrightarrow\!| \; 1 \; |\!\longleftarrow S_F$$

(b)

$$\dot{x} = 0\,x - \tfrac{1}{m}p + 1[V(t)] + 0[F(t)]$$
$$\dot{p} = k\,x - \tfrac{b}{m}p + b[V(t)] + 1[F(t)]$$

(c)

Figure 6.3. Example system. (a) Schematic diagram; (b) bond graph; (c) state equations.

$u_1 = V$, $u_2 = F$, and one output, $y_1 = f$. The parameters of the system are

$$a_{11} = 0, \qquad a_{12} = \frac{-1}{m},$$

$$a_{21} = k, \qquad a_{22} = \frac{-b}{m},$$

$$b_{11} = 1, \qquad b_{12} = 0,$$

$$b_{21} = b, \qquad b_{22} = 1, \qquad (6.25)$$

$$c_{11} = k, \qquad c_{12} = \frac{-b}{m},$$

$$d_{11} = b, \qquad d_{12} = 0,$$

which correspond to those defined by Eqs. (6.1a) and (6.2a). Note that in Figure 6.3c the notation $V(t)$ and $F(t)$ is used as a reminder that these input quantities are specified as time functions that must be known *before* the state equations can be solved.

For the free response, we consider that $u_1(t) = u_2(t) = \cdots u_r(t) = 0$, or in the example, $V(t) = F(t) = 0$. Thus the equations to be studied are simply

$$\dot{x} = 0x - \frac{1}{m} p,$$

$$\dot{p} = kx - \frac{b}{m} p. \qquad (6.26)$$

One trivial solution to Eq. (6.10) is evident. It is

$$x \equiv 0, \qquad p \equiv 0. \qquad (6.27)$$

This solution corresponds to the case in which the spring is unstretched and the mass has no momentum (or velocity). This solution is again not very useful and is of no help in fitting given initial conditions. Suppose a solution satisfying

$$x(t_0) = x_0, \qquad p(t_0) = p_0, \qquad (6.28)$$

where x_0 is the initial stretch in the spring and $p_0 = mv_0$ is the initial momentum of the mass, is desired.

To find a solution, assume that both energy state variables may be represented by an amplitude and an exponential time function. In the example, we assume

$$x(t) = Xe^{st},$$

$$p(t) = Pe^{st}, \tag{6.29}$$

where X and P are constant amplitudes and s is a complex or real number with the dimension of inverse time. If the assumption of Eq. (6.29) is substituted into Eq. (6.26) the result is

$$sXe^{st} = \left(\frac{-1}{m}\right)Pe^{st},$$

$$sPe^{st} = kXe^{st} - \frac{b}{m}Pe^{st},$$

or $\qquad\qquad\qquad\qquad\qquad\qquad\qquad\qquad\qquad\qquad\qquad$ (6.30)

$$\left[sX + \frac{1}{m}P\right]e^{st} = 0,$$

$$\left[-kX + \left(s + \frac{b}{m}\right)P\right]e^{st} = 0.$$

Now Eq. (6.30) shows that Eq. (6.29) results in a transformation of a differential equation problem into a problem in algebra. One way to solve Eq. (6.30) would be for e^{st} to vanish. But this is not a useful solution since this would again imply the trivial solution of (6.27). Rejecting this possibility, we concentrate on solving for X and P which satisfy

$$sX + \frac{1}{m}P = 0,$$

$$-kX + \left(s + \frac{b}{m}\right)P = 0. \tag{6.30a}$$

Once again, the trivial solution appears since clearly $X = 0$, $P = 0$ is a solution to Eq. (6.30a). In fact, the elementary theory of linear algebra contains the result that there is only one solution to a set of linear simultaneous equations if the determinant of the coefficients does not vanish. Cramer's rule furnishes a recipe for expressing the solution in terms of ratios of determinants, the denominator of each being the determinant of the equation coefficients. Since we have one trivial solution by inspection, our assumed solution will prove useless unless the solution to Eq. (6.30a) turns out *not* to be unique. This will only occur if

$$\det \begin{vmatrix} s & \dfrac{1}{m} \\ -k & s+\dfrac{b}{m} \end{vmatrix} = 0,$$

or

$$s^2 + \frac{b}{m}s + \frac{k}{m} = 0.$$

(6.31)

The requirement that the determinant vanish leads to the requirement that a polynomial in s must vanish. This equation is called the *characteristic equation* and leads to a set of *characteristic values* for the parameters. Equation (6.19) was the characteristic equation for our first example. (In German a characteristic value is called an *eigenwert*, and in English the mongrel term, *eigenvalue*, is commonly used. Why "characteristic wert" never caught on remains a mystery.)

Solving Eq. (6.31), the result is

$$s_1 = \frac{-b/m + [(b/m)^2 - 4k/m]^{1/2}}{2},$$

$$s_2 = \frac{-b/m - [(b/m)^2 - 4k/m]^{1/2}}{2}.$$

(6.32)

Note that s_1 and s_2 will be complex for small values of b. For each value of s that is determined from an equation such as Eq. (6.31) a solution satisfying Eq. (6.30a) may be found. For an nth-order system, there will generally be n special values of s; in our second-order example, there are two. For high-order systems it will generally not be possible to solve for the characteristic values without resorting to numerical methods, but the existence of n roots of an nth-order characteristic equation is assured by the theory of algebra.

Much useful information is contained in the characteristic values of Eq. (6.32) themselves, but before discussing what can be attained by simply solving the characteristic equation, let us complete the program of finding a complete free response satisfying given initial conditions. For each value of s, s_1, and s_2, let us attempt to determine X and P values from Eq. (6.30a). For s_1 we will attempt to find X_1 and P_1.

$$s_1 X_1 + \frac{1}{m} P_1 = 0,$$

$$-kX_1 + \left(s_1 + \frac{b}{m}\right)P_1 = 0,$$

or

$$\left(\frac{-b/m+[(b/m)^2-4k/m]^{1/2}}{2}\right)X_1+\left(\frac{1}{m}\right)P_1=0, \qquad (6.33)$$

$$-kX_1+\left(\frac{-b/m+[(b/m)^2-4k/m]^{1/2}}{2}+\frac{b}{m}\right)P_1=0. \qquad (6.34)$$

Of course, one should not expect to be able to solve uniquely for X_1 and P_1 from Eqs. (6.33) and (6.34) since s_1 was chosen to make the determinant of the coefficients vanish. In fact, after some algebraic manipulation, it can be shown that both the equations are equivalent to the single equation,

$$X_1-\left(\frac{b+(b^2-4km)^{1/2}}{2km}\right)P_1=0, \qquad (6.35)$$

which means that the ratio of X_1 to P_1 is determined, but not the magnitudes of either one. Whenever a characteristic value is substituted into equations for amplitudes such as Eq. (6.30a), the equations will be found not to be independent. The most one can expect to solve for is the ratios of $n-1$ amplitudes in terms of one arbitrary amplitude.

Substituting s_2 from Eq. (6.32) into Eq. (6.30a) and solving for amplitudes X_2 and P_2, one again finds that both equations are equivalent to the single equation:

$$X_2+\left(\frac{-b+(b^2-4km)^{1/2}}{2mk}\right)P_2=0. \qquad (6.36)$$

Now whenever X_1 and P_1 are related according to Eq. (6.35), then Eq. (6.30a) will be satisfied, and the original differential equation Eq. (6.26) will be satisfied by

$$x(t)=X_1e^{s_1t},$$
$$P(t)=P_1e^{s_1t}.$$

Similarly, another solution is

$$x(t)=X_2e^{s_2t},$$
$$p(t)=P_2e^{s_2t},$$

whenever X_2 and P_2 satisfy Eq. (6.36). It is also easy to verify that another valid solution is obtained by adding the two solutions together:

$$x(t) = X_1 e^{s_1 t} + X_2 e^{s_2 t},$$
$$p(t) = P_1 e^{s_1 t} + P_2 e^{s_2 t}.$$

(6.37)

This is an example of the principle of superposition for linear systems and is readily verified by simply substituting Eq. (6.37) into Eq. (6.26) and noting that as long as the components of the sum satisfy the differential equations so does the sum.

Although s_1 and s_2 are completely determined by Eq. (6.32), the four quantities, X_1, X_2, P_1, and P_2, are so far related only by the two relations, Eqs. (6.35) and (6.36). Two more conditions are necessary to completely specify the free response, and these are provided by the two initial conditions of Eq. (6.28).

$$x_0 = x(t_0) = X_1 e^{s_1 t_0} + X_2 e^{s_2 t_0},$$

(6.38)

$$p_0 = p(t_0) = P_1 e^{s_1 t_0} + P_2 e^{s_2 t_0}.$$

(6.39)

Thus, the four equations, (6.35), (6.36), (6.38), and (6.39), completely specify the free response of the system satisfying the given initial conditions. Although in this simple case one could solve these equations completely, the solution is too complex for easy understanding, so we will discuss some useful special cases.

6.3.3   Example: The Undamped Oscillator

If the dashpot parameter, b, is set to zero, or if the dashpot is removed from the system, the equations of the system are somewhat simplified. Equation (6.23) becomes

$$\dot{x} = 0x - \frac{1p}{m} + 1V + 0F,$$
$$\dot{p} = kx + 0p + 0V + 1F.$$

(6.40)

Equation (6.24) becomes

$$f = kx + 0p + 0V + 0F.$$

For the free response we need study only a simple version of Eq. (6.26):

$$\dot{x} = \frac{-p}{m},$$
$$\dot{p} = kx.$$

Following the development leading to Eq. (6.31), the following characteristic equation is obtained:

$$s^2 + \frac{k}{m} = 0, \tag{6.40}$$

which yields characteristic values

$$s_1 = +j\left(\frac{k}{m}\right)^{1/2} = j\omega_n,$$
$$s_2 = -j\left(\frac{k}{m}\right)^{1/2} = -j\omega_n, \tag{6.41}$$

where $j = (-1)^{1/2}$ and the *undamped natural frequency*, ω_n, is defined as $(k/m)^{1/2}$. In distinction to the first-order example, we now have encountered purely complex eigenvalues.

In attempting to solve for X_1 and P_1, it is found that only the single relation, Eq. (6.35) is available:

$$X_1 - \frac{jP_1}{(km)^{1/2}} = 0. \tag{6.42}$$

In solving for X_2 and P_2, one encounters a simplified version of Eq. (6.36):

$$X_2 + \frac{jP_2}{(km)^{1/2}} = 0. \tag{6.43}$$

Finally, we must satisfy the initial conditions in Eqs. (6.38) and (6.39).

Using Eqs. (6.42) and (6.43) to eliminate P_1 and P_2 in Eqs. (6.38) and (6.39), the result is

$$X_1 e^{s_1 t_0} + X_2 e^{s_2 t_0} = x_0, \tag{6.44}$$

$$-j(km)^{1/2} X_1 e^{s_1 t_0} + j(km)^{1/2} X_2 e^{s_2 t_0} = p_0. \tag{6.45}$$

These equations are readily solved, the result being

$$X_1 = \frac{1}{2}\left(x_0 + \frac{jp_0}{(mk)^{1/2}}\right) e^{-j\omega_n t_0}, \tag{6.46}$$

$$X_2 = \frac{1}{2}\left(x_0 - \frac{jp_0}{(mk)^{1/2}}\right) e^{+j\omega_n t_0}, \tag{6.47}$$

in which Eq. (6.42) has been used. For completeness, P_1 and P_2 may also be found:

$$P_1 = -j(mk)^{1/2}X_1 = \tfrac{1}{2}[-j(mk)^{1/2}x_0 + p_0]e^{-j\omega_n t_0}, \tag{6.48}$$

$$P_2 = +j(mk)^{1/2}X_2 = \tfrac{1}{2}[j(mk)^{1/2}x_0 + p_0]e^{+j\omega_n t_0}. \tag{6.49}$$

We have now achieved our objective. The free response is given by Eq. (6.37), in which we now know s_1 and s_2, and we can determine X_1, X_2, P_1, and P_2 from arbitrary initial conditions. It remains only to interpret the solutions.

The solution just determined is of the form,

$$x(t) = X_1 e^{j\omega_n t} + X_2 e^{-j\omega_n t}, \tag{6.50}$$

$$p(t) = P_1 e^{j\omega_n t} + P_2 e^{-j\omega_n t}, \tag{6.51}$$

and is to be interpreted as valid for $t \geq t_0$. The time-dependent parts, $e^{j\omega_n t}$ and $e^{-j\omega_n t}$, may be represented in the complex plane as rotating vectors or, as they are sometimes called, *phasors* (see Figure 6.4). The vectors rotate at constant angular rate, ω_n, around the unit circle. The sketch of Figure 6.4 provides an easy way to remember the relations,

$$e^{+j\omega_n t} = \cos \omega_n t + j \sin \omega_n t, \tag{6.52}$$

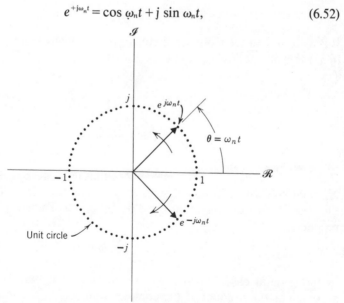

Figure 6.4. Complex plane representation of $e^{+j\omega_n t}$ and $e^{-j\omega_n t}$.

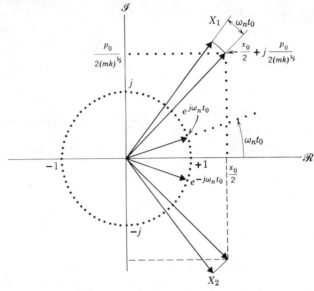

Figure 6.5. Representation in the complex plane of X_1 and X_2. See Eqs. (6.46) and (6.47).

and

$$e^{-j\omega_n t} = \cos \omega_n t - j \sin \omega_n t. \tag{6.53}$$

It is also easy to see from the sketch or algebraically from Eqs. (6.52) and (6.53) that the following relations are true:

$$\cos \omega_n t = \frac{e^{j\omega_n t} + e^{-j\omega_n t}}{2}, \tag{6.54}$$

$$\sin \omega_n t = \frac{e^{j\omega_n t} - e^{-j\omega_n t}}{2j}. \tag{6.55}$$

From the form of Eq. (6.50) one can see that if X_1 were real and equal to X_2, then $x(t)$ would be a cosine wave with amplitude $2X_1$. This would be the case if, in Eqs. (6.46) and (6.47), $t_0 = p_0 = 0$. Then Eq. (6.50) would reduce to

$$x(t) = x_0 \cos \omega_n t.$$

In the general case, X_1 and X_2 are complex numbers that are easily sketched as the product of the complex numbers, $x_0/2 \pm j p_0/2(mk)^{1/2}$, and the complex exponentials, $e^{\pm j\omega_n t_0}$. See Figure 6.5 and recall that in

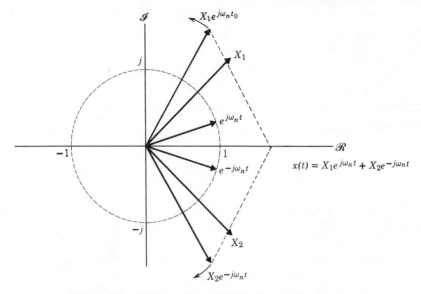

Figure 6.6. Sketch of Eq. (6.50) in the complex plane.

multiplying two complex numbers, the magnitudes multiply and the angles add. It is no accident that X_1 and X_2 (and P_1 and P_2) turn out to be complex conjugates. According to Eq. (6.50), the real quantity, $x(t)$, must be the sum of X_1 and X_2 multiplied by the complex conjugate quantities, $e^{j\omega_n t}$ and $e^{-j\omega_n t}$, respectively. Figure 6.6 shows how the sum of complex quantities adds up to a real variable, $x(t)$.

Another way to represent Eq. (6.50) is as a cosine wave with a phase angle, φ, and an amplitude, A,

$$x(t) = A \cos(\omega_n t + \varphi). \tag{6.56}$$

From Figure 6.6 it is easy to verify that

$$A = 2\,|X_1| = 2\,|X_2| = 2\left(\frac{x_0^2}{4} + \frac{p_0^2}{4mk}\right)^{1/2} \tag{6.57}$$

and

$$\varphi = \angle X_1 = \left[\tan^{-1}\frac{p_0}{x_0(mk)^{1/2}}\right] + \omega_n t_0 \tag{6.58}$$

where $|\;\;|$ stands for "magnitude of" and $\angle$ stands for "angle of."

Another commonly used notation involves the "real part of" operator, Re( ), and the "imaginary part of" operator, Im( ). Sine and cosine functions are readily expressed in terms of these functions as follows:

$$\cos \omega_n t = \text{Re}(e^{j\omega_n t}),\tag{6.59}$$

$$\sin \omega_n t = \text{Im}(e^{j\omega_n t}).\tag{6.60}$$

Then $x(t)$ may be expressed in yet another way as

$$x(t) = A \text{ Re}(e^{(j\omega_n t + \varphi)}) = \text{Re}(2X_1 e^{j\omega_n t}).\tag{6.61}$$

Equation (6.61) is sometimes simpler to use than some other equivalent representations since it deals with only one of the pair X_1, X_2 and one of the pair $e^{+j\omega_n t}$, $e^{-j\omega_n t}$. Both pairs must be complex conjugates in order for x to be real; therefore a knowledge of one member of each pair is sufficient.

Finally, it should be noted that the discussion of $p(t)$ in Eq. (6.51) exactly parallels that of $x(t)$ just given. We may note from Eqs. (6.48) and (6.49) that P_1 and P_2 are proportional to X_1 and X_2, but are 90° away due to the j factors. (Remember that multiplication of a complex number by j results in a number with identical magnitude, but with an angle advanced by $\pi/2$ rad or 90°.) In fact, if one were to sketch $P_1 e^{j\omega_n t}$ and $P_2 e^{-j\omega_n t}$ as rotating vectors as was done for the components of x in Figure 6.6, the vectors for P_1 and P_2 would lag the vectors for X_1 and X_2 by 90°. This corresponds with the fact that $p(t)$ is proportional to either the negative of the derivative of $x(t)$, or the integral of $x(t)$, Eq. (6.40). (The reader should verify that the time derivative of $X_1 e^{+j\omega_n t}$ is a vector advanced by 90° and the integral of $X_1 e^{+j\omega_n t}$ is a vector retarded by 90°, and that similar statements can be made for the counter-rotating vector $X_2 e^{-j\omega_n t}$.)

The 90° phase difference between x and p means that if x is a cosine wave, p is a sine wave. Then, in the state space, x, p, the trajectories are closed ellipses that can be made to be circles upon proper scaling. The direction that the representative point travels on the circles depends on the sign convention used in setting up the original differential equations, but, in any case, the representative point moves around the center of the circular trajectory at an angular rate of ω_n rad per unit time.

6.3.4 Example: The Damped Oscillator

When the dashpot coefficient, b, is positive and not too large, the free response of the system will be oscillatory but will eventually damp out. The characteristic values of Eq. (6.32) clearly change character depending on whether the expression under the radical is positive or negative. If

$$\left(\frac{b}{m}\right)^2 < 4\frac{k}{m}, \tag{6.62}$$

the system is said to be *underdamped*, if

$$\left(\frac{b}{m}\right)^2 > 4\frac{k}{m} \tag{6.63}$$

the system is said to be *overdamped*, and if

$$\left(\frac{b}{m}\right)^2 = 4\frac{k}{m} \tag{6.64}$$

the system is said to be *critically damped*.

Often the characteristic equation, Eq. (6.31), is written as follows:

$$s^2 + 2\zeta\omega_n s + \omega_n^2 = 0, \tag{6.65}$$

in which the undamped natural frequency, ω_n, is as defined in Eq. (6.41) and the *damping ratio*, ζ, is

$$\zeta \equiv \frac{b}{2(mk)^{1/2}}. \tag{6.66}$$

Then the system is underdamped, overdamped, or critically damped according to whether $\zeta < 1$, $\zeta > 1$, or $\zeta = 1$, respectively. With these definitions, the characteristic values for the underdamped case of Eq. (6.32) may be written:

$$\begin{aligned} s_1 &= -\zeta\omega_n + j\omega_n(1-\zeta^2)^{1/2}, \\ s_2 &= -\zeta\omega_n - j\omega_n(1-\zeta^2)^{1/2}. \end{aligned} \tag{6.67}$$

Equations (6.35) and (6.36) become

$$X_1 - \left(\frac{\zeta\omega_n + j\omega_n(1-\zeta^2)^{1/2}}{k}\right)P_1 = 0 \tag{6.68}$$

and

$$X_2 + \left(\frac{-\zeta\omega_n + j\omega_n(1-\zeta^2)^{1/2}}{k}\right)P_2 = 0. \tag{6.69}$$

The free response solution, Eq. (6.37), becomes a bit more complex than in the undamped case because the characteristic values are now complex numbers rather than pure imaginary numbers. A typical term is

$$X_1 e^{s_1 t} = X_1 e^{[-\zeta \omega_n t + j\omega_n(1-\zeta^2)^{1/2}t]} = X_1 e^{-\zeta \omega_n t} e^{j\omega_n(1-\zeta^2)^{1/2}t}. \tag{6.70}$$

As before, X_1 may be a complex number determined partly by Eq. (6.68) and Eq. (6.69) and partly by the necessity to fit initial conditions, Eqs. (6.38) and (6.39). The term, $e^{-\zeta \omega_n t}$ is real and decreases exponentially with time. (In the previous example, ζ was equal to zero, and this factor remained unity.) The last factor in Eq. (6.70) is a complex exponential form which represents a sinusoidal wave of frequency, ω_d, the *damped natural frequency*,

$$\omega_d \equiv \omega_n(1-\zeta^2)^{1/2}, \tag{6.71}$$

or

$$\left(\frac{\omega_d}{\omega_n}\right)^2 + \zeta^2 = 1.$$

When there is damping, the sinusoidal part of the system response has a frequency somewhat different from the undamped natural frequency, and, as Eq. (6.71) shows, the variation of ω_d with ζ may be plotted as a segment of a circle. When ζ is small compared to unity, $\omega_d \simeq \omega_n$, but $\omega_d \rightarrow 0$ as $\zeta \rightarrow 1$ (see Figure 6.7).

Although it is possible to solve for X_1, X_2, P_1, P_2 in terms of general initial conditions, x_0, p_0, the result is not very useful since it is usually easier simply to solve the equations for some particular initial conditions rather than to specialize the general result. What is important is to understand the nature of the solution. Figure 6.8 is the equivalent of Figure 6.6 for the underdamped case. The major differences between the

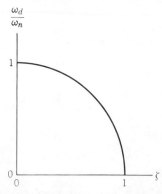

Figure 6.7.   Variation of damped natural frequency with damping ratio.

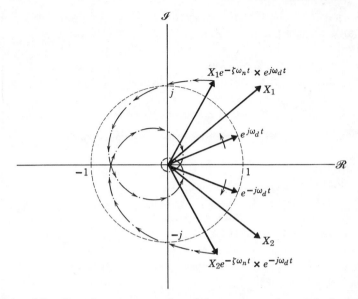

Figure 6.8. Complex representation of the response of an underdamped system.

two figures are that both components of the x response in Figure 6.8 rotate at an angular speed ω_d instead of ω_n and that there is an attenuation factor, $e^{-\zeta\omega_n t}$, that reduces the amplitudes exponentially so that the complex components spiral toward the origin of the complex plane. The corresponding response in time is sketched in Figure 6.9. Note that the response passes through zero whenever the sinusoidal component does; thus the zeros occur periodically. The envelope of the time response is the exponentially decreasing amplitude shown. When the sinusoidal component is unity, the $x(t)$ wave form is locally tangent to the exponential. As a result, the actual maxima of the wave are not quite periodic.

Since so much understanding of the general nature of the free response is contained in the roots of the characteristic equation, it is useful to show how the damping ratio and the damped and undamped natural frequencies are correlated with the locations in the imaginary plane of the characteristic values, s_1 and s_2. This is shown in Figure 6.10. Note that for any system in which the state variables are real quantities, any complex characteristics will occur in complex conjugate pairs, and it is easy to read off natural frequencies and the damping ratio corresponding to each pair.

Stable systems result in characteristic values which lies in the *left* half of the complex plane; that is, for stable systems the real component of e^{st}

is a *decreasing* exponential. *Unstable* systems correspond to s values in the right half plane, and the intermediate case of the undamped oscillator corresponds to purely imaginary values of s on the j axis.

For the overdamped oscillator, the character of the roots of Eq. (6.32) changes, and it is convenient to rewrite Eq. (6.67) as

$$s_1 = -\zeta\omega_n + \omega_n(\zeta^2-1)^{1/2},$$
$$s_2 = -\zeta\omega_n - \omega_n(\zeta^2-1)^{1/2}. \tag{6.72}$$

These roots are plotted in Figure 6.11. Note that for $1 < \zeta < \infty$ the roots, s_1 and s_2, lie in the range $-\infty < s_1, s_2 < 0$; that is, the roots are negative and real. In such a case, terms of the form e^{st} are conveniently written in terms of a time constant, τ, as was done for the first-order case. Suppose, for example, s_1 were equal to $-a$. Then,

$$e^{s_1 t} = e^{-at} = e^{-t/\tau}, \tag{6.73}$$

where

$$\tau = \frac{1}{a} = \frac{-1}{s_1}.$$

The $e^{-t/\tau}$ factor in the second-order system solution has the same effect as was shown in Figure 6.2 for the first-order system. Thus, in Figure 6.11 s values far out on the negative real axis correspond to quickly decreasing exponentials. Also, it is clear that the real part of s_1 and s_2 in Figure 6.10, $\zeta\omega_n$, corresponds to a time constant of $1/\zeta\omega_n$. Characteristic values on the

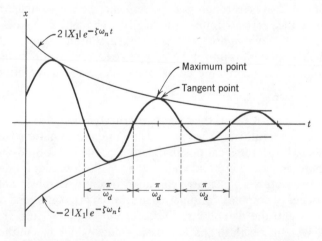

Figure 6.9.   Time response for underdamped system.

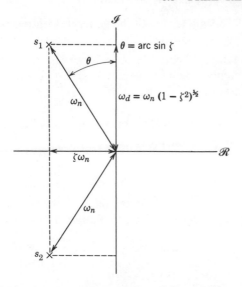

Figure 6.10. Correlation of natural frequencies and damping ratio with location of characteristic values in the complex plane for underdamped case.

positive real axis also possess time constants but represent increasing exponentials. A value of $s = 0$ implies a time response component of $e^{0t} = 1$ or a constant component: an infinite time constant.

6.3.5 General Comments

Although only first- and second-order systems have been studied above, the characteristics of the free response of all linear systems are not much more complicated. The procedure for finding the free response is summarized below.

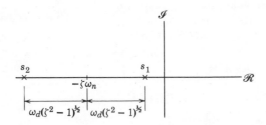

Figure 6.11. Characteristic values for overdamped case.

1. Neglect all forcing terms. In the general equations of Eq. (6.1a), let $u_1, u_2, \ldots, u_r$ all vanish.

2. Assume each state variable, x_i, will possess solutions of the form, $X_i e^{st}$. After canceling out the e^{st} terms, Eq. (6.1a) becomes a purely algebraic equation:

$$(s - a_{11})X_1 - a_{12}X_2 - \cdots - a_{1n}X_n = 0,$$
$$-a_{21}X_1 + (s - a_{22})X_2 - \cdots - a_{2n}X_n = 0, \qquad (6.74)$$
$$\cdots$$
$$-a_{n1}X_1 - a_{n2}X_2 - \cdots + (s - a_{nn})X_n = 0.$$

This set of equations always possesses a trivial solution, $X_1 = X_2 = \cdots = X_n = 0$, which is the only solution unless

$$\det \begin{vmatrix} s - a_{11} & -a_{12} & \cdots & -a_{1n} \\ -a_{21} & s - a_{22} & & \\ \cdot & & & \\ \cdot & & & \\ \cdot & & & \\ -a_{n1} & -a_{n2} & & s - a_{nn} \end{vmatrix} = 0. \qquad (6.75)$$

When the determinant is expanded out into an nth-order polynomial, the characteristic equation results.

3. Solve the characteristic equation for the n characteristic values, $s_1, s_2, \ldots, s_n$. The s values will either be real or, if complex, will occur in complex conjugate pairs. The location of the characteristic values in the complex plane indicates the type of component of the free response that is associated with the characteristic values. Figure 6.12 indicates how root location may be correlated with time response. Stable systems possess characteristic values with negative real parts.

4. For each value of s corresponding to a characteristic value, Eq. (6.74) yields a partial solution. When the characteristic values are distinct, that is, when no two of the n characteristic values are equal, then $n - 1$ of the n equations in Eq. (6.74) will be found to be independent when a characteristic value is substituted for s.* One could solve then for $n - 1$ of the Xs in terms of the last X, for example.

---

* The special case when roots are repeated is treated in all books on differential equations. It is rarely important in practice, so we will not treat it here.

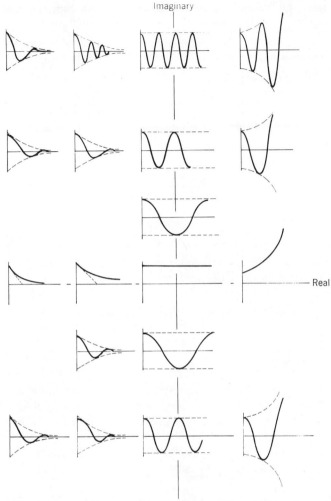

Figure 6.12. Correlation of time response with complex characteristic values.

The total solution for the free response is

$$x_1(t) = X_{11}e^{s_1 t} + X_{12}e^{s_2 t} + \cdots + X_{1n}e^{s_n t},$$
$$x_2(t) = X_{21}e^{s_1 t} + X_{22}e^{s_2 t} + \cdots + X_{2n}e^{s_n t}, \qquad (6.76)$$
$$\cdots$$
$$x_n(t) = X_{n1}e^{s_1 t} + X_{n2}e^{s_2 t} + \cdots + X_{nn}e^{s_n t},$$

where $X_{11}, X_{21}, X_{31}, \ldots, X_{n1}$ appear in Eq. (6.74) when $s = s_1$, $X_{12}, X_{22}, X_{32}, \ldots, X_{n2}$ appear in Eq. (6.74), when $s = s_2$, and so on.

The n^2 values of X in Eq. (6.76) cannot be completely determined by writing Eq. (6.74) for each characteristic value of s, since in this manner only $n(n-1)$ independent equations will be obtained. The remaining n conditions are represented by the n arbitrary initial conditions:

$$x_1(t_0) = x_{10} = X_{11}e^{s_1t_0} + X_{12}e^{s_2t_0} + \cdots + X_{1n}e^{s_nt_0},$$

$$x_2(t_0) = x_{20} = X_{21}e^{s_1t_0} + X_{22}e^{s_2t_0} + \cdots + X_{2n}e^{s_nt_0},$$

$$\cdot$$
$$\cdot \qquad\qquad\qquad\qquad\qquad\qquad\qquad (6.77)$$
$$\cdot$$

$$x_n(t_0) = x_{n0} = X_{n1}e^{s_1t_0} + X_{n2}e^{s_2t_0} + \cdots + X_{nn}e^{s_nt_0},$$

where the $x_{10}, x_{20}, \ldots, x_{n0}$ are the given initial conditions.

For very low-order systems it may be possible to solve the characteristic equation and actually determine all the Xs using Eq. (6.74) and the initial conditions. For systems of higher order it is generally necessary to resort to machine computation, in which case, if the system is specified as linear state equations, digital computer programs are available that will provide analytical expressions for the system response or numerical results for the time histories of the state variables or output variables. See, for example, Reference [1]. Also, ENPORT will provide state equations and eigenvalues directly from bond-graph data as well as time-response plots. Thus, an understanding of how various eigenvalues contribute to system response is more important than an ability to carry out the algebraic manipulations required to find an analytic solution for system free response.

6.4 FORCED RESPONSE

We now turn to the problem of finding solutions of Eq. (6.1a) that are compatible with given time functions for $u_1(t), u_2(t), \ldots, u_r(t)$. Again, the principle of superposition will prove useful for the linear system since it will allow us to find solutions for each input separately and then permit the simple addition of the solutions to find a solution for all the inputs acting simultaneously.

In principle, any time function might be studied as an input, but, in practice, sinusoidal forcing is of overwhelming importance. This is partly because periodic forcings may be decomposed into sinusoidal components using the Fourier series and then the effects of the component forcing functions superposed using the principle of superposition, and

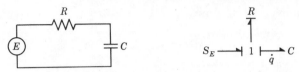

Figure 6.13. An electrical circuit analogous to the thermal system of Figure 6.1.

partly because the testing of real devices is conveniently accomplished in many cases by using sinusoidal inputs at frequencies varied over a range of interest.

For linear systems, it is convenient to represent sinusoidal waves in terms of complex exponentials. A straightforward way to represent sines and cosines may be developed from Eqs. (6.54) and (6.55), but it turns out to be simpler to use the real and imaginary part operators of Eqs. (6.59) and (6.60). Here we will illustrate the use of Eq. (6.59).

Initially, let us return to a first-order system such as the thermal system of Figure 6.1. Since it is hard to imagine the temperature of an oil bath cycling sinusoidally at very high frequencies, let us consider an electrical R–C circuit attached to an oscillatory voltage source as shown in Figure 6.13. The equation for the charge in the capacitor is readily found.

$$\dot{q} = \frac{-1}{RC}\, q + \frac{E(t)}{R}.\tag{6.78}$$

If we assume that $E(t)$ is sinusoidal with adjustable forcing frequency, ω_f,

$$E(t) = E_0 \cos \omega_f t = \mathrm{Re}(E_0 e^{j\omega_f t}),\tag{6.79}$$

then we can search for the *steady-state response*, $q(t)$, that results after any transient response due to the free response and the initial conditions has died away. (Since this system is stable, whatever free response is needed to fit initial conditions will indeed decay in time, leaving only the steady-state or forced response.)

For the linear operations of multiplication by a constant and differentiation with respect to time, or integration, it can be shown that it is immaterial whether one first operates on a complex function and then takes the real part of the result, or first takes the real part and then performs the operation. Thus it is possible to suppress the Re operator in Eq. (6.79) until all operations are completed and then take the real part of the final result. To find the forced response of the system, it is only necessary to assume the system variables are all sinusoidal with frequency, ω_f. (This assumption is not valid for nonlinear systems.) In the present example, we assume

$$q(t) = \text{Re}(Qe^{i\omega_f t}). \tag{6.80}$$

Substituting Eq. (6.79) and Eq. (6.80) into Eq. (6.78) and suppressing Re, we have

$$j\omega_f Qe^{i\omega_f t} = \frac{-Qe^{i\omega_f t}}{RC} + \frac{E_0}{R} e^{i\omega_f t}, \tag{6.81}$$

which leads immediately to

$$Q = \frac{E_0/R}{j\omega_f + 1/RC} = \frac{CE_0}{RCj\omega_f + 1}. \tag{6.82}$$

Although we assumed that E_0 was a real sinusoidal amplitude in Eq. (6.79), it turns out that Q in Eq. (6.80) needs to be complex. At very low forcing frequencies, Q is almost purely real,

$$Q \to CE_0 \quad \text{as} \quad \omega_f \to 0, \tag{6.83}$$

but for high frequencies, Q is almost purely imaginary,

$$Q \to \frac{E_0}{Rj\omega_f} = \frac{-jE_0}{R\omega_f} \quad \text{as} \quad \omega_f \to \infty. \tag{6.84}$$

We also note that for high forcing frequencies, the magnitude of Q is proportional to $1/\omega_f$ and thus approaches zero.

Let us now plot Q in Eq. (6.82) as a complex number for $0 < \omega_f < \infty$. As an intermediate step, consider the denominator of Q, $j\omega_f + 1/RC$. This is shown in Figure 6.14a. Now Q itself is just E_0/R times the reciprocal of this complex number.

Recall, if you can, that the inverse of a complex number has a magnitude that is the inverse of the magnitude of the number and an angle that is the negative of the angle of the number. Using this fact, one can easily sketch the behavior of $(j\omega_f + 1/RC)^{-1}$ as ω_f varies as in Figure 6.14b. (Perhaps you can prove to yourself that the curve in Figure 6.14b should be a semicircle.)

Since Q is just E_0/R times the complex number shown in Figure 6.14b, we see that the angle of Q varies from 0 at $\omega_f = 0$ to $-90°$ or $-\pi/2$ as $\omega_f \to \infty$. What this means is that if we think of $E(t)$ in Eq. (6.79) as the real part of a phasor or rotating vector, then $q(t)$ in Eq. (6.80) is represented

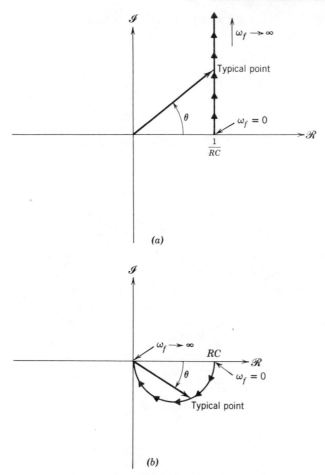

Figure 6.14. Plots of (a) $j\omega_f + 1/RC$ and (b) $(j\omega_f + 1/RC)^{-1}$ as ω_f varies.

by a rotating vector that lags behind the vector, $E(t)$, by an angle, θ (dependent on ω_f), that is somewhere between zero and 90°. Also, the magnitude of the vector representing q varies from CE_0 to zero as ω_f varies from zero to infinity.

A sketch of the rotating vectors is shown in Figure 6.15 for some particular ω_f. Since Q is shown as having an angle of about $-45°$, when Q is multiplied by $e^{j\omega_f t}$, the result is a vector rotating with speed, ω_f, but lagging about 45° behind $e^{j\omega_f t}$ since when one multiplies two complex numbers, one must multiply the magnitudes and add the phase angles. The vector representing $E(t)$, on the other hand, rotates with $e^{j\omega_f t}$ since E_0 is just a real number (zero angle).

By direct manipulation of Eq. (6.82), or by considering the sketches in Figure 6.14, we can write an expression for the angle of Q

$$\theta = \angle Q = -\tan^{-1} \frac{\omega_f}{1/RC} = -\tan^{-1} RC\omega_f \qquad (6.85)$$

The angle, Q, is often called the *phase angle* and is *negative* or *lagging* in this case. This notation is related to the rotating vectors of Figure 6.15. The amplitude or magnitude of Q is

$$|Q| = \frac{E_0/R}{[\omega_f^2 + (1/RC)^2]^{1/2}} = \frac{CE_0}{[(RC\omega_f)^2 + 1]^{1/2}} . \qquad (6.86)$$

Note that, except for the real factor, E_0/R, both $\angle Q$ and $|Q|$ are essentially plotted for all ω_f in Figure 6.14b.

It is common also to plot $\angle Q$ and $|Q|$ (or $\angle Q/E_0$ and $|Q/E_0|$) separately as a function of ω_f. These plots are called *Bode plots* when $|Q|$ versus ω_f is plotted on log-log scales and when $\angle Q$ is plotted versus log ω_f. Such plots are shown in Figure 6.16. Bode plots are convenient because they can often be quickly sketched using asymptotic behavior of the functions at low and high frequencies.

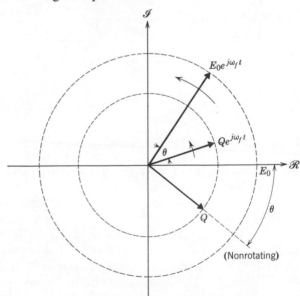

Figure 6.15.   Rotating vectors representing $E(t)$ and $q(t)$ in Eqs. (6.79) and (6.80).

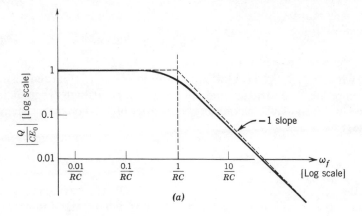

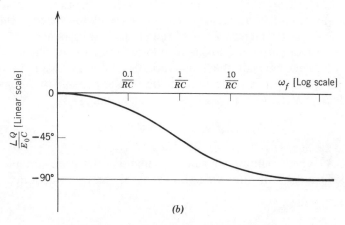

Figure 6.16. Bode plots for $Q/CE_0 = 1/(RCj\omega_f + 1)$.

In our case, we see that Q/CE_0 approaches unity for $\omega_f \rightarrow 0$ from Eq. (6.83). This means unit amplitude and zero phase angle. At high frequencies Eq. (6.84) tells us that

$$\frac{Q}{E_0 C} \rightarrow -j\frac{1}{(RC)\omega_f} \qquad \text{as} \qquad \omega_f \rightarrow \infty.$$

This means

$$\frac{\angle Q}{E_0 C} \rightarrow -90°,$$

$$\left|\frac{Q}{E_0 C}\right| \rightarrow \left(\frac{1}{RC}\right)\omega_f^{-1},$$

(6.87)

or

$$\log \left| \frac{Q}{E_0 C} \right| \rightarrow \log \left(\frac{1}{RC} \right) - \log \omega_f. \qquad (6.88)$$

Thus, when $\log |Q/E_0 C|$ is plotted versus $\log \omega_f$, we see that the result will be asymptotic to a straight line with a -1 slope at high frequencies.

The low-frequency asymptote and the high-frequency asymptote intersect at

$$RC\omega_f = 1 \qquad \text{or} \qquad \tau\omega_f = 1, \qquad (6.89)$$

using the definition of time constant in Eq. (6.20). The frequency, $\omega_f = 1/\tau$, is called the *break frequency* since this is where the low-frequency asymptote breaks into the high-frequency asymptote. At this frequency, the actual value of $|Q/E_0 C|$ is $1/\sqrt{2}$ as one may see from Eq. (6.86), and $\angle Q/EC$ is $-45°$. For a first-order system, the time constant and its reciprocal, the break frequency, characterize both the speed of the free response and the frequency scale of the sinusoidal response.

We now turn our attention to our second-order example, which has two state variables and two possible input variables. Suppose that the velocity source of Figure 6.3 is an electromagnetic shaker capable of forcing the input velocity, $V(t)$, in Eq. (6.23) to be sinusoidal:

$$V(t) = V_0 \cos \omega_f t = \text{Re}(V_0 e^{j\omega_f t}). \qquad (6.90)$$

To find a forced solution, we again assume that the state variables will also be sinusoidal with the same frequency as the input quantity:

$$x(t) = \text{Re}(X e^{j\omega_f t}), \qquad (6.91)$$

$$p(t) = \text{Re}(P e^{j\omega_f t}). \qquad (6.92)$$

Substituting Eqs. (6.90), (6.91), and (6.92) into Eq. (6.23) and neglecting F temporarily, we have

$$j\omega_f X e^{j\omega_f t} = \frac{-1}{m} P e^{j\omega_f t} + V_0 e^{j\omega_f t}, \qquad (6.93)$$

$$j\omega_f P e^{j\omega_f t} = kX e^{j\omega_f t} - \frac{b}{m} P e^{j\omega_f t} + bV_0 e^{j\omega_f t}, \qquad (6.94)$$

in which the Re operator has been suppressed. The problem now is an algebraic problem in which X and P are to be determined. The equations could be satisfied if $e^{j\omega_f t}$ should vanish, but this is not possible [and even $\text{Re}(e^{j\omega_f t})$ only vanishes at particular instants of time], so one attempts to satisfy the equations independent of $e^{j\omega_f t}$:

$$j\omega_f X + \frac{1}{m} P = V_0, \tag{6.95}$$

$$-kX + \left(j\omega_f + \frac{b}{m}\right)P = bV_0. \tag{6.96}$$

Note that there is some similarity between Eqs. (6.95) and (6.96) and Eq. (6.30a), but in the present case (1) $j\omega_f$ is given and (2) the right-hand sides of the equations are nonzero. These equations are readily solved as long as the determinant of the coefficients is not zero. For example, using Cramer's rule, we have

$$X = \frac{\begin{vmatrix} V_0 & \dfrac{1}{m} \\[2ex] bV_0 & \left(j\omega_f + \dfrac{b}{m}\right) \end{vmatrix}}{\begin{vmatrix} j\omega_f & \dfrac{1}{m} \\[2ex] -k & \left(j\omega_f + \dfrac{b}{m}\right) \end{vmatrix}} = \frac{(j\omega_f)V_0}{-\omega_f^2 + \dfrac{b}{m}j\omega_f + \dfrac{k}{m}}, \tag{6.97}$$

$$P = \frac{\begin{vmatrix} j\omega_f & V_0 \\[1ex] -k & bV_0 \end{vmatrix}}{\begin{vmatrix} j\omega_f & \dfrac{1}{m} \\[2ex] -k & \left(j\omega_f + \dfrac{b}{m}\right) \end{vmatrix}} = \frac{(bj\omega_f + k)V_0}{-\omega_f^2 + \dfrac{b}{m}j\omega_f + \dfrac{k}{m}}. \tag{6.98}$$

Note that X and P, the amplitudes of the complex exponential representation of the state variable forced response, are ratios of two complex numbers that vary with ω_f. The denominator is always the same; it is the same determinant used to generate the characteristic equation, Eq. (6.31), but with s replaced by $j\omega_f$. It is not difficult to evaluate X and P for any given ω_f, but it is of even more interest to determine how X and P vary as ω_f varies from zero to infinity. In Figure 6.17a the denominator expression for Eqs. (6.97) and (6.98) is plotted as a complex number as ω_f varies. In Figure 6.17b the reciprocal complex number is plotted. These plots are the second-order analogs of the plots in Figure 6.14 for the

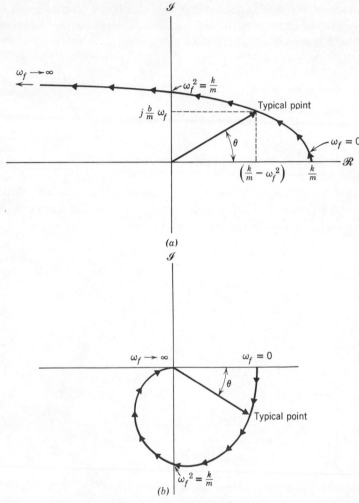

Figure 6.17. Plots of (a) $k/m - \omega_f^2 + j(b/m)\omega_f$ and (b) $[k/m - \omega_f^2 + j(b/m)\omega_f]^{-1}$ in the complex plane as ω_f varies from $\omega_f = 0$ to $\omega_f \to \infty$.

first-order system. Bode plots for amplitude and phase of the denominator factor are shown in Figure 6.18. The general shape of the Bode plots may be seen by studying Figure 6.17b, but it is useful to consider some asymptotic cases in order to determine the plots more precisely.

For a start, consider very low frequencies:

$$\lim_{\omega_f \to 0} \left(\frac{1}{k/m - \omega_f^2 + j\omega_f(b/m)} \right) = \frac{1}{k/m} = \frac{m}{k}. \tag{6.99}$$

Therefore, a horizontal asymptote may be drawn on the left side of Figure 6.18a, and the phase angle in Figure 6.18b for very low frequencies is zero. (Note that because of the logarithmic scale $\omega_f = 0$ cannot appear on the sketches, but the far left on the diagrams corresponds to very small values of ω_f.)

At the *undamped* natural frequency the expression also simplifies:

$$\frac{1}{[(k/m) - \omega_f^2] + j\omega_f(b/m)} \bigg|_{\omega_f = (k/m)^{1/2}} = \frac{1}{j(k/m)^{1/2}(b/m)} = -j\left(\frac{m}{k}\right)\frac{(mk)^{1/2}}{b}$$

$$= \frac{m}{k}\frac{(mk)^{1/2}}{b} \angle \frac{-\pi}{2}. \quad (6.100)$$

At the undamped natural frequency the magnitude is as given in Eq. (6.100), and the phase angle is exactly $-90°$ or $-\pi/2$ radius.

Finally, for very high frequencies, we have

$$\lim_{\omega_f \to \infty} \frac{1}{k/m - \omega_f^2 + j\omega_f(b/m)} = \frac{1}{-\omega_f^2} = \frac{1}{\omega_f^2} \angle -\pi. \quad (6.101)$$

Equation (6.101) indicates that the magnitude of the function varies as ω_f^{-2} and its phase is $-\pi$ radius or $-180°$ at high frequencies. This indicates that the logarithm of the amplitude is -2 times the logarithm of ω_f—hence the two-to-one negative slope shown on the amplitude plot. The reader should verify that the high- and low-frequency asymptotes intersect as shown in Figure 6.18a when extended to the undamped natural frequency. The use of asymptotes that plot as straight lines on a log-log plot, corresponding to terms of the form ω_f^m where m is usually a positive or negative integer, proves to be of general usefulness.

Now the plots of Figure 6.18 may be used to plot the frequency behavior of X and P from Eqs. (6.97) and (6.98). The numerator functions for X and P (apart from the constant factor, V_0) have been plotted in Figure 6.19 using the same concepts as in Figure 6.18. To sketch X and P as a function of ω_f, we must multiply the complex numbers plotted in Figure 6.18 with those in Figure 6.19. This can be done by *adding* the phase angles and multiplying the amplitudes or equivalently, *adding the logarithms of the magnitudes* at each ω_f. This process is illustrated in Figures 6.20 and 6.21.

It is also possible to construct asymptotic expressions for X and P directly from their defining relations, Eqs. (6.97) and (6.98), but the method just illustrated indicates how the common denominator expression interacts with different numerators to produce the frequency response of all the state variables.

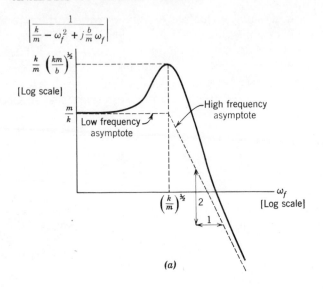

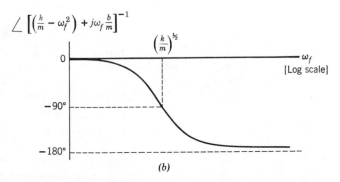

Figure 6.18. Bode plots of (a) amplitude and (b) phase of $[(k/m - \omega_i^2) + j(b/m)\omega_i]^{-1}$.

Finally, it remains to interpret the significance of $|X|$, $|P|$, $\angle X$, $\angle P$ in light of Eqs. (6.91) and (6.92). In Figure 6.22 a rotating vector representing $V(t) = V_0 e^{j\omega_i t}$ is shown, and the complex number, X, constant for any particular ω_i, is also shown. The rotating vector representing $x(t) = X e^{j\omega_i t}$ has amplitude $|X|$ and an angle $\angle X$ relative to V, (shown negative). Thus, in this case, the $x(t)$ vector lags the $V(t)$ vector by the phase angle shown in Figure 6.20. Actually, we consider that $V(t)$ is the real part of a rotating vector (a cosine wave in this case) and $x(t)$ is the real part of $X e^{j\omega_i t}$ (a

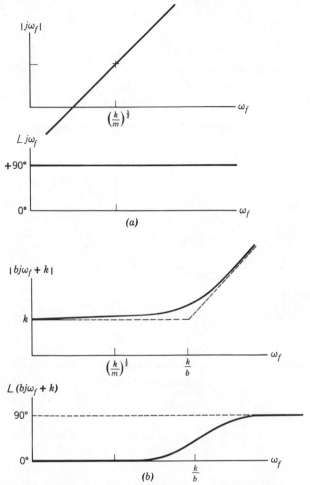

Figure 6.19. Bode plots for the numerators of (a) Eq. (6.97) and (b) Eq. (6.98).

cosine wave with amplitude $|X|$ and a lagging phase angle given by $\angle X$).

$$x(t) = |X| \cos(\omega_f t - \angle X). \qquad (6.102)$$

Of course, it is possible to encounter leading phase angles also.

6.4.1 Normalization of Response Curves

Although, up to now, only the physical parameters, m, b, and k, have been used in studying the frequency response, it sometimes proves

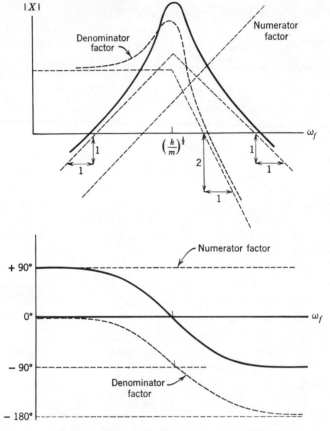

Figure 6.20. Bode plot construction for X.

convenient and insightful to nondimensionalize expressions as much as possible. As a first step in this, consider rewriting the response quantities, X and P, in Eqs. (6.97) and (6.98) in terms of damping ratio and undamped natural frequency, Eq. (6.41) and Eq. (6.66):

$$\frac{X}{V_0} = \frac{j\omega_f}{\omega_n^2 - \omega_f^2 + j2\zeta\omega_n\omega_f},$$

$$\frac{P}{V_0} = \frac{m(j2\zeta\omega_n\omega_f + \omega_n^2)}{\omega_n^2 - \omega_f^2 + j2\zeta\omega_n\omega_f}.$$

By defining a frequency ratio,

$$\beta \equiv \frac{\omega_f}{\omega_n}, \tag{6.103}$$

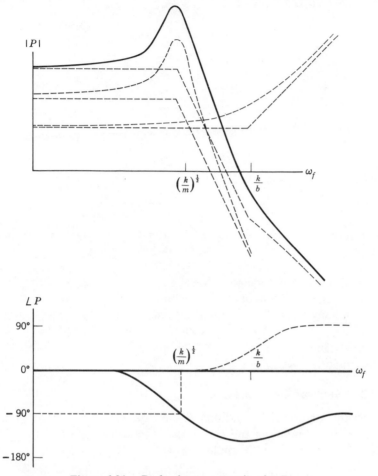

Figure 6.21. Bode plot construction for P.

further simplification is achieved:

$$\frac{X}{V_0} = \frac{(m/k)^{1/2}j\beta}{(1-\beta^2)+j2\zeta\beta},\qquad(6.104)$$

$$\frac{P}{V_0} = \frac{m(j2\zeta\beta+1)}{(1-\beta^2)+j2\zeta\beta}.\qquad(6.105)$$

Since all aspects of the sinusoidal forced response of the system contain the function,

$$H(\beta) = [(1-\beta^2)+j2\zeta\beta]^{-1},\qquad(6.106)$$

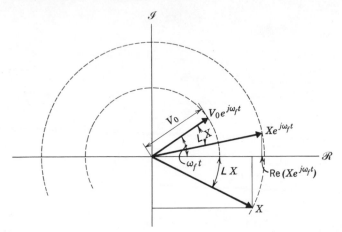

Figure 6.22. Representation of $V(t)$ and $x(t)$ in terms of complex exponentials.

it is useful to show Bode plots for H as in Figure 6.23. Note that by using β, which essentially involves the definition of nondimensional frequency (or nondimensional time), only ζ remains as a parameter in H. In computing actual frequency response quantities such as X and P, however, some physical parameters are necessary in general, as illustrated in Eqs. (6.104) and (6.105).

6.4.2   General Comments

For any state-space equations of the form of Eq. (6.1a) the pattern of analysis for computing the sinusoidal response is just the same as has been illustrated in the example.

1. Recognizing that if there are multiple inputs, the responses may be computed individually and later simply summed in order to find the total forced response, pick one input, say $u_1(t)$, and if it is sinusoidal, represent it in complex exponential form, for example,

$$u_1(t) = \text{Re}(U_1 e^{j\omega_f t}).$$

2. Let each x variable be represented as a complex exponential, for example,

$$x_1(t) = \text{Re}(X_1 e^{j\omega_f t})$$

Since only linear operations will be required, one may temporarily suppress the Re operator and simply apply it to the final result.

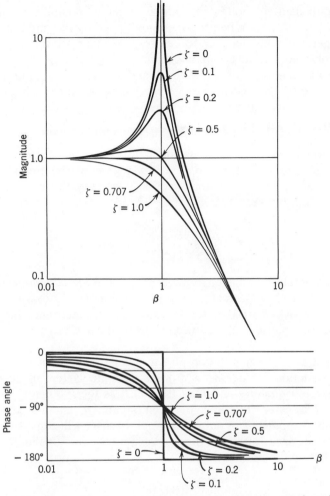

Figure 6.23. Bode plots for normalized second-order response function, Eq. (6.106).

3. Substituting into Eq. (6.1a) one is faced with the algebraic problem below:

$$(j\omega_f - a_{11})X_1 + (-a_{12})X_2 + \cdots + (-a_{1n})X_n = b_{11}U_1,$$

$$(-a_{21})X_1 + (j\omega_f - a_{22})X_2 + \cdots + (-a_{2n})X_n = b_{21}U_1,$$

$$\cdots \qquad \cdots \qquad \cdots \qquad \cdots$$

$$(-a_{n1})X_1 + (-a_{n2})X_2 + \cdots + (j\omega_f - a_{nn})X_n = b_{n1}U_1.$$

4. When this equation set is solved, using Cramer's rule or any

equivalent method, the result is a set of complex numbers for $X_1, X_2, \ldots, X_n$. When these are multiplied by $e^{j\omega_f t}$ and the real-part operator is applied, the forced response computation is complete. Polar plots in the complex plane as in Figures 6.14 or 6.17 or Bode plots such as Figures 6.16, 6.18, and 6.21 may aid in interpretation of the forced response as a function of the forcing frequency.

6.5 TRANSFER FUNCTIONS

The term *transfer function* is much used in connection with linear systems and usually refers to a generalization of the frequency concepts presented in the previous section. One way to define a transfer function, $H_{y/x}$, between two quantities, x and y, is

$$H_{y/x} \equiv \frac{y}{x}\bigg|_{x=e^{st}} = H_{y/x}(s), \qquad (6.107)$$

where the notation is meant to imply that if the *forced response* of y is computed when $x = e^{st}$, then the ratio y/x will be $H_{y/x}$ and will depend on s. When $s = j\omega_f$, $H_{y/x}(j\omega_f)$ is the function which determines the relative amplitude and phase angle between sinusoidal waveforms representing x and y.

To illustrate the computation of a transfer function, let us compute a transfer function for our second-order example system using f in Eq. (6.24) as the y quantity and F in Eq. (6.23) as the x quantity. For state equations, transfer functions between inputs, u, and outputs, y, are commonly computed. For our present purposes we will consider that V in Eq. (6.23) vanishes although there is also a transfer function between f and V that could be computed.

If

$$F(t) = e^{st} \qquad (6.108)$$

for forced response, it may be assumed that all variables have the same time dependence, but with amplitude factors to be determined.

$$x(t) = Xe^{st}, \qquad (6.109)$$

$$p(t) = Pe^{st}. \qquad (6.110)$$

Then, from Eq. (6.24)

$$f(t) = kXe^{st} - \frac{b}{m}Pe^{st} = \left(kX - \frac{b}{m}P\right)e^{st}. \qquad (6.111)$$

To find X and P, one again substitutes the assumed forms into the state equations,

$$sXe^{st} = -\frac{1}{m} Pe^{st},$$

$$sPe^{st} = kXe^{st} - \frac{b}{m}Pe^{st} + e^{st},$$

or, upon suppressing the e^{st} terms, the following algebraic equations emerge:

$$sX + \frac{P}{m} = 0, \tag{6.112}$$

$$-kX + \left(s + \frac{b}{m}\right)P = 1. \tag{6.113}$$

This problem closely resembles the problem of Eqs. (6.95) and (6.96) except that s has replaced $j\omega_f$ and the input is F instead of V.

Solving for X and P using Cramer's rule, we have

$$X = \frac{\begin{vmatrix} 0 & 1/m \\ 1 & (s+b/m) \end{vmatrix}}{\begin{vmatrix} s & 1/m \\ -k & (s+b/m) \end{vmatrix}} = \frac{-1/m}{s^2 + (b/m)s + k/m}, \tag{6.114}$$

$$P = \frac{\begin{vmatrix} s & 0 \\ -k & 1 \end{vmatrix}}{\begin{vmatrix} s & 1/m \\ -k & (s+b/m) \end{vmatrix}} = \frac{s}{s^2 + (b/m)s + k/m}. \tag{6.115}$$

Now, using Eq. (6.111), the result is

$$f(t) = \left(\frac{-(k/m)-(b/m)s}{s^2 + (b/m)s + k/m}\right)e^{st}, \tag{6.116}$$

and the desired transfer function, $H_{f/F}(s)$, is

$$H_{f/F}(s) = \frac{f(t)}{F(t)}\bigg|_{F(t)=e^{st}} = \frac{-[(b/m)s + k/m]}{s^2 + (b/m)s + k/m} = \frac{-(2\zeta\omega_n s + \omega_n^2)}{s^2 + 2\zeta\omega_n s + \omega_n^2}. \tag{6.117}$$

Clearly, a transfer function immediately yields a frequency response function if $j\omega_f$ is substituted for s. For example, if

$$F(t) = A \cos \omega_f t = \mathrm{Re}(Ae^{j\omega_f t}), \tag{6.118}$$

then

$$f(t) = \mathrm{Re}(H_{f/F}(j\omega_f)Ae^{j\omega_f t}). \tag{6.119}$$

Block Diagrams

Block diagrams are commonly used to represent transfer functions, and the block shown in Figure 6.24a indicates that the output signal, f, is the product of the input signal, F, and the transfer function, $H_{f/F}$.

For signals with the time dependence e^{st}, differentiation is equivalent to multiplication by s, integration to dividing by s. Thus the transfer function for an integrator is $1/s$. Using this notion, the block diagram of Figure 6.24b represents the state equations (6.23) and (6.24) directly and shows that causally F is an input and f is an output. It is possible to show that Figure 6.24b is equivalent to Figure 6.24a using essentially graphical means. The major tool is the well-known relation shown in Figure 6.25 for combining feedback loops. Although much of classical

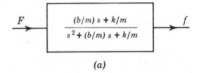

(a)

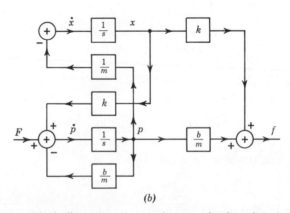

(b)

Figure 6.24. Block diagrams representing transfer functions between F and f. (a) Combined transfer function; (b) direct representation of state and output equations.

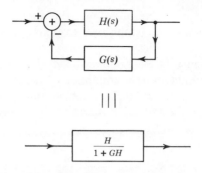

Figure 6.25.   Block diagram reduction for feedback loops.

control theory and linear system analysis is concerned with such techniques it will suffice for present purposes to point out that since bond graphs generate highly organized state equations only the simple algebra needed to solve equations such as Eqs. (6.112) and (6.113) is required to generate transfer functions between any input and any output. There are some cases where block-diagram-reduction techniques provide convenient guides to accomplishing algebraic manipulation, but for large systems computer-automated algebraic methods presented in Ref. [1], for example, are far superior to graphical techniques in avoiding human error.

6.6   TOTAL RESPONSE

The total response of a system is determined by initial conditions and given input time functions and, for linear systems, is a sum of a free response and a forced response. It is possible to compute total response using the methods of Sections 6.3 and 6.4, although for most systems of practical interest, it is more practical simply to simulate the system using analog- or digital-computer techniques. It is, however, important to understand the nature of the total response to be expected from a simulation.

To understand the role of the initial conditions and the forcing on total response, let us consider the problem of determining the response of the second-order example system to a suddenly applied cosine force. Let

$$F(t) = F_0 \cos \omega_f t = \mathrm{Re}(F_0 e^{j\omega_f t}), \qquad (6.120)$$

$$t \geq 0,$$

and

$$x(0) = p(0) = 0. \qquad (6.121)$$

This example will serve to point out the fact that vanishing initial conditions do not imply vanishing free response when a forcing term is present.

The forced response is easily found since in computing the transfer function in the previous section, amplitudes X and P were found when F was e^{st}. Equations (6.114) and (6.115) may be modified by multiplying by the amplitude factor, F_0, and substituting $s = j\omega_f$ in order to construct the forced solution, x_{forced} and p_{forced}:

$$x_{\text{forced}} = \text{Re}\left(\frac{-F_0/m}{-\omega_f^2 + (b/m)j\omega_f + k/m} \, e^{j\omega_f t}\right), \tag{6.122}$$

$$p_{\text{forced}} = \text{Re}\left(\frac{j\omega_f F_0}{-\omega_f^2 + (b/m)j\omega_f + k/m} \, e^{j\omega_f t}\right). \tag{6.123}$$

Suppose $\omega_f < \omega_n$ so that the phase angle sketched in Figure 6.18 is between $0°$ and $90°$ and $\zeta < 1$ so that there is some amplitude magnification over the amplitude at low frequency. Then

$$x_{\text{forced}} = -A_x \cos(\omega_f t - \phi), \tag{6.122a}$$

$$p_{\text{forced}} = -A_p \sin(\omega_f t - \phi). \tag{6.123a}$$

It is easy to see in Eq. (6.122) that x_{forced} is a negative cosine wave since ϕ accounts for the phase lag and the numerator contributes a negative sign. In Eq. (6.123) one must first imagine the rotating vector associated with $j\omega_f e^{j\omega_f t}$. This vector is rotated $90°$ *ahead* of the vector, $e^{j\omega_f t}$, and its real part is found to describe a negative sine wave. Another way to see this is to note that in a transfer function sense, multiplication by $j\omega t$ is equivalent to differentiation, and the derivative of $\cos \omega_f t$ is $-\omega_f \sin \omega_f t$. The functions x_{forced} and p_{forced} are sketched in Figure 6.26.

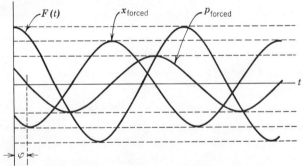

Figure 6.26. Sketches of forced response quantities.

Note that the forced response satisfies (or balances) the forced state equations at all times (even $t < 0$), and its only defect as a complete solution is that at $t = 0$ the forced response does not satisfy the initial conditions of Eq. (6.121). One may think of x_{forced} and p_{forced} as the system response that would have occurred if the cosine force had been applied a long time in the past and if the system had settled down to a steady-state type of response. (This interpretation is useful for stable systems which really do settle down.)

Even though x_{forced} and p_{forced} satisfy the forced state equations at all times, the free response satisfies the state equations without F, and, thus, if we write

$$x_{total} = x_{free} + x_{forced}, \tag{6.124}$$

$$p_{total} = p_{free} + p_{forced}, \tag{6.125}$$

and substitute into the state equations we find that the equations are still satisfied. Now the amplitudes of the free responses may be adjusted so that the *total response satisfies* the initial conditions. For the given zero initial conditions, Eq. (6.121), this results in

$$x_{free}(0) = -x_{forced}(0), \tag{6.126}$$

$$p_{free}(0) = -p_{forced}(0). \tag{6.127}$$

Equations (6.126) and (6.127) now may be used in Eqs. (6.38) and (6.39) after the forced response at $t = 0$ is extracted from Eqs. (6.122) and (6.123). Then, using Eqs. (6.35) and (6.36), enough conditions have been found to completely determine the free response. Although the determination of the free response in detail for a particular case is straightforward, even for a simple second-order system the process is rather tedious. The solutions are not hard to sketch, as shown in Figure 6.27. Note that the free responses start at $t = 0$ at values that are just the negative of the values of the forced response and then decay as damped oscillations at the system *damped natural frequency* and with the system damping ratio.

When the free and forced responses are summed, the result is as shown in Figure 6.28. Note that at the start of the transient, the solution appears confused because the free response at the system frequency interacts with the forced response at the forcing frequency. After some time, however, the free response becomes very small for this stable system due to the factor $e^{-\zeta \omega_n t}$, and the forced response dominates. If the system were unstable, of course, the free response would blow up as time increased and the forced response would not be important after some time.

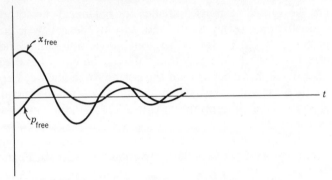

Figure 6.27. Sketches of free response quantities.

The major characteristics worth keeping in mind for the total response of linear systems are as follows:

1. The free-response components of all state variables (and hence, output variables) always have a time behavior dependent on the system eigenvalues; that is, oscillatory response components will be associated with each pair of complex eigenvalues, and a natural frequency and damping ratio may be associated with each pair; exponential response components will be associated with each real eigenvalue, and a time constant may be associated with each such eigenvalue. Changing the initial conditions changes the amplitudes of the free-response components of the total response.

2. The forced-response components of all state variables all have a time dependence determined by the given inputs. If an input is sinusoidal, all state variables will contain a forced component at the same frequency as the input. The forced components of response are directly proportional to the amplitude of the input and are not influenced by initial conditions.

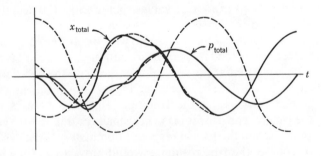

Figure 6.28. Sketches of total response.

When simulation is used to find system response, only the total response will be found, and it is a mental exercise to imagine the splitting of the response into the free and forced parts. If the input is itself transient in contrast to the case of sinusoidal waves or steps which go on forever, then the valid concept of splitting the response into two types of response may not be very useful.

It is also important to remember that nonlinear systems may react in much more complicated ways than linear systems.

Nonlinear systems excited by sinusoidal inputs can, in contrast to linear systems, respond with periodic solutions containing frequencies other than the exciting frequency, and initial conditions can have a drastic effect on nature of the response of a nonlinear system to a given forcing function.

6.7   OTHER REMARKS

Although only physically motivated state variables have been used in this book, it is possible and sometimes analytically convenient to consider other equivalent state variables. For example, the state equations for the undamped oscillator,

$$\dot{x} = \frac{-1}{m} p + V,$$

$$\dot{p} = kx + F,$$

$$(6.40)$$

may be transformed into a new set of state equations in x_1 and x_2 by defining the linear transformation

$$x = x_1 + x_2,$$

$$p = -j(km)^{1/2}x_1 + j(km)^{1/2}x_2,$$

$$(6.128)$$

and the corresponding inverse transformation,

$$x_1 = \tfrac{1}{2}x + \frac{jp}{2(km)^{1/2}},$$

$$x_2 = \tfrac{1}{2}x - \frac{jp}{2(km)^{1/2}}.$$

$$(6.129)$$

Using the transformation and substituting into the original state equation, state equations for x_1 and x_2 may be derived.

$$\dot{x}_1 = j\omega_n x_1 + \frac{V}{2} + \frac{kF}{2(km)^{1/2}},$$

$$\dot{x}_2 = -j\omega_n x_2 + \frac{V}{2} - \frac{jF}{2(km)^{1/2}}.$$

(6.130)

The interesting feature of this set of state equations is that the state variables are not coupled to each other and the system eigenvalues are in clear evidence. When the eigenvalues are distinct, state equations may be transformed as in the example, and even if some of the roots of the characteristic equation are repeated a slightly more coupled normal form, called the Jordan normal form, can be found.

Although Eq. (6.130) is useful for theoretical considerations (in control theory, for example), it is hard to interpret the physical significance of x_1 and x_2 except through the formalism of the transformation. Clearly, x_1 and x_2 are complex even though the system variables, x and p, are not. For this reason, even when the x_1, x_2 state equations are used to design a controller or to derive system characteristics, it is generally necessary to retransform back into physically meaningful variables for interpretation of results or implementation of a control scheme. This type of transformation is conveniently discussed using matrix notation, and the study of such transformations forms an important part of modern control theory.

Another type of transformation is between the n first-order state space equations that are the basic output of bond-graph techniques and nth-order differential equations which are equivalent. One way in which differential equations are sometimes found is by interpreting the s variable in transfer functions as a d/dt operator. If one were given the transfer function in Figure 6.24a or Eq. (6.117), for example, one could write

$$\left(s^2 + \frac{b}{m}s + \frac{k}{m}\right)f = \left(\frac{b}{m}s + \frac{k}{m}\right)F(t),$$

(6.131)

and hence a differential equation,

$$\ddot{f} + \frac{b}{m}\dot{f} + \frac{k}{m}f = \frac{b}{m}\dot{F}(t) + \frac{k}{m}F(t).$$

(6.132)

It is certainly true that Eq. (6.132) would generate the same transfer function that was found from the original state equations. It is also true that f and $\dot{f}$ may be regarded as state variables in some sense. For example, if one calls $\dot{f}$ by a new name, say, g, then Eq. (6.132) may be rewritten as follows:

$$\dot{f} = g,$$

$$\dot{g} = -\frac{b}{m} g - \frac{k}{m} f + \frac{b}{m} \dot{F}(t) + \frac{k}{m} F(t). \tag{6.133}$$

This is almost in the standard form for state space equation—only the necessity to differentiate $F(t)$ seems bothersome. But this too can be avoided by considering the basic input to be $\dot{F}(t) \equiv H(t)$ and defining an extra state variable, X, so that the integral of H will be a state variable.

$$\dot{f} = g,$$

$$\dot{g} = \frac{b}{m} g - \frac{k}{m} f + \frac{b}{m} H(t) + \frac{k}{m} X, \tag{6.134}$$

$$\dot{X} = H(t).$$

Now it is clear that the third-order system of Eq. (6.134) is somewhat artifical since the block diagram of Figure 6.24b or the original state space equations provided a second-order system that yielded the given transfer function. Although there are techniques to convert Eq. (6.132) from its second-order form into two state equations analogous to the original state equations without introducing superfluous state variables, it is again true that the state variables may be hard to interpret physically. For this reason, it is generally preferable to start with physically motivated state equations, and, if desired, to derive nth-order equations rather than the reverse. Although any nth-order equation may be found in the time domain by differentiating and combining the first-order state equations, it is generally simpler to do the algebra in the s domain, as in deriving transfer functions, and then simply to interpret s as the derivative operator in order to write the final differential equation.

Finally, it should be mentioned that several analytical methods of importance for linear system analysis have not been mentioned in this chapter since they are not required for the purposes of this book. Fourier and Laplace transformation techniques, for example, form some of the heavy artillery of analytical methods for linear systems, but their discussion would take too long for a single chapter and there are numerous excellent presentations of these topics. The simpler techniques of this chapter will suffice for the purposes of this book, and since we deal with general systems, many of which are not linear, techniques limited to linear systems will not be presented in depth here. In later chapters, it will be assumed that the reader has some familiarity with more advanced techniques for linear systems, but such a knowledge will not be essential to understanding the system dynamics discussed, and references for self-study will be included where appropriate.

REFERENCE

1. J. L. Melsa and S. K. Jones, *Computer Programs for Computational Assistance in the Study of Linear Control Theory*, 2nd ed., N.Y.: McGraw-Hill, 1973.

PROBLEMS

6-1 Consider Eq. (6.12) with time constant $\tau = RC = 1.5$ sec. At $t = 0$ let $T_s = 1500°F$ and let T_0 be $100°F$ and constant. Compute the initial behavior of T_s by using Euler's method, Eq. (6.6). Choose the time step, Δt, for reasonable accuracy based on τ. Plot T_s versus t for several steps and show that when the ratio $\Delta t / \tau$ is too large, the integration scheme is very inaccurate. Find an analytic expression for T_s by considering the exponential decay of T_s to T_0.

6-2 Consider the undamped oscillator with natural frequency of 10 rad/sec. See Eq. (6.40). Determine the free response starting from zero initial position and an initial velocity. Apply Euler's integration method, Eq. (6.6), for a few steps. What would be a reasonable way to pick Δt so that the digital solution would be accurate?

6-3

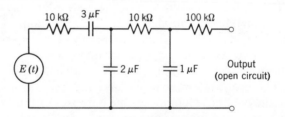

For the filter circuit shown, write state and output equations with the voltage, $E(t)$, as the input and the terminal voltage as the output. Write the characteristic equation from which the eigenvalues could be determined.

6-4 Consider the bond graph shown below:

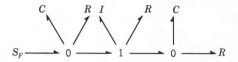

Assuming that all elements are linear, write state equations. Pick two variables that could be output variables, and write output equations for them. Identify the coefficients in Eqs. (6.1a) and (6.2a) with your equation coefficients as was done in Eq. (6.25). Sketch mechanical, electrical, and hydraulic systems that could be modeled with this bond graph.

6-5 Using equations such as Eq. (6.59) or sketches in the complex plane, demonstrate that

$$\frac{d}{dt} \text{Re}(e^{j\omega t}) = \text{Re}\left(\frac{d}{dt} e^{j\omega t}\right).$$

6-6 If an automobile weighs 3000 lb and has an undamped natural frequency of 1.0 Hz when it oscillates on its main suspension springs, what should the total stiffness of the four shock absorbers be if a damping ratio of 0.707 is to be achieved? Consider the shock absorbers to be linear dampers even though, in practice, shock absorbers are distinctly nonlinear.

6-7 Sketch a frequency-response plot for the output voltage divided by the input voltage for the system of Problem 6.3.

6-8 In the system shown, consider $V(t)$ to be a velocity input. Set up state equations for the system, and let the acceleration of the top mass be a system output. (Note that the acceleration is proportional to the total force on the mass.) Find the ratio of the acceleration to

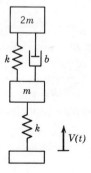

the input velocity when the velocity is sinusoidal, and sketch the frequency-response plot.

6-9 If a first-order system is excited by a cosine wave starting from a zero initial condition at $t = 0$, sketch the forced response, free response, and total response as was done in Figures 6.26–6.28.

6-10 By labeling signals and writing equations implied by the blocks, demonstrate the validity of the block diagram identity of Figure 6.25.

6-11 Using Figure 6.25, reduce Figure 6.24b to Figure 6.24a.

6-12

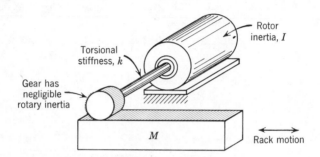

The rotor of the motor shown is connected through a relatively flexible shaft to the pinion which drives the rack and load mass, M. Assuming the pinion may be modeled as having no rotary inertia, construct a bond graph for the free motion of the system. Find the eigenvalues of the system and interpret them.

6-13

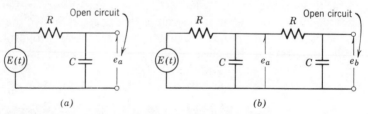

Figures a and b depict two linear filters which are supposed to remove unwanted high-frequency noise from a voltage signal, $E(t)$. In Figure a the output voltage measured when no current is allowed to flow is e_a; in Figure b the output is e_b. All resistors have the same resistance, R, and all capacitors have the capacitance C.

(a) For a derive state equations, and relate the output, e_a, to the state variables. For b find state equations, and relate both e_a and e_b to the state variables.

(b) For a find the transfer function relating e_a and E, $H_{e_a/E}(s)$. For b find transfer functions for e_a and e_b, $H_{e_a/E}(s)$ and $H_{e_b/E}(s)$.

(c) Shouldn't the two transfer functions for e_a in Figure a and Figure b be equal? Shouldn't the transfer function for e_b be the square of the transfer function for e_a? If not, why not?

6-14

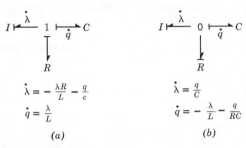

$$\dot{\lambda} = -\frac{\lambda R}{L} - \frac{q}{c}$$

$$\dot{q} = \frac{\lambda}{L}$$

(a)

$$\dot{\lambda} = \frac{q}{C}$$

$$\dot{q} = -\frac{\lambda}{L} - \frac{q}{RC}$$

(b)

(a) Draw the circuits that correspond to the bond graphs and state equations.

(b) Describe qualitatively the effect in both cases when $R \to 0$ and $R \to \infty$.

(c) Find relations for the system eigenvalues in both cases using the following procedure:

  i. Let $\dot{x} = Ax$ be the state equation.

  ii. If $x = Xe^{st}$ then $sX = AX$ or $[sI - A][X] = [0]$ if $[I]$ is the unit matrix.

  iii. The eigenvalues are values of s for which $\det[sI - A] = 0$ since only when the determinant vanishes can X be nontrivial.

Note that A is given so all that is required is to form $[sI - A]$ and set is determinant equal to zero. The results should square with (b).

6-15

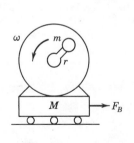

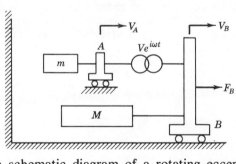

The sketches show a schematic diagram of a rotating eccentric vibrator and a "mechanical network" representation of the device. The assumption is that the motor produces a simple harmonic relative horizontal velocity of

$$V \cos \omega t = r\omega \cos \omega t = R_e\{Ve^{j\omega t}\}$$

between the rotating mass, m, and the main mass, M. Because of the assumed constraints, we do not discuss vertical forces or vertical motion.

(a) Construct a bond graph for the system, and fully augment it. Assume F_B is a known force applied to the system.

(b) Write a state equation for the system, and write output equations for V_A and V_B.

(c) Verify, using the results of (b), that the following terms sometimes used to describe the exciter are valid when $Ve^{j\omega t}$ is the complex representation of $V \cos \omega t$ for the velocity source:

   i. The *internal mobility* looking back at B is the ratio of velocity to force when the velocity generator is stopped and locked in position.

$$\frac{V_B}{F_B} = \frac{1}{j\omega(M+m)}.$$

   ii. The *blocked force* output at B is F_B when some external agent prevents motion of B.

$$F_{B\,(\text{blocked})} = j\omega m V$$

   iii. The *free velocity* output at B is the velocity at B when there is no force F_B.

$$V_{B\,(\text{free})} = \frac{m}{M+m}\, V.$$

   iv. A general network theorem states that the internal mobility is the quotient of the free velocity and the blocked force. See H. H. Skilling, *Electrical Engineering Circuits*, N.Y.: Wiley, 1957, p. 339. Verify the theorem for this example.

7

MULTIPORT FIELDS AND
JUNCTION STRUCTURES

In the first part of this book, dynamic models for a variety of physical systems were constructed using the 1-port elements, $-R$, $-C$, $-I$, S_e-, S_f-; the 2-port elements, $-TF-$ and $-GY-$, and the 3-port, 0- and 1-junctions. In this chapter, *fields*, which are multiport generalizations of $-R$, $-C$, and $-I$ elements, are introduced, and the concept of a *junction structure*, which is an assemblage of the power-conserving elements,

$-0-$, $-1-$, $-TF-$, and $-GY-$, is presented. Using fields and junction structures, one may conveniently study systems containing complex multiport components using bond graphs. In fact, bond graphs with fields and junction structures prove to be a most effective way to handle the modeling of complex multiport systems, combining both structural detail and clarity with ease of visualization. Succeeding chapters demonstrate the application of the elements introduced here to the modeling of systems containing multiport devices.

7.1 ENERGY-STORING FIELDS

In Chapter 3 it was shown that the elements $-C$ and $-I$ can store energy and can give back stored energy without loss. This is true no matter what the constitutive laws for these elements happen to be. Any single-valued functional relationship between effort and displacement defines an energy-conservative capacitor, and any single-valued relationship between flow and momentum defines an energy-conservative inertia element. The multiport generalizations of $-C$ and $-I$, which we will call C-fields and I-fields, respectively, will also be conservative. As we shall see, the conservation of energy in the multiport cases imposes a constraint on the constitutive laws for C- and I-fields.

7.1.1   C-Fields

The symbol for a C-field is simply the letter, C, with as many bonds as the C-field has ports. An n-port C-field is shown in Figure 7.1. Note that the flow at the ith port has been labeled $\dot{q}_i$ and that an inward sign convention is assumed. The energy, $\mathbf{E}$, stored in the C-field can then be expressed as

$$\mathbf{E} = \int_{t_0}^{t} \sum_{i=1}^{n} (e_i f_i)\, dt = \int_{t_0}^{t} \sum_{i=1}^{n} e_i \dot{q}_i\, dt$$

$$= \int_{q_0}^{q} \sum_{i=1}^{n} e_i(\mathbf{q})\, dq_i = \int_{q_0}^{q} \mathbf{e}(\mathbf{q})\, d\mathbf{q} = \mathbf{E}(\mathbf{q}), \tag{7.1}$$

in which the identity, $f_i\, dt = dq_i$, has been used, and column vectors of efforts and flows have been defined thus:

$$\mathbf{q} \equiv \begin{bmatrix} q_1 \\ q_2 \\ q_3 \\ \cdot \\ \cdot \\ \cdot \\ q_n \end{bmatrix}, \qquad \mathbf{e} \equiv \begin{bmatrix} e_1 \\ e_2 \\ e_3 \\ \cdot \\ \cdot \\ \cdot \\ e_n \end{bmatrix}, \tag{7.2}$$

and $\mathbf{e}\, d\mathbf{q}$ represents the scalar or inner product, which could also be represented as $\mathbf{e}^t\, d\mathbf{q}$ using the transpose of the $\mathbf{e}$ column matrix.

At time t_0, $\mathbf{q} = \mathbf{q}_0$, and at time t, $\mathbf{e} = \mathbf{e}(\mathbf{q})$ represents the constitutive laws of the C-field evaluated at $\mathbf{q}(t)$, the instantaneous value of the vector of displacements.

Before going on to study the properties of general C-fields as defined by Eq. (7.1), it may be worthwhile to exhibit some relatively simple C-fields as they arise in practice. Some components of systems are described

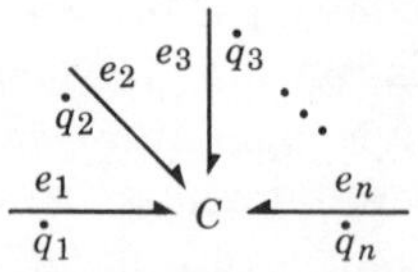

Figure 7.1.   The symbol for an n-port C-field.

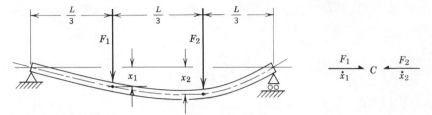

Figure 7.2. An elastic beam deformed by the action of two forces and represented as a 2-port C-field. Uniform beam: elastic modulus, E; area moment of inertia, I.

naturally by C-fields. The beam of Figure 7.2, for example, might be part of a vibratory system with masses or other elements attached at the locations indicated by the F_1, $\dot{x}_1$ and F_2, $\dot{x}_2$ ports. If one neglects the mass of the beam (which is permissible if the system frequencies are sufficiently low) and if one neglects at least temporarily the power losses associated with inelastic material behavior and the support pivots, then the beam is a purely elastic structure and can be represented as a 2-port C-field. The beam then may be represented by constitutive relations among F_1, F_2, x_1, and x_2.

Virtually all engineering undergraduates could find the constitutive laws for this C-field at some point in their careers, although when the skills involved are not exercised by an individual, the computation of beam deflections can rapidly become a lost art. A convenient way to represent the beam is through the use of *superposition* to add up the effects on x_1 and x_2 of the forces, F_1 and F_2. This procedure, (see Reference [1], for example), works only because the beam is represented for a *linear* range of strains and deflections. The result is conveniently expressed in a matrix form:

$$\begin{bmatrix} x_1 \\ \overline{} \\ x_2 \end{bmatrix} = \frac{L^3}{243EI} \begin{bmatrix} 4 & \frac{7}{2} \\ \hline \frac{7}{2} & 4 \end{bmatrix} \begin{bmatrix} F_1 \\ \overline{} \\ F_2 \end{bmatrix}. \tag{7.3}$$

This form of the constitutive laws, which gives displacements as a function of efforts, is the *compliance form*. As long as the matrix in Eq. (7.3) is invertible, one could solve the equation for F_1 and F_2 in terms of x_1 and x_2. This would be the *stiffness form* for the constitutive laws.

When an element is described from the beginning as a set of effort-displacement relations at n ports, as in the beam example, we will call the model of the element an *explicit field*. Generally, it is much more convenient to treat explicit field elements as fields, although in some cases it would be possible to find an equivalent system of 1-port energy-storing

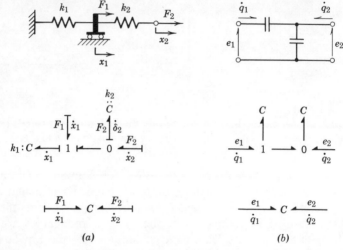

Figure 7.3. A C-field composed of 1-port C elements and junction elements. (a) A mechanical example; (b) an electrical example.

elements bonded together with junction structure elements that would have the same port constitutive laws as the field. On the other hand, when systems are assembled, it often happens that a group of 1-port energy-storing elements, say —Cs, are interconnected with —0—' —1—, and —TF— elements. Such a subsystem often can be usefully treated as a field. Such a field will be called an *implicit field* since the field constitutive laws at the external ports must be deduced from the constitutive laws of the elements comprising the field. An example is shown in Figure 7.3. Except for the sign convention on one internal bond, the electrical and mechanical systems shown in Figure 7.3 have the same bond graph. If one imposes integral causality on the 1-port C elements as in Figure 7.3a, then it is clear that the state variables are x_1 and δ_2, namely, the deflections of the two springs. Thus it is not immediately clear that the C-field representation in terms of x_1 and x_2 is possible. Using the mechanical example, let us work out the C-field constitutive laws.

The causal strokes indicate that we can find F_1 and F_2 as output variables in terms of the state variables x_1 and δ_2:

$$F_1 = k_1 x_1 - k_2 \delta_2,$$

$$F_2 = 0 x_1 + k_2 \delta_2,$$

(7.4)

where, for the sake of preserving simplicity in the example, the springs have been assumed to be linear. The state equations are

$$\dot{x}_1 = \dot{x}_1(t), \qquad \text{(given)}, \tag{7.5}$$

$$\dot{\delta}_2 = \dot{x}_2(t) - \dot{x}_1(t), \tag{7.6}$$

in which $\dot{x}_1$ and $\dot{x}_2$ are determined by some system external to the C-field and thus function as input variables.

Using Eq. (7.6), we see that δ_2 may be expressed in terms of the C-field variables, x_1 and x_2, by means of a time integration.

$$\delta_2 = x_2 - x_1 + \text{const.} \tag{7.7}$$

Referring to Figure 7.3a, it can be seen that x_1 and x_2 can be defined such that when $x_1 = x_2 = 0$, $F_1 = F_2 = \delta_2 = 0$. Thus the integration constant in Eq. (7.7) can be set to zero when x_1 and x_2 represent deviations from the positions at which the springs are unstretched. Using Eq. (7.7), we can eliminate δ_2 from Eq. (7.4). Then

$$\begin{bmatrix} F_1 \\ \hline F_2 \end{bmatrix} = \begin{bmatrix} k_1 + k_2 & -k_2 \\ \hline -k_2 & k_2 \end{bmatrix} \begin{bmatrix} x_1 \\ \hline x_2 \end{bmatrix}, \tag{7.8}$$

which represents the constitutive laws of a 2-port C-field in stiffness matrix form. [The reader may benefit from carrying out the steps required to express the constitutive laws of the electrical C-field of Figure 7.3b in a form analogous to Eq. (7.8).]

An inspection of the constitutive laws of Eqs. (7.3) and (7.8) shows that for these linear C-fields, the *compliance matrix* [Eq. (7.3)] and the *stiffness matrix* [Eq. (7.8)] are symmetric. In general, compliance matrices and stiffness matrices, which are inverses of each other, are always symmetric. This can be proven in a form valid for even nonlinear C-fields by studying the stored energy function of Eq. (7.1).

A change in any component of $\mathbf{q}$, say Δq_i, will produce a change in E, say ΔE. By a direct inspection of Eq. (7.1), one can deduce that the coefficient relating $\Delta \mathbf{E}$ to Δq_i is, in fact, e_i. Thus, the partial derivative of $\mathbf{E}$ with respect to q_i is just e_i, or

$$\frac{\partial \mathbf{E}}{\partial q_i} = e_i(\mathbf{q}), \qquad i = 1, 2, \ldots, n. \tag{7.9}$$

Because $\mathbf{E} = \mathbf{E}(\mathbf{q})$ is supposed to represent the stored-energy function, which we assume to be a single-valued scalar function of the $\mathbf{q}$ vector, if $\mathbf{E}$

is smooth enough to have second derivatives, then

$$\frac{\partial e_i}{\partial q_j} = \frac{\partial^2 \mathbf{E}}{\partial q_j \, \partial q_i} = \frac{\partial e_j}{\partial q_i}, \qquad i, j = 1, 2, 3, \ldots, n, \tag{7.10}$$

which shows how $e_i(\mathbf{q})$ and $e_j(\mathbf{q})$ are constrained by the existence of the stored energy function, E. In other words, not every set of constitutive laws expressing efforts in terms of displacements could come from an energy-storing, or conservative, C-field. Physically, it is clear that elastic structures, capacitor networks, and the like cannot supply more energy than that amount previously stored, so Eq. (7.10) will apply to such devices.

In the linear case, Eq. (7.10) implies that stiffness matrices are necessarily symmetrical for any conservative C-field. If we use k_{ij} for stiffness matrix coefficients and $\mathbf{k}$ for the matrix itself, then the C-field constitutive laws for an n-port can be written thus:

$$e_i = \sum_{i=1}^{n} k_{ij} q_j,$$

or

$$\mathbf{e} = \mathbf{kq},$$

and Eq. (7.8) is a particular example. The stored energy is

$$\mathbf{E}(\mathbf{q}) = \tfrac{1}{2} \sum_{i=1}^{n} \sum_{j=1}^{n} k_{ij} q_i q_j = \tfrac{1}{2} \mathbf{q}^t \mathbf{kq}. \tag{7.11}$$

Application of Eq. (7.10) to Eq. (7.11) yields

$$k_{ij} = k_{ji} \qquad \text{or} \qquad \mathbf{k}^t = \mathbf{k}, \tag{7.12}$$

and since compliance matrices such as that in Eq. (7.3) are inverses of stiffness matrices and the inverse of a symmetric matrix is also symmetric, we conclude that compliance matrices must also be symmetrical. Equations (7.12), which are often called Maxwell's reciprocal relations, are most easily derived from the more general nonlinear relations of Eqs. (7.10), but graphical derivations based on special integration paths are often used for the linear case. See Reference [2], for example.

The reciprocal relations for C-fields and other energy-storing fields are more useful and unexpected in some cases than it might at first appear. In later chapters dealing with a variety of applications these relations will prove to be very important. For now, we show only a few examples that

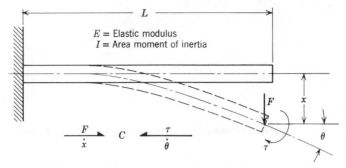

Figure 7.4. Cantilever beam represented as a C-field.

are less obvious than our first ones. The system of Figure 7.4 shows how the end of a cantilever beam can interact with a rotational and a translational port. Using beam superposition tables or any other method to find the beam deflections, the compliance matrix may be found and inverted to find the stiffness matrix:

$$
\begin{bmatrix} F \\ - \\ \tau \end{bmatrix} = EI \left[\begin{array}{c|c} 12/L^3 & -6/L^2 \\ \hline -6/L^2 & 4/L \end{array} \right] \begin{bmatrix} x \\ - \\ \theta \end{bmatrix}.
\tag{7.13}
$$

The symmetry of this matrix serves as a useful check on the correctness of the computations leading up to it. Furthermore, we can make statements such as the following: The number of newtons (the force) required to keep x at zero when $\theta = 0.1$ rad is equal to the number of newton-meters (the torque) required to keep θ at zero when $x = 0.1$ m. Or, if $F = 1$ N and $\tau = 0$, then θ in radians will be numerically equal to x in meters when $F = 0$ and $\tau = 1$ N-m. (The latter statement is based on the symmetry of the compliance matrix.)

Nonlinear C-fields are common in certain application areas. Here we show how a mechanical element that behaves as a linear 1-port in some restricted cases can be modeled as a nonlinear 2-port C-field when two-dimensional motion is permitted. The element, shown schematically in Figure 7.5, is a simple linear spring attached to ground at one end and deflecting under the action of forces, F_x and F_y, at the other end. It is possible to express F_x and F_y in terms of x and y and, thus, to consider this device as a C-field. The constitutive relations are

$$
\begin{aligned}
F_x &= kx - k(x^2 + y^2)^{-1/2} L_0 x, \\
F_y &= ky - k(x^2 + y^2)^{-1/2} L_0 y,
\end{aligned}
\tag{7.14}
$$

where k is the spring constant and L_0 is the free length of the spring.

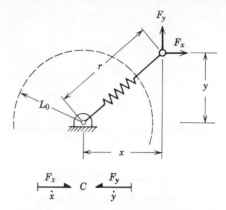

Figure 7.5. A nonlinear C-field representation of a linear spring, one end of which can move in plane motion.

A check on Eqs. (7.14) is given by Eq. (7.10).

$$\frac{\partial F_x}{\partial y} = -k(x^2 + y^2)^{-3/2}L_0 x 2y, \tag{7.15}$$

$$\frac{\partial F_y}{\partial x} = -k(x^2 + y^2)^{-3/2}L_0 y 2x \tag{7.16}$$

Because the expressions on the right-hand sides in Eqs. (7.15) and (7.16) are identical, the constitutive laws of Eqs. (7.14) do represent a nonlinear conservative C-field.

Actually, it is easiest to derive Eqs. (7.14) by using energy methods. The stored energy, $\mathbf{E}$, is easily expressed in terms of the radius, r.

$$\mathbf{E} = \tfrac{1}{2}k(r - l_0)^2 = \tfrac{1}{2}k(r^2 - 2rl_0 + l_0^2),$$

and, using $r^2 = x^2 + y^2$, we have

$$\mathbf{E} = \tfrac{1}{2}k(x^2 + y^2 - 2l_0(x^2 + y^2)^{1/2} + l_0^2). \tag{7.17}$$

When Eq (7.17) is used according to Eq. (7.9) to derive Eq. (7.14), then the reciprocal relations, Eqs. (7.15) and (7.16), must follow immediately.

7.1.2 Causal Considerations for C-Fields

As in the case of 1-port C elements, we may distinguish between integral and derivative causality for multiport C-fields, but multiport fields also admit of mixed integral-derivative causalities. The completely

integral causality form is

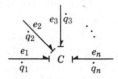

and the constitutive laws may be stated thus:

$$e_i = \Phi_{c_i}^{-1}(q_1, q_2, \ldots, q_n),$$
$$i = 1, 2, \ldots, n$$

(7.18)

The completely derivative causality form is

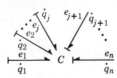

and the constitutive laws may be restated in the form:

$$q_i = \Phi_{c_i}(e_1, e_2, \ldots, e_n),$$
$$i = 1, 2, \ldots, n.$$

(7.19)

In Eqs. (7.18) and (7.19), the generally nonlinear functions, $\Phi_{c_i}^{-1}$ and Φ_{c_i}, are named in analogy with their 1-port counterparts from Chapter 3. In mechanical systems the compliance form corresponds to derivative causality and the stiffness form to integral causality. When the ports are numbered such that the first j ports have integral causality and the remainder have derivative causality, then a mixed causal form appears as follows:

and the constitutive laws are

$$e_1 = \Phi_i(q_1, q_2, \ldots, q_j, e_{j+1}, \ldots, e_n),$$
$$i = 1, 2, \ldots, j;$$

(7.20)

$$q_k = \Phi_k(q_1, q_2, \ldots, q_j, e_{j+1}, e_n),$$
$$k = j + 1, \ldots, n.$$

(7.21)

The question of which causal forms are possible for an energy-storing field is an interesting one. For an explicit field that is specified by a set of constitutive laws in a particular causal form, the question is an algebraic one in which each causal form involves solving the given relationships again in terms of different input and output variables. For nonlinear fields, it can be very hard to decide whether this will be possible, and even for linear fields the problems are not trivial. Of course, to convert from completely integral to completely derivative causality or the reverse, one needs only to invert a matrix in the linear case. Since there are straightforward tests of matrices to determine whether or not the inverse exists, it is not hard to decide, for example, whether given a field in derivative causality the integral causality form is permissible. It is harder, however, to establish a universal rule for testing all the mixed causal forms that an n-port field can exhibit.

For implicit fields, on the other hand, the rules of causality can often be used on the elements comprising the field to deduce which causal forms of the field are possible. As a very simple example, consider the implicit field formed by a 1-port C- and a 0-junction as shown in Figure 7.6. This

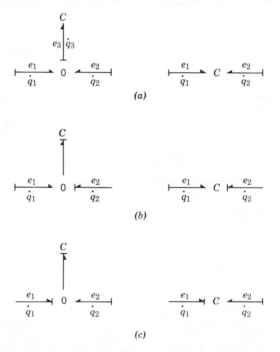

Figure 7.6. A simple implicit field. (a) Integral causality; (b) and (c) mixed causality. (Derivative causality for this field is not possible.)

system is readily treated as a 2-port C-field in integral causality (see Figure 8.6a). Reading the implicit field causality, we may find e_1 and e_3, assuming that the 1-port C is linear.

$$e_1 = e_3 = \frac{q_3}{C_3}. \tag{7.22}$$

$$e_2 = e_3 = \frac{q_3}{C_3}. \tag{7.23}$$

Now we wish to convert from q_3 as a state variable to the field state variables, q_1 and q_2. Reading the bond graph, we have

$$\dot{q}_3 = \dot{q}_1 + \dot{q}_2 \tag{7.24}$$

or

$$q_3 = q_1 + q_2 + \text{const.} \tag{7.25}$$

If for simplicity we define q_1, q_2, and q_3 so that the integration constant vanishes in Eq. (7.25), then the field equations follow by substitution in Eqs. (7.22) and (7.23).

$$\begin{bmatrix} e_1 \\ \hline e_2 \end{bmatrix} = \begin{bmatrix} 1/C_3 & 1/C_3 \\ \hline 1/C_3 & 1/C_3 \end{bmatrix} \begin{bmatrix} q_1 \\ \hline q_2 \end{bmatrix}. \tag{7.26}$$

There are two ways to see that the completely derivative causal form is not possible: (1) the rules of causality when applied to the implicit field of Figure 7.6 do not allow the imposition of both e_1 and e_2 as inputs to the 0-junction, and (2) the determinant of the matrix in Eq. (7.26) clearly vanishes, indicating that Eq. (7.26) cannot be solved for q_1, q_2 in terms of e_1 and e_2.

On the other hand, two mixed causal forms are possible as shown in Figure 7.6b and c. For the form of Figure 7.6b, the relations are derived by reading the implicit field bond graph.

$$e_1 = e_2, \tag{7.27}$$

$$\dot{q}_2 = -\dot{q}_1 + \dot{q}_3 = -\dot{q}_1 + \frac{d}{dt}(C_3 e_3) = -\dot{q}_1 + \frac{d}{dt}(C_3 e_2) \tag{7.28}$$

Equation 7.28 must be integrated in time and, if q_1 and q_2 are properly defined, the integration constant can be made to vanish.

$$q_2 = -q_1 + C_3 e_2. \tag{7.29}$$

In matrix form, the mixed causal form is

$$
\begin{bmatrix} e_1 \\ \hline q_2 \end{bmatrix} = \begin{bmatrix} 0 & 1 \\ \hline -1 & C_3 \end{bmatrix} \begin{bmatrix} q_1 \\ \hline e_2 \end{bmatrix}. \tag{7.30}
$$

Note that the integral form, Eq. (7.26), is symmetric, as it must be according to the energy argument, but mixed forms such as Eq. (7.30) will generally have antisymmetric terms.

Although almost any nontrivial system may be considered to contain an energy-storing field (as we have just seen, even a single $-C$ and a 0-junction can be treated as a 2-port field), it is often not worthwhile to consider the manipulation of a bond graph into a form in which an implicit field is obvious. An exception to this general rule occurs when, in the course of assigning causality, it is found that one or more elements must

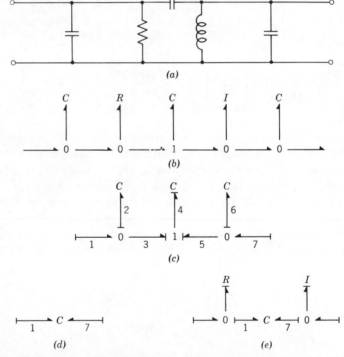

Figure 7.7. System containing an implicit C-field. (a) Section of an electrical network; (b) bond graph; (c) implicit field identified; (d) explicit field representation; (e) bond graph of (a) using explicit field representation.

be assigned derivative causality. Often this occurrence signals the presence of an implicit field that can usefully be converted into explicit form, thus simplifying further equation formulation. Of course, the methods of equation formulation presented in Chapter 5 for the cases involving derivative causality permit equation formulation without identifying fields. Although there is no less labor involved in converting an implicit field with derivative causality into explicit form, as we do here, this process may increase one's insight into the causes and nature of derivative causality. Also, the explicit field representation can be used repeatedly, thus avoiding the necessity of handling derivative causality in each system of which the field is a part.

As an example, consider the electrical subsystem shown in Figure 7.7a. The bond graph of Figure 7.7b does not appear to have any unusual characteristics, but a little experimentation with causal assignments will show that the three $-C$ elements cannot all have integral causality. What this means is that the three displacements, in this case, three electrical charge variables, cannot be independent. In Figure 7.7c, the C-field is singled out for study. In this form, the charges, q_2 and q_6, for example, can play the roles of state variables and the charge, q_4, will be statically related to q_2 and q_6. Let us now find the explicit field representation shown in Figure 7.7d in which charge variables, q_1 and q_7, will play the roles of state variables.

Reading the graph of Figure 7.7c, we find that the port output variables, e_1 and e_7, can be expressed in terms of the state variables, q_2 and q_6. (For simplicity, we assume all the $-C$ elements are linear, although our operations could be carried out for nonlinear $-C$ elements as well.)

$$e_1 = e_2 = \frac{q_2}{C_2} \; ; \qquad e_7 = e_6 = \frac{q_6}{C_6} . \tag{7.31}$$

The state equations may be used to relate q_2 and q_6 to q_1 and q_7.

$$\dot{q}_2 = f_1 - f_3 = f_1 - f_4 = f_1 - \frac{d}{dt} q_4$$

$$= f_1 - \frac{d}{dt} C_4 e_4 = f_1 - \frac{d}{dt} [C_4(e_3 + e_5)]$$

$$= f_1 - \frac{d}{dt} C_4(e_2 + e_6) = f_1 - \frac{d}{dt} \left[C_4 \left(\frac{q_2}{C_2} + \frac{q_6}{C_6} \right) \right]. \tag{7.32}$$

Similarly,

$$\dot{q}_6 = f_7 - \frac{d}{dt} \left[C_4 \left(\frac{q_2}{C_2} + \frac{q_6}{C_6} \right) \right]. \tag{7.33}$$

These equations may be rearranged in matrix form:

$$
\left[
\begin{array}{c|c}
\dfrac{C_2 + C_4}{C_2} & \dfrac{C_4}{C_6} \\
\hline
\dfrac{C_4}{C_2} & \dfrac{C_6 + C_4}{C_6}
\end{array}
\right]
\left[
\begin{array}{c}
\dot{q}_2 \\
- \\
\dot{q}_6
\end{array}
\right]
=
\left[
\begin{array}{c}
f_1 \\
- \\
f_7
\end{array}
\right]
\tag{7.34}
$$

Basically, we wish to integrate Eq. (7.34) in order to relate q_2 and q_6 to q_1 and q_7, but, because of the differential causality, a matrix inversion will be required. The result is

$$
\left[
\begin{array}{c}
\dot{q}_2 \\
- \\
\dot{q}_6
\end{array}
\right]
=
\frac{1}{C_2 C_4 + C_2 C_6 + C_4 C_6}
\left[
\begin{array}{c|c}
C_2 C_6 + C_2 C_4 & -C_2 C_4 \\
\hline
-C_4 C_6 & C_2 C_6 + C_4 C_6
\end{array}
\right]
\left[
\begin{array}{c}
\dot{q}_1 \\
- \\
\dot{q}_7
\end{array}
\right],
\tag{7.35}
$$

where $\dot{q}_1 = f_1$, $\dot{q}_7 = f_7$.

Since Eq. (7.35) is a relation between derivatives of charge variables, it can be integrated in time to yield the desired relations between the charges. Again, one must consider the charges at some initial time in order to evaluate integration constants. If the system is assembled out of initially uncharged capacitors, then q_2, q_6, q_1, and q_7 can all vanish initially and the integration constants also vanish. Then Eq. (7.35) can be integrated by simply removing the dots over the qs. When this is done, a substitution into Eq. (7.31) yields the explicit field equations:

$$
\left[
\begin{array}{c}
e_1 \\
- \\
e_7
\end{array}
\right]
=
\frac{1}{C_2 C_4 + C_2 C_6 + C_4 C_6}
\left[
\begin{array}{c|c}
C_6 + C_4 & -C_4 \\
\hline
-C_4 & C_2 + C_4
\end{array}
\right]
\left[
\begin{array}{c}
q_1 \\
- \\
q_7
\end{array}
\right].
\tag{7.36}
$$

This relation is the constitutive law for the 2-port C-field of Figure 7.7d in integral causality form. Note that we have arranged an inward sign convention for the C-field. This means that Eq. (7.36) should have a symmetric matrix, and it does. Now, the explicit field representation, which does not have derivative causality problems, can be used in the original system as shown in Figure 7.7e. The same field might appear in many systems, and the explicit representation would eliminate the need to perform algebraic manipulations associated with derivative causality in each system.

Finally, it should be mentioned that it is sometimes convenient to reduce implicit fields to an explicit form, regardless of whether or not derivative causality is involved. Figure 7.8 shows two examples in which

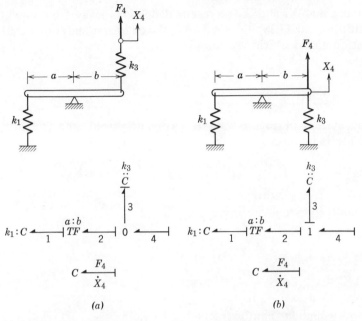

Figure 7.8. Two implicit fields. (*a*) A 1-port field with internal derivative causality; (*b*) A 1-port field without internal derivative causality.

1-ports, 2-ports, and 3-ports are connected and form implicit fields. In Figure 7.8a, the presence of the field is signaled by the derivative causality required on one of the $-C$ elements. In Figure 7.8b, no derivative causality is required, but by defining an explicit field one can reduce the number of state variables required.

Reading the bond graph of Figure 7.8a, one finds

$$F_4 = F_2 = \frac{a}{b} F_1 = \frac{a}{b} k_1 X_1,$$

$$\dot{X}_1 = \frac{a}{b} V_2 = \frac{a}{b} (V_4 - V_3) = \frac{a}{b}\left(V_4 - \frac{d}{dt}\left(\frac{F_3}{k_3}\right)\right) = \frac{a}{b}\left(V_4 - \frac{d}{dt}\frac{F_2}{k_3}\right)$$

$$= \frac{a}{b}\left(V_4 - \frac{d}{dt}\left(\frac{a}{b}\frac{F_1}{k_3}\right)\right) = \frac{a}{b}\left(V_4 - \frac{d}{dt}\left(\frac{a}{b}\frac{k_1 X_1}{k_3}\right)\right),$$

(7.37)

or

$$\dot{X}_1 = \frac{b^2 k_3}{a^2 k_1 + b^2 k_3}\frac{a}{b} V_4,$$

or

$$X_1 = \frac{b^2 k_3}{a^2 k_1 + b^2 k_3}\frac{a}{b} X_4,$$

(7.38)

assuming that X_1 and X_4 represent deflections away from equilibrium. Substituting Eq. (7.38) into Eq. (7.37), the characteristics of an equivalent 1-port compliance field are found.

$$F_4 = \frac{a^2 k_1 k_3}{a^2 k_1 + b^2 k_3} X_4. \tag{7.39}$$

The system of Figure 7.8b is easily described since no derivative causality is involved.

$$F_4 = F_2 + F_3 = \frac{a}{b} F_1 + F_3 = \frac{a}{b} k_1 X_1 + k_3 X_3. \tag{7.40}$$

Two state equations are involved:

$$\dot{X}_1 = \frac{a}{b} V_2 = \frac{a}{b} V_4, \tag{7.41}$$

$$\dot{X}_3 = V_4, \tag{7.42}$$

but both equations can be integrated to yield

$$X_1 = \frac{a}{b} X_4, \qquad X_3 = X_4, \tag{7.43}$$

again assuming that integration constants vanish, that is, that $X_1 = X_3 = X_4 = 0$ represents the condition when all springs are unstretched. Substituting Eqs. (7.43) into Eq. (7.40), the equivalent 1-port C-field constitutive law is found.

$$F_4 = \left(\frac{a^2}{b^2} k_1 + k_3\right) X_4. \tag{7.44}$$

Both the C-field representations of Eqs. (7.39) and (7.44) are practically useful, in the first case because the algebra associated with differential causality is solved once and for all, and in the second case because an essentially trivial extra state equation has been integrated once and for all.

7.1.3 I-Fields

Inertial fields are strictly analogous to the capacitive fields just discussed. Instead of constitutive laws relating efforts to displacements, inertial elements have constitutive laws relating flows to momenta. All the results

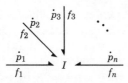

Figure 7.9. The symbol for an n-port I-field.

for C-fields will hold for I-fields if flows are substituted for efforts and momenta are substituted for displacements.

For example, the energy stored in the n-port I-field shown in Figure 7.9 is just

$$
\begin{aligned}
\mathbf{E} &= \int_{t_0}^{t} \sum_{i=1}^{n} f_i e_i \, dt \\
&= \int_{t_0}^{t} \sum_{i=1}^{n} f_i \dot{p}_i \, dt \\
&= \int_{\mathbf{p}_0}^{\mathbf{p}} \sum_{i=1}^{n} f_i(\mathbf{p}) \, dp_i \\
&= \int_{\mathbf{p}_0}^{\mathbf{p}} \mathbf{f}(\mathbf{p}) \, d\mathbf{p} = \mathbf{E}(\mathbf{p}),
\end{aligned}
\tag{7.45}
$$

in which, as in Eq. (7.1), column vectors for the flows and momenta have been defined:

$$
\mathbf{p} \equiv \begin{bmatrix} p_1 \\ p_2 \\ \cdot \\ \cdot \\ \cdot \\ p_n \end{bmatrix}, \qquad \mathbf{f} \equiv \begin{bmatrix} f_1 \\ f_2 \\ \cdot \\ \cdot \\ \cdot \\ f_n \end{bmatrix}
\tag{7.46}
$$

The analogs of Eq. (7.9) are

$$
\frac{\partial \mathbf{E}}{\partial p_i} = f_i(\mathbf{p}), \qquad i = 1, 2, \ldots, n,
\tag{7.47}
$$

and the reciprocity relations are

$$
\frac{\partial f_i}{\partial p_j} = \frac{\partial^2 \mathbf{E}}{\partial p_j \, \partial p_i} = \frac{\partial f_j}{\partial p_i}.
\tag{7.48}
$$

Equation (7.48), which is valid for nonlinear fields, carries the implication for the linear cases that mass matrices, inductance matrices, and the like must be symmetric because energy must be conserved.

In common with C-fields, I-fields occur in both implicit and explicit forms. Often the form which an I-field takes depends on the system modeler's point of view, and it is important to realize that manipulations of fields may result in practically useful, simplified means of representing system components. In mechanics, for example, the concept of a rigid body implies that elemental masses are constrained to move in ways such that distances between mass elements do not vary. This means that all rigid bodies are I-fields, and in Chapter 9, the general means of describing mechanical systems containing rigid bodies will be discussed in some detail. For now, a single example may suffice to show how an analyst could construct either an implicit or an explicit field representation of a rigid body.

The long, thin, rigid bar of Figure 7.10a has area, A, mass density, ρ, and length, L. Thus its total mass, m, is

$$m = \rho AL, \tag{7.49}$$

and its centroidal moment of inertia, J, about an axis perpendicular to the long dimension is

$$J = \frac{mL^2}{12}. \tag{7.50}$$

If we consider plane motion of the bar and allow only vertical motion of the center of mass and a small angular rotation relative to a horizontal axis, and if we consider two ports at the end of the bar with forces, F_1 and F_2, and velocities, V_1 and V_2, then this rigid body can be described by a linear 2-port I-field.

One way to find the constitutive laws for the I-field is to describe the motion first in terms of the velocity of the center of mass, V_c, and the angular velocity, $\omega = \dot{\theta}$. The net force on the bar is then the rate of change of the linear momentum, and the momentum is related to V_c by the mass ($V_c = p_c/m$). Similarly, the net torque about the center of mass is the rate of change of the angular momentum, and the angular momentum is related to $\omega = \dot{\theta}$ by the moment of inertia ($\omega = p_\theta/J$). In Figure 7.10b, these constitutive laws are represented by the two 1-port inertial elements. The remainder of the graph serves to relate V_1 and V_2 to V_c and ω, and also to relate the net force and net torque to F_1 and F_2. (The part of the graph involving 0- and 1-junctions and transformers is a special kind of junction

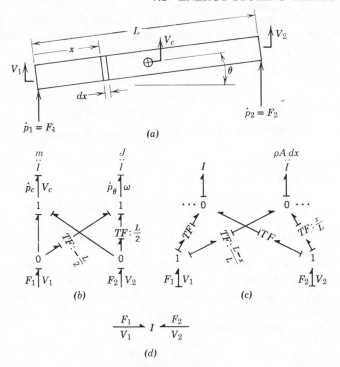

Figure 7.10. *I*-Field representation of a rigid body in plane motion. (*a*) Schematic diagram; (*b*) implicit field using total mass and centroidal moment of inertia; (*c*) implicit field using differential elements; (*d*) explicit field.

structure that appears frequently in mechanics and that will be discussed further in Chapter 9.)

From the bond graph (or the schematic diagram) we may now write the output equations:

$$V_1 = V_c - \frac{L}{2}\,\omega = \frac{p_c}{m} - \frac{L}{2}\frac{p_\theta}{J}, \tag{7.51}$$

$$V_2 = V_c + \frac{L}{2}\,\omega = \frac{p_c}{m} + \frac{L}{2}\frac{p_\theta}{J}, \tag{7.52}$$

where p_c and p_θ designate the linear and angular momentum variables, respectively. The state equations are

$$\dot{p}_c = F_1 + F_2, \tag{7.53}$$

$$\dot{p}_\theta = -\frac{L}{2}\,F_1 + \frac{L}{2}\,F_2. \tag{7.54}$$

In order to have an explicit field at the external ports, these equations must be integrated in time. If p_c and p_θ (and hence, V_c and ω) vanish at the initial time in the integration, then the possible integration constants vanish, and the results are

$$p_c = p_1 + p_2, \tag{7.55}$$

$$p_\theta = -\frac{L}{2} p_1 + \frac{L}{2} p_2. \tag{7.56}$$

Substituting Eqs. (7.55) and (7.56) into Eqs. (7.51) and (7.52), the result is an explicit I-field constitutive law:

$$\begin{bmatrix} V_1 \\ \overline{} \\ V_2 \end{bmatrix} = \begin{bmatrix} (1/m)+(L^2/4J) & (1/m)-(L^2/4J) \\ \hline (1/m)-(L^2/4J) & (1/m)+(L^2/4J) \end{bmatrix} \begin{bmatrix} p_1 \\ \overline{} \\ p_2 \end{bmatrix} \tag{7.57}$$

A different, more fundamental, but less convenient approach to this problem will generate an explicit I-field representation directly. We consider the bar to be a rigid massless rod on which are attached an infinite number of masses of value $\rho A\, dx$ at the generic position, x. The velocity, V_x, of such a mass is

$$V_x = \left(\frac{L-x}{L}\right) V_1 + \frac{x}{L} V_2. \tag{7.58}$$

The bond graph of Figure 7.10c shows how these elemental masses are related to the external ports. Since the forces, F_1 and F_2, are the sum of the forces generated by the elemental masses, the derivative causality shown is convenient. Reading the bond graph we have F_1 and F_2 expressed as the sum (integral) of the components due to the elemental masses:

$$F_1 = \int_0^L \left(\frac{L-x}{L}\right)\frac{d}{dt}\left((\rho A\, dx)V_x\right) = \int_0^L \rho A\left(\frac{L-x}{L}\right)\left[\left(\frac{L-x}{L}\right)\dot{V}_1 + \left(\frac{x}{L}\right)\dot{V}_2\right] dx,$$
$$\tag{7.59}$$

$$F_2 = \int_0^L \left(\frac{x}{L}\right)\frac{d}{dt}\left((\rho A\, dx)V_x\right) = \int_0^L \rho A\left(\frac{x}{L}\right)\left[\left(\frac{L-x}{L}\right)\dot{V}_1 + \left(\frac{x}{L}\right)\dot{V}_2\right] dx, \tag{7.60}$$

where Eq. (7.58) has been used. The result of the integration in x is

$$\begin{bmatrix} F_1 \\ \overline{} \\ F_2 \end{bmatrix} = \begin{bmatrix} \dot{p}_1 \\ \overline{} \\ \dot{p}_2 \end{bmatrix} = \frac{\rho AL}{6} \begin{bmatrix} 2 & 1 \\ \hline 1 & 2 \end{bmatrix} \begin{bmatrix} \dot{V}_1 \\ \overline{} \\ \dot{V}_2 \end{bmatrix}, \tag{7.61}$$

or, if we agree to define velocities such that p_1 and p_2 vanish when V_1 and V_2 vanish, then

$$\begin{bmatrix} p_1 \\ \overline{} \\ p_2 \end{bmatrix} = \frac{\rho A L}{6} \begin{bmatrix} 2 & 1 \\ \hline 1 & 2 \end{bmatrix} \begin{bmatrix} V_1 \\ \overline{} \\ V_2 \end{bmatrix}. \tag{7.62}$$

This is the constitutive law for the explicit field of Figure 7.10d in derivative causality (or mass matrix) form. It is left as an exercise for the reader to show that, using Eqs. (7.49) and (7.50), the constitutive laws of Eqs. (7.57) and (7.62) are the same except for the different causality.

Electrical circuits containing mutually interacting coils can be conveniently represented using explicit I-fields. In order to write the constitutive laws for such fields, one requires not only a sign convention for currents and voltages, but also a convention dealing with the relative orientation of the coils. One such convention is shown in the circuit diagrams of Figure 7.11a and b as dots placed near the ends of the coils. The idea is that if the currents, i_1 and i_2, are defined such that when positive, they both enter or both leave dotted ends of their respective coils, then the mutual-inductance effects will be positive. If, on the other hand, the dots are placed as in Figure 7.11b so that one of the currents enters through a dotted end and the other leaves through a dotted end, then the mutual-inductance effects are negative. The convention is most readily illustrated for the linear case, but it applies equally well for the nonlinear case. For example, if one denotes self-inductance coefficients by the letter L, and mutual-inductance coefficients by the letter M, as is the usual convention, then the system of Figure 7.11a has the constitutive law:

$$\begin{bmatrix} \lambda_1 \\ \overline{} \\ \lambda_2 \end{bmatrix} = \begin{bmatrix} L_1 & M_{12} \\ \hline M_{12} & L_2 \end{bmatrix} \begin{bmatrix} i_1 \\ \overline{} \\ i_2 \end{bmatrix}. \tag{7.63}$$

Note that this field representation is in derivative causality form and is symmetric, as it must be. The field of Figure 7.11b has the constitutive law:

$$\begin{bmatrix} \lambda_1 \\ \overline{} \\ \lambda_2 \end{bmatrix} = \begin{bmatrix} L_1 & -M_{12} \\ \hline -M_{12} & L_2 \end{bmatrix} \begin{bmatrix} i_1 \\ \overline{} \\ i_2 \end{bmatrix}. \tag{7.64}$$

For three or more interacting coils, it is rather awkward to show the orientations. In Figure 7.11c, the dots have become circles, squares, and

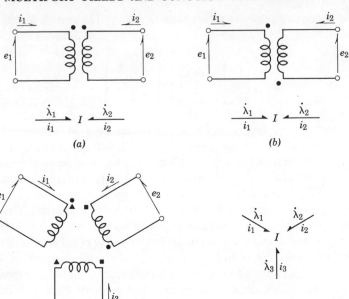

Figure 7.11. Mutual inductance in electrical systems. (a) and (b) 2-port I-fields with different coil orientation; (c) 3-port I-field.

triangles to indicate the signs of the three mutual-inductance coefficients. The field laws are readily written:

$$\begin{bmatrix} \lambda_1 \\ \lambda_2 \\ \lambda_3 \end{bmatrix} = \begin{bmatrix} +L_1 & -M_{12} & -M_{13} \\ -M_{12} & +L_2 & +M_{23} \\ -M_{13} & +M_{23} & +L_3 \end{bmatrix} \begin{bmatrix} i_1 \\ i_2 \\ i_3 \end{bmatrix}. \tag{7.65}$$

In the bond-graph I-field representation, the convention indicated by the circles, squares, and triangles is evident in the sign pattern of the mutual-inductance coefficients. Note that with the inward sign convention shown, the self-inductance terms must be positive for physical coils. Also, independently of the coil orientation convention, the inductance matrix and its inverse will be symmetric if the I-field is conservative.

When energy-storing fields are used in systems, it sometimes happens that explicit representations such as Eq. (7.65) must be algebraically manipulated. For example, if the coils of Figure 7.11c are interconnected as shown in Figure 7.12a, then the rules of causality indicate that integral causality cannot be applied to all three I-field ports. If there were no mutual inductance, the system would be as shown in Figure 7.12b. This system has one element in differential causality, but the formulation methods of Chapter 5 would handle the system without trouble. In fact, this system is just the dual of the C-field of Figure 7.7. On the other hand, when mutual-inductance effects are considered, the system appears as shown in Figure 7.12c. If one assigns integral causality on bonds 1 and 3, then bond 2 must have derivative causality.

It is not particularly difficult to switch the constitutive law of Eq. (7.65) into the mixed causal form of Figure 7.12c. Rewriting parts of Eq. (7.65), we have

$$
\begin{bmatrix} \lambda_1 \\ \overline{} \\ \lambda_3 \end{bmatrix} = \left[\begin{array}{c|c} L_1 & -M_{13} \\ \hline -M_{13} & L_3 \end{array} \right] \begin{bmatrix} i_1 \\ \overline{} \\ i_3 \end{bmatrix} + \begin{bmatrix} -M_{12} \\ \overline{} \\ M_{23} \end{bmatrix} [i_2].
\tag{7.66}
$$

Inverting this relation to yield i_1 and i_3 in terms of λ_1, λ_3, and i_2 gives

$$
\begin{bmatrix} i_1 \\ \overline{} \\ i_3 \end{bmatrix} = \frac{1}{L_1 L_3 - M_{13}^2} \left[\begin{array}{c|c|c} L_3 & M_{13} & L_3 M_{12} - M_{13} M_{23} \\ \hline M_{13} & L_1 & M_{13} M_{12} - L_1 M_{23} \end{array} \right] \begin{bmatrix} \lambda_1 \\ \overline{} \\ \lambda_3 \\ \overline{} \\ i_2 \end{bmatrix}.
\tag{7.67}
$$

Substituting Eq. (7.67) into the second of Eqs. (7.65), we can also obtain the third equation for λ_2 in terms of λ_1, λ_3, and i_2:

$$
\lambda_2 = \lambda_2(\lambda_1, \lambda_3, i_2).
\tag{7.68}
$$

When mutual-inductance effects are present, the writing of state equations is not quite as straightforward as when there is no mutual inductance. Basically, this is because of the role of i_2 in Eq. (7.67). Starting with the equation for λ_1,

$$
\dot{\lambda}_1 = e_4 - e_2 = e_4 - \frac{d}{dt} \lambda_2(\lambda_1, \lambda_3, i_2).
\tag{7.69}
$$

Now,

$$
i_2 = i_1 + i_3,
\tag{7.70}
$$

and we can use Eq. (7.67) to find i_1 and i_3 (and hence i_2) in terms of λ_1, λ_3, and i_2. But now we are in an algebraic loop reminiscent of those discussed in Chapter 5 when the imposition of integral causality did not suffice to

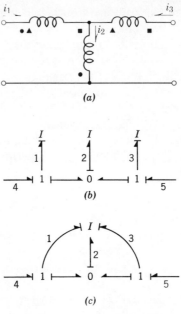

Figure 7.12. A network of interacting coils. (*a*) The coils of Figure 8.11 interconnected; (*b*) system bond graph if mutual inductance is absent; (*c*) system bond graph with *I*-field.

determine the causality of all bonds, Using Eqs. (7.70) and (7.67) we can express i_2 in terms of itself and then solve this algebraic equation for i_2 in terms of λ_1, λ_3 in order to compute the state equation, Eq. (7.69). Failure to do this will result in an endless looping through the equations in an attempt to eliminate i_2.

Although one could write state equations for the system in the manner outlined above, there is a simpler way to handle the system. In Figure 7.13, it is shown that the imposition of complete differential causality [which corresponds to Eq. (7.65) anyway] allows one to compute the explicit 2-port field without solving any algebraic equations. Reading the

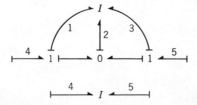

Figure 7.13. Reduction of system of Figure 7.12 to explicit 2-port *I*-field form.

graph, and using Eq. (7.65), we have

$$e_4 = e_1 + e_2 = \frac{d}{dt}(\lambda_1 + \lambda_2) = \frac{d}{dt}[(L_1 - M_{12})i_1 + (L_2 - M_{12})i_2 + (M_{23} - M_{13})i_3]$$

$$= \frac{d}{dt}[(L_1 - M_{12})i_4 + (L_2 - M_{12})(i_4 + i_5) + (M_{23} - M_{13})i_5],$$

or

$$\lambda_4 = (L_1 + L_2 - 2M_{12})i_4 + (L_2 - M_{12} - M_{13} + M_{23})i_5 \qquad (7.71)$$

upon integration. Similarly,

$$\lambda_5 = (L_2 - M_{12} - M_{13} + M_{23})i_4 + (L_2 + L_3 + 2M_{23})i_5. \qquad (7.72)$$

These equations represent the constitutive laws for the 2-port I-field in derivative-causality form. The all-integral-causality form follows by inverting the equations. Thus, λ_4 and λ_5 can readily be used as the two independent state variables. Clearly, the 2-port I-field is much more convenient to use in equation formulation than the 3-port field and its associated junction structure. Such studies of fields are, of course, as useful for C-field problems as they are for I-fields.

7.1.4 Mixed Energy-Storing Fields

There are occasions when an energy-storing device cannot be described as a C-field or an I-field, but rather acts as a C-field at some ports and an I-field at others. Within a single energy domain, the need for such an element is far from evident, but, when transducers are studied in Chapter 8, it will be seen that many transducers are essentially energy conservative, but are not pure I-fields or C-fields. Even in the case of transducers, it has been a common practice to arrange the analogies between variables in the two energy domains linked by the transducer in such a way that the transducer is a pure field. There is no real need to do this, however, and in bond graphs we prefer to retain our identification of effort, flow, displacement, and momentum quantities for all energy domains. We need, then, to discuss what will be called IC-fields. Not only is this policy just as convenient as the policy of switching analogies to suit the problem at hand; it is, in principle, required for some systems. There is no way, for example, to pick an effort-flow identification for an electromechanical system containing both a movable-plate capacitor and a solenoid transducer such that both transducers would be described as pure I- or C-fields.

Figure 7.14 shows the general n-port IC-field. The ports are numbered so that the first j-ports ($1 \le j \le n$) are inertial in character and the ports

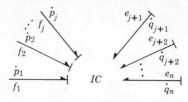

Figure 7.14.   The general mixed energy-storage field.

from $j + 1$ to n are capacitive in character. In integral-causality form as shown, the first j state variables are momenta and the remainder are displacements. The stored energy, $\mathbf{E}$, is a function of this mixed set of state variables:

$$\mathbf{E} = \int^t \sum_{i=1}^n e_i f_i \, dt = \int^p \sum_{i=1}^j f_i \, dp_i + \int^q \sum_{k=j+1}^n e_k \, dq_k, \tag{7.73}$$

where $\mathbf{p}$ represents the vector of momenta and $\mathbf{q}$ represents the vector of displacements. From Eq. (7.73), we see that

$$f_i = \frac{\partial \mathbf{E}}{\partial p_i}, \qquad i = 1, 2, \ldots, j, \tag{7.74}$$

$$e_k = \frac{\partial \mathbf{E}}{\partial q_k}, \qquad k = j+1, j+2, \ldots, n. \tag{7.75}$$

The reciprocity conditions for the constitutive laws can be easily derived by computing second partial derivatives of $\mathbf{E}$. For example, the second partial derivatives of $\mathbf{E}$ with respect to a momentum from the first set of ports and a displacement from the second set of ports yields

$$\frac{\partial f_i}{\partial q_k} = \frac{\partial^2 \mathbf{E}}{\partial q_k \, \partial p_i} = \frac{\partial e_k}{\partial p_i}, \tag{7.76}$$

where $1 \le i \le j$, $j + 1 \le k \le n$. These results should be compared to the results for pure C-fields and I-field [Eqs. (7.1), (7.10), (7.45), and (7.46)]. Physical examples involving IC-fields will be found in Chapter 8.

7.2   RESISTIVE FIELDS

An R-field is an n-port, the constitutive laws of which relate the n-port efforts and the n-port flows by means of static (or algebraic) functions. This definition includes power-conservative elements such as 0- and 1-junctions, and elements containing sources as R-fields, but in practice

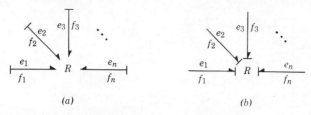

(a)   *(b)*

Figure 7.15. The general R-field. (a) Resistance form; (b) conductance form.

most R-fields studied dissipate power. Both explicit and implicit R-fields exist. Explicit multiport R-fields arise frequently in modeling of nonlinear devices. Implicit fields arising from the interconnection of 1-port resistances, transformers, gyrators, and junction 3-ports may be conveniently represented in R-field form as relations between external port efforts and flows.

The causality for R-fields usually is determined by the source and energy-storing elements in the system. Two fundamental causal patterns are shown in Figure 7.15. The constitutive laws for the *resistance* causality shown in Figure 7.15a may be written:

$$e_i = \Phi_{Ri}(f_1, f_2, \ldots, f_n),$$
$$i = 1, 2, \ldots, n. \tag{7.77}$$

The *conductance* causality shown in Figure 7.15b has constitutive laws that may be represented by the following:

$$f_i = \Phi_{R_i}^{-1}(e_1, e_2, \ldots, e_n),$$
$$i = 1, 2, \ldots, n. \tag{7.78}$$

In addition to these two fundamental causal forms, a large number of other forms are possible in which some of the bonds are causally oriented as in Figure 7.15a and the remainder are oriented as in Figure 7.15b. Although R-fields are basically neutral with respect to causality, we will encounter some constitutive laws for R-fields that cannot be unique for certain causalities (this happens, of course, only in the nonlinear case).

Although there is no stored energy function for R-fields as there is for energy-storing fields, and hence there is no simple way to show that R-field constitutive laws are constrained, still, for special classes of R-fields, some useful properties of the constitutive laws may be found. We will illustate some of these properties by example.

First, let us consider linear R-fields which contain no sources and no gyrators. A typical example is shown in Figure 7.16. This R-field will

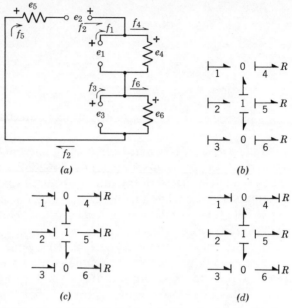

Figure 7.16. An implicit R-field. (a) Circuit graph; (b) bond graph showing resistance causality; (c) bond graph showing conductance causality; (d) bond graph showing mixed causality.

accept resistance causality on the three external ports as shown in Figure 7.16b. Reading the bond graph, the constitutive laws are readily derived:

$$
\begin{bmatrix} e_1 \\ \hline e_2 \\ \hline e_3 \end{bmatrix} = \begin{bmatrix} R_4 & R_4 & 0 \\ \hline R_4 & R_4+R_5+R_6 & R_6 \\ \hline 0 & R_6 & R_6 \end{bmatrix} \begin{bmatrix} f_1 \\ \hline f_2 \\ \hline f_3 \end{bmatrix}. \tag{7.79}
$$

As the bond graph of Figure 7.16c demonstrates, this R-field will also accept conductance causality. So, by inverting Eq. (7.79) or, more simply, by reading the bond graph of Figure 7.15c, the constitutive laws in conductance form may be found:

$$
\begin{bmatrix} f_1 \\ \hline f_2 \\ \hline f_3 \end{bmatrix} = \begin{bmatrix} 1/R_4+1/R_5 & -1/R_5 & 1/R_5 \\ \hline -1/R_5 & 1/R_5 & -1/R_5 \\ \hline 1/R_5 & -1/R_5 & 1/R_5+1/R_6 \end{bmatrix} \begin{bmatrix} e_1 \\ \hline e_2 \\ \hline e_3 \end{bmatrix}. \tag{7.80}
$$

Notice that both Eqs. (7.79) and (7.80) are symmetric. These forms may be called "Onsager" forms in analogy to the well-known Onsager reciprocal relations of irreversible thermodynamics [3]. Onsager proposed his reciprocity conditions for variables called "affinities" and "fluxes" that are analogous to the efforts and flows of bond graphs. In general, implicit R-fields composed of linear 1-port resistances, 0- and 1-junctions, and transformers will obey Onsager reciprocity; that is, when expressed in resistance or conductance form, the matrix in the constitutive laws will be symmetric. On the other hand, for mixed causality, the matrix will have antisymmetrical terms, and if gyrators are present, the Onsager reciprocity relations do not hold. Thus the reciprocity of energy-storing fields, which may be called Maxwell reciprocity, is more general than Onsager reciprocity. A perfectly reasonable explicit R-field characterization expressed in resistance or conductance form can be unsymmetrical. One may think of the unsymmetrical part of the matrices as arising from gyrational effects that introduce antisymmetrical terms, although in an explicit field one cannot identify a gyrator unless one can find some implicit field that has the same port constitutive laws as the explicit field.

Using our example, let us first show that an R-field in mixed causality has antisymmetric terms. A possible mixed causal pattern is shown in Figure 7.16d. Reading the bond graph, the constitutive laws are as follows:

$$
\begin{bmatrix} f_1 \\ e_2 \\ f_2 \end{bmatrix} = \begin{bmatrix} 1/R_4 & -1 & 0 \\ 1 & R_5 & 1 \\ 0 & -1 & 1/R_6 \end{bmatrix} \begin{bmatrix} e_1 \\ f_2 \\ e_3 \end{bmatrix},
\tag{7.81}
$$

in which the antisymmetric terms are evident. Such a form, which results when some of the input and output variables are interchanged for an R-field that obeys the Onsager reciprocal relations in resistance or conductance causality, is sometimes called a Casimir form [4].

It may be tempting to conclude that if R-fields are described with either resistance or conductance causality, they will exhibit Onsager reciprocity, and if they are described with a mixed causality, they will show a Casimir form. But it is easy to show that if gyrators are allowed in an implicit R-field, then this conclusion is false. Consider, for example, the R-field of Figure 7.17 that is formed from the field of Figure 7.16 by adding a single gyrator. In the conductance causality shown in Figure 7.17a, the field exhibits a Casimir form rather than an Onsager form,

$$
\begin{bmatrix} f_1 \\ f_2 \\ f_3 \end{bmatrix} = \begin{bmatrix} 1/R_4 & -1/r & 0 \\ 1/r & R_5/r^2 & 1/r \\ 0 & -1/r & 1/R_6 \end{bmatrix} \begin{bmatrix} e_1 \\ e_2 \\ e_3 \end{bmatrix},
\tag{7.82}
$$

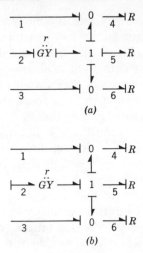

Figure 7.17.   Modified version of the R-field of Figure 7.15. (a) Conductance causality; (b) mixed causality.

in which r is the gyrator parameter. Similarly, for the mixed causality of Figure 7.17b, the R-field is symmetric:

$$\begin{bmatrix} f_1 \\ \hline e_2 \\ \hline f_3 \end{bmatrix} = \left[\begin{array}{c|c|c} 1/R_4 + 1/R_5 & -r/R_5 & 1/R_5 \\ \hline -r/R_5 & r^2/R_5 & -r/R_5 \\ \hline 1/R_5 & -r/R_5 & 1/R_5 + 1/R_6 \end{array} \right] \begin{bmatrix} e_1 \\ \hline f_2 \\ \hline e_3 \end{bmatrix}. \qquad (7.83)$$

A general R-field is not representable in an Onsager or a Casimir form. A simple example is shown in Figure 7.18. Its constitutive laws in resistance form are

$$\begin{bmatrix} e_1 \\ \hline e_2 \end{bmatrix} = \left[\begin{array}{c|c} R_3 + R_4 & r + R_4 \\ \hline -r + R_4 & R_4 + R_5 \end{array} \right] \begin{bmatrix} f_1 \\ \hline f_2 \end{bmatrix}, \qquad (7.84)$$

in which a symmetrical part due to the elements which obey Onsager reciprocity and an antisymmetrical part due to the gyrator can be recognized. If Eq. (7.84) were presented with numerical values for the parameters, it would have a rather undistinguished appearance, being neither an Onsager nor a Casimir form.

Note that if positive resistance parameters are used, then all of the example fields will dissipate energy for any possible port conditions. This is true because the 1-port resistors can only dissipate power, and the

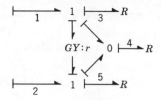

Figure 7.18. An R-field containing a gyrator.

junction elements and the 2-ports, —TF— and —GY—, conserve power. A way to check about power dissipation given an explicit linear field is to check whether the matrix of coefficients is positive definite or not. In the resistance form, for example, if **e** is the column vector of efforts, **f** the column vector of flows, and **R** the resistance matrix, then the power, P, is

$$P = f^t e,$$

and the constitutive law for the field is

$$e = Rf,$$

so that

$$P = f^t R f, \tag{7.85}$$

where ( )^t stands for the transpose of a vector or matrix.

Equation (7.85) shows that power will be dissipated for any **f** if **R** is positive definite. If one can only say that no power should be created, then **R** must be positive semidefinite; that is, there may be some finite flows for which zero power is dissipated. Using the rules for checking a matrix for positive definiteness, the reader may demonstrate that all the examples of R-fields given above are at least positive semidefinite. In this regard, it is useful to remember that the anitsymmetric part of any matrix contributes nothing to the definiteness of the matrix. This correlates with the idea that the antisymmetric terms due to the presence of a gyrator represent no power generation or dissipation in a field such as that associated with Eq. (7.84).

7.3 MODULATED 2-PORT ELEMENTS

The 2-port elements, —TF— and —GY—, are linear, power-conserving elements, the usefulness of which has been demonstrated in previous chapters. Here we discuss the modulated transformer, —MTF—, and modulated gyrator, —MGY—, which are nonlinear, power-conserving generalizations of —TF— and —GY—. Basically, the parameters of the

—TF— and —GY— elements are allowed to be a function of some parameter, say, ξ, in the modulated 2-ports. The usual symbols and constitutive laws for these elements are shown below:

$$m(\xi) \qquad\qquad r(\xi)$$

$$\xrightarrow[1]{} MTF \xrightarrow[2]{} \qquad\qquad \xrightarrow[1]{} MGY \xrightarrow[2]{}$$

$$m(\xi)e_1 = e_2 \qquad\qquad e_1 = r(\xi)f_2 \qquad\qquad (7.86)$$

$$f_1 = m(\xi)f_2 \qquad\qquad r(\xi)f_1 = e_2$$

Notice that for both elements, the power, $e_1 f_1$, is always equal to the power, $e_2 f_2$, no matter what the value of $m(\xi)$ or $r(\xi)$ may be. Also, the parameters, m or r, are shown as changing by means of a signal or active bond rather than by means of a power bond. Thus, it is characteristic of these modulated-parameter elements that the parameters change value without a directly associated power flow.

The modulated elements will prove useful for modeling certain classes of systems in later chapters. In Figure 7.19, two examples are shown that illustrate typical uses of the modulated elements. The modulated trans-former is particularly useful in describing mechanical systems moving through large angles. The rigid, pivoted bar of Figure 7.19a may be described thus:

$$\tau = (l \cos \theta)F, \qquad\qquad (7.87)$$

$$(l \cos \theta)\omega = V, \qquad\qquad (7.88)$$

where the variables are defined in the figure.

These constitutive relations may be derived by applying the laws of mechanics to the linkage. Equation (7.87) is a torque equilibrium state-ment and Eq. (7.88) is an angular velocity–velocity relation which could be derived by time-differentiating the equation,

$$l \sin \theta = x. \qquad\qquad (7.89)$$

The bond graph of Figure 7.19b represents Eqs. (7.87) and (7.88), where the transformer modulus is $l \cos \theta$, and θ plays the role of ξ in the general equations, Eqs. (7.86). It is a peculiarity of the MTFs of mechanics that the modulus varies with the displacements at the ports. For this reason, these MTFs are sometimes called displacement-modulated transformers.

Figure 7.19c shows an electromechanical system in which a field

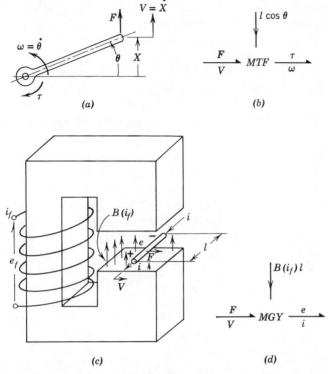

Figure 7.19. Examples of modulated 2-port elements. (*a*) Mechanical linkage; (*b*) modulated transformer for (*a*); (*c*) electromechanical system; (*d*) modulated gyrator for (*c*).

current, i_f, is responsible for establishing a magnetic field, $B(i_f)$, in a gap. If then a current-carrying conductor of length l moves with velocity V, as shown, it is a consequence of Faraday's law that a voltage, e, will be induced along the conductor according to the following relation:

$$e = B(i_f)lV. \tag{7.90}$$

The force, F, required to move the conductor is found from the Lorentz force law to be

$$B(i_f)li = F. \tag{7.91}$$

The modulated gyrator of Figure 7.19d represents these laws, with $B(i_f)l$ representing the modulus, r, and i_f the parameter, ξ, in the general form of Eq. (7.86). Note that $ei = FV$ (as long as the electrical and mechanical

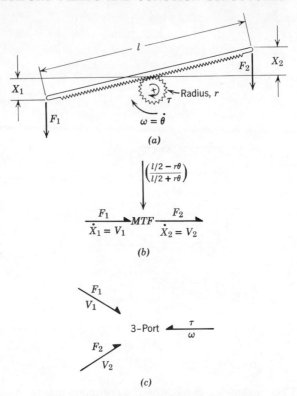

Figure 7.20. Rack and pinion system. (*a*) Schematic diagram; (*b*) incorrect bond graph; (*c*) correct bond graph.

power is measured in the same units), and i_f can be changed with no power associated with the *MGY*. On the other hand, i_f is also associated with a self-inductance effect and coil resistance so that there is power flow and energy storage associated with i_f. The point is that no energy is associated with changes in Bl from the point of view of the *MGY*. The modulated gyrator will be used in modeling voice coils, electric motors, and similar devices in Chapter 8.

Since the sign conventions and causal restrictions for the modulated elements are the same as for the —*TF*— and —*GY*— elements studied earlier, there is little new to be said here. A few words of caution are in order, however. Since the modulated elements incorporate a pure signal interaction, it is quite possible to construct bond graphs that do not have physical interpretations by imagining that the modulus of a transformer or gyrator can be a function of any variable at all. Just as it is easy to make incorrect block diagrams, signal flow graphs, or other signal descriptions which violate power and energy constraints that exist in real physical systems, so too is it rather easy to assume that an element can be modeled using a modulated 2-port when, in fact, the element may really be a true 3-port, or when the physical system may not allow the modulus to be a function of the signal assumed.

To show just one example of the trap into which one may sometimes fall, consider the rack and pinion system shown in Figure 7.20. Clearly, if the pinion is small ($r \ll l$), then the rack is nearly a lever with a lever ratio of $(l/2 - r\theta)/(l/2 + r\theta)$ if the rack is centered when $\theta = 0$. Furthermore, if we rotate the pinion to some position and then hold θ fixed, the rack will function almost as a —TF— with the transformer modulus set by the value of θ. Thus, it seems reasonable to represent the system as in Figure 7.20b, but this representation is false. The reason is that, whenever F_1 and F_2 are not zero, θ cannot be changed with no power. In fact, the torque, τ, is proportional to F_1 or F_2. Thus, the device is a true 3-port as shown in Figure 7.20c. It can be modeled using MTF elements and 0- and 1-junctions as a multiport modulated transformer, as will be shown in Chapter 9, but the representation of Figure 7.20b is fundamentally incorrect.

7.4 JUNCTION STRUCTURES

Junction structures, which are assemblages of 0- and 1-junctions, transformers, and gyrators, are the energy switchyards which enforce the constraints among parts of dynamic systems. No power is dissipated or generated in a junction structure, so the *net* power into a junction structure at its ports always vanishes. Since junction structures do provide relations between efforts and flows at their ports, junction structures are special types of R-fields that never dissipate power. One might expect, for linear junction structures, that the Onsager and Casimir forms that were found for general R-fields would also be found for junction structures. These forms are indeed found, but junction structures typically cannot accept all possible causal assignments at their ports, so some forms for the constitutive laws simply cannot exist.

Some example junction structures are shown in Figure 7.21. Note that all these structures are shown with an inward sign convention so that the sum of the port powers must vanish. This means that, in each case, the matrix relating inputs to outputs must be antisymmetric; that is, zeros must appear on the main diagonal and the ijth component must be the negative of the jith component.

For systems without gyrators and in mixed resistance-conductance causality, such as Figures 7.21a, c, and d, the relations are Casimir forms

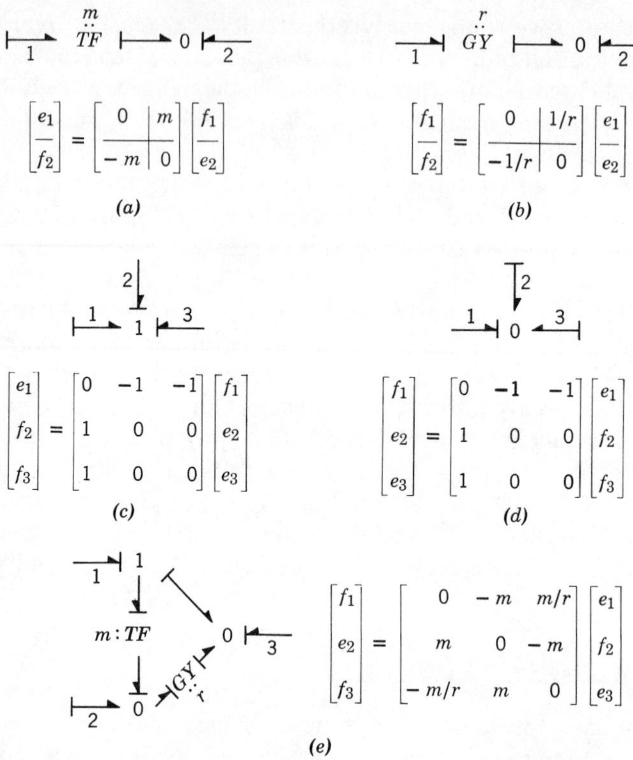

$$\begin{bmatrix} e_1 \\ \hline f_2 \end{bmatrix} = \begin{bmatrix} 0 & m \\ \hline -m & 0 \end{bmatrix} \begin{bmatrix} f_1 \\ \hline e_2 \end{bmatrix}$$

(a)

$$\begin{bmatrix} f_1 \\ \hline f_2 \end{bmatrix} = \begin{bmatrix} 0 & 1/r \\ \hline -1/r & 0 \end{bmatrix} \begin{bmatrix} e_1 \\ \hline e_2 \end{bmatrix}$$

(b)

$$\begin{bmatrix} e_1 \\ f_2 \\ f_3 \end{bmatrix} = \begin{bmatrix} 0 & -1 & -1 \\ 1 & 0 & 0 \\ 1 & 0 & 0 \end{bmatrix} \begin{bmatrix} f_1 \\ e_2 \\ e_3 \end{bmatrix}$$

(c)

$$\begin{bmatrix} f_1 \\ e_2 \\ e_3 \end{bmatrix} = \begin{bmatrix} 0 & -1 & -1 \\ 1 & 0 & 0 \\ 1 & 0 & 0 \end{bmatrix} \begin{bmatrix} e_1 \\ f_2 \\ f_3 \end{bmatrix}$$

(d)

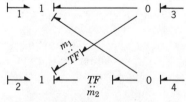

$$\begin{bmatrix} f_1 \\ e_2 \\ f_3 \end{bmatrix} = \begin{bmatrix} 0 & -m & m/r \\ m & 0 & -m \\ -m/r & m & 0 \end{bmatrix} \begin{bmatrix} e_1 \\ f_2 \\ e_3 \end{bmatrix}$$

(e)

Figure 7.21.   Constitutive laws for some simple junction structures.

(but with zeros on the main diagonal). The system of Figure 7.20*b* is in conductance form, but because it contains a gyrator, does not obey Onsager reciprocity and is instead antireciprocal. Figure 7.21*e* is antisymmetrical partly because of mixed causality (which contributes two antisymmetrical terms) and partly because of the gyrator (which contributes the other two antisymmetrical terms).

Since an *R*-field that contains no gyrators must obey Onsager reciprocity in resistance or conductance form, and any junction structure must

Figure 7.22.   A junction structure that is a 2×2 bilaterial transformation of variables.

have antisymmetric constitutive laws, we conclude that junction structures without gyrators cannot accept resistance or conductance causality on all ports.

Another interesting viewpoint about junction structures involves the transformations of variables that they provide. Consider, for example, the junction structure of Figure 7.22. It transforms the variables, e_3 and e_4, into the variables, e_1 and e_2:

$$\begin{bmatrix} e_1 \\ \hline e_2 \end{bmatrix} = \left[\begin{array}{c|c} -1 & -1 \\ \hline -m_1 & -m_2 \end{array} \right] \begin{bmatrix} e_3 \\ \hline e_4 \end{bmatrix}. \tag{7.92}$$

But this transformation of efforts is accompanied by a transformation of flows:

$$\begin{bmatrix} f_3 \\ \hline f_4 \end{bmatrix} = \left[\begin{array}{c|c} 1 & m_1 \\ \hline 1 & m_2 \end{array} \right] \begin{bmatrix} f_1 \\ \hline f_2 \end{bmatrix}, \tag{7.93}$$

and, since the junction structure is accomplishing the transformations, the transformations must conserve power. Rearranging the equations, the antisymmetric form characteristic of power conservation appears:

$$\begin{bmatrix} e_1 \\ \hline e_2 \\ \hline f_3 \\ \hline f_4 \end{bmatrix} = \left[\begin{array}{cc|cc} 0 & 0 & -1 & -1 \\ 0 & 0 & -m_1 & -m_2 \\ \hline 1 & m_1 & 0 & 0 \\ 1 & m_2 & 0 & 0 \end{array} \right] \begin{bmatrix} f_1 \\ \hline f_2 \\ \hline e_3 \\ \hline e_4 \end{bmatrix}. \tag{7.94}$$

Although this example illustrates the transformation of two efforts into two other efforts (and a concommitant transformation of two flows into two flows), junction structures can transform m variables into n variables, where these variables may be mixed sets of efforts and flows. In all cases, there is an invariant power flow involved. The power at one set of junction structure ports is always equal to the power at the remainder of the ports with proper sign considerations, so that the net power entering the junction structure always vanishes. This power invariance with the transformations enforced by junction structures was the basis of the tensorial analysis of networks by Kron [5] and is related to the use of generalized coordinates in mechanics, as will be seen in Chapter 9.

Junction structures composed of the linear elements —0—, —1—,

—*TF*—, and —*GY*— are, not surprisingly, linear, and their constitutive laws [e.g., Eqs. (7.92), (7.93), or (7.94)] represent linear transformations of variables. Such junction structures embody Kirchoff's laws for electric circuits, equations of continuity and dynamic equilibrium in fluid systems, and Newton's laws and geometric compatibility constraints for a restricted class of mechanical systems. In other cases, notably for certain transducer systems and mechanical systems in which geometric non-linearity is present, the nonlinear, modulated elements, —*MTF*— and —*MGY*—, are part of system junction structures. These junction structures are nonlinear or, at least, time-varying because the transformer moduli, m, and gyrator moduli, r, vary. Such junction structures still conserve power, however, and, in fact, these structures can be described by the same constitutive laws as their linear counterparts, except that the moduli are functions of some variables. Thus, the equations of Figure 7.21*a*, *b*, and *e* and Eqs. (7.92), (7.93), and (7.94) could represent nonlinear junction structures with varying parameters. The fact that electric circuits contain only linear junction structures and mechanical systems, in general, contain nonlinear junction structures means that the analogy between electric circuits and mechanical systems is strictly limited. Although nonlinearity in junction structures may not seem to add significantly to the difficulty of a problem, equation formulation can be quite complex when the modulated elements are present. An electric circuit containing linear 1-ports is itself linear. A mechanical system containing linear 1-ports but a nonlinear junction structure can exhibit very complicated nonlinear behavior. Examples of this will appear in subsequent chapters.

7.5 CLASSIFICATION OF BONDS AND FIELDS

Let us now examine multiport systems as a whole and look at them as interconnected fields. For some types of systems, at least, looking at them as interconnected fields gives insight into the dynamic structure, and formulation of equations can be quite straightforward.

This way of treating systems is most useful for large systems in which generation, manipulation, and reduction of many equations requires a systematic approach for successful implementation. Computer programs can be developed to perform reduction and formulation operations, and, in fact, the ENPORT-4 program organizes bond graphs into interconnected fields, as discussed below. (General references for this section are Reference [6] on the classification of fields and the general reduction procedure, Reference [7] on the transformation of energy-storing fields, and References [8] and [9] on the ENPORT program.)

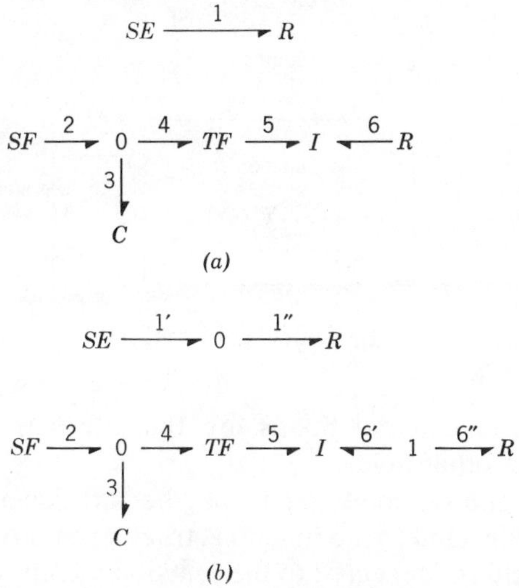

$$SE \xrightarrow{\quad 1 \quad} R$$

$$SF \xrightarrow{\quad 2 \quad} 0 \xrightarrow{\quad 4 \quad} TF \xrightarrow{\quad 5 \quad} I \xleftarrow{\quad 6 \quad} R$$

$$3 \downarrow$$

$$C$$

(a)

$$SE \xrightarrow{\quad 1' \quad} 0 \xrightarrow{\quad 1'' \quad} R$$

$$SF \xrightarrow{\quad 2 \quad} 0 \xrightarrow{\quad 4 \quad} TF \xrightarrow{\quad 5 \quad} I \xleftarrow{\quad 6' \quad} 1 \xrightarrow{\quad 6'' \quad} R$$

$$3 \downarrow$$

$$C$$

(b)

Figure 7.23. Examples of a standard form of bond graph. (a) Nonstandard form; (b) standard form.

In order to partition a multiport system into distinct, interconnected fields, we require a standard form of graph, so that our methods can be described succinctly. Therefore, in this section we shall assume that every bond in a bond graph is joined to a junction element, 0, 1, TF, or GY, at least at one end, if not at both. To achieve such a bond graph it is only necessary to insert a (dummy) junction of either the 0 or 1 type into each bond that violates the assumption.

For example, in Figure 7.23a, bond graphs that violate the assumption are shown. In part b of that figure a set of junctions has been introduced such that no change in the system relations has been made. That is, the bond variables on bonds 1' and 1'' are identical with those of bond 1, and the bond variables on bonds 6' and 6'' are identical with those of bond 6 except that $e_{6'} = -e_{6''}$. In both cases, either type of junction could have been introduced.

With a bond graph in standard form, every bond may now be classified as either an external or internal bond, according to whether the bond adjoins an R, C, I, SE, or SF at one end or not. That is, *external bonds* adjoin one of the set (R, C, I, SE, SF) at one end; *internal bonds* adjoin only members of the set (0, 1, TF, GY).

External bonds may be classified further according to the type of element they adjoin; namely, *storage-field bonds* adjoin C or I elements; *dissipative-field bonds* adjoin R elements; and *source-field bonds* adjoin SE or SF elements. Internal bonds are internal to the junction structure

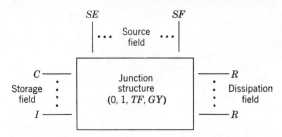

Figure 7.24.   Basic fields of a multiport system.

of the system, and external bonds are the interface of the junction structure with the other fields.

As Figure 7.24 shows, bond graphs may be partitioned as follows: the storage field is connected to the junction structure by storage-field bonds, the dissipation field is connected to the junction structure by dissipation-field bonds, and the source field is connected to the junction structure by source-field bonds. In the example of the bond graph of Figure 7.23b, bond 4 is *internal*, bonds 1' and 2 are source-field bonds, bonds 1" and 6" are dissipative-field bonds, and bonds 3, 5, and 6' are storage-field bonds.

Significant Vectors and Field Relations
in Multiport Systems

It is now possible to identify vectors of bond variables associated with particular fields, but this is generally not useful until causality has been assigned to the graph. Then the vectors may be identified not only by fields, but also by the input-output role they play in the causal system.

Referring to Figures 7.25a, we see that for a multiport system with completely integral causality on the storage elements, the SE elements impose their efforts on the junction structure, the SF elements impose their flows, the C elements impose their efforts, the I elements impose their flows, and the R elements impose a mixture of efforts and flows as dictated by the particular case. The identification of key vectors is shown in part (b) of Figure 7.25 as follows: U is the input to the junction structure from the source field, V is the output from the junction structure to the source field, Z is the co-energy variables vector, $\dot{X}$ is the time derivative of the energy variables vector, D_0 is the input to the junction structure from the dissipation field (generally a mixture of efforts and flows); and D_i is the input to the dissipation field from the junction structure (generally a mixture of efforts and flows).

The field equations relate some of the vectors directly. The storage field relations are

$$Z = \phi_F(X) \tag{7.95}$$

if the storage field is nonlinear, or

$$Z = FX \tag{7.96}$$

if it is linear. Furthermore, we may write that

$$X(t) = \int_{t_0}^{t} \dot{X} \, dt + X(t_0), \tag{7.97}$$

to show how the storage field behaves in terms of power variables, Z and $\dot{X}$.

For the dissipative field we have in the nonlinear case,

$$D_o = \phi_L(D_i), \tag{7.98}$$

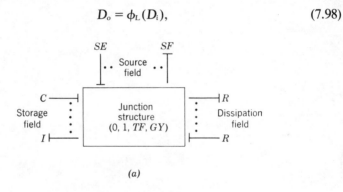

(a)

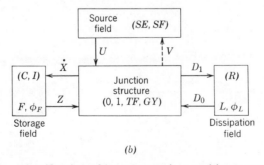

(b)

Figure 7.25. Identification of key vectors in a multiport system having integration causality. (a) Bond graph with integration causality; (b) key vectors identified.

and in the linear case,

$$D_o = LD_i. \tag{7.99}$$

There is no direct relation between U and V for the source field, but we note that U is a vector of input functions of time, the independent variable.

$$U = U(t). \tag{7.100}$$

It should be apparent at this point that we must turn our attention to the junction structure in order to continue the development of state equations. This has been done in developing the ENPORT digital simulation programs (see References [8] and [9]). Here we now show how the field equations are modified when a mixed integral and derivative causal pattern exists.

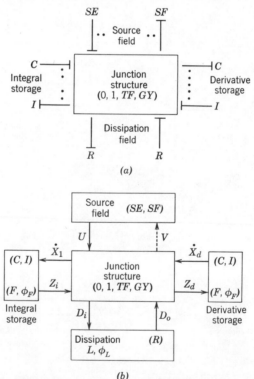

Figure 7.26. Identification of key vectors in a mixed causality system. (a) Mixed causality; (b) key vectors.

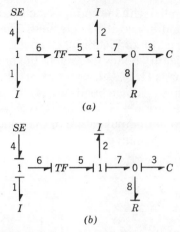

Figure 7.27. Bond-graph model of a lever mechanism. (a) Powers assigned; (b) causality assigned.

Figure 7.26 is similar to Figure 7.25 except for the addition of a derivative energy-storage field and the partitioning of the Z and X vectors into integral (Z_i and X_i) and derivative (Z_d and X_d) parts. The field relations may now be written as

$$Z_i = \phi_{Fi}(X_i, X_d) \qquad (7.101a)$$

and

$$Z_d = \phi_{Fd}(X_i, X_d) \qquad (7.101b)$$

in the nonlinear case, and

$$Z_i = F_{11}X_i + F_{12}X_d \qquad (7.102a)$$

and

$$Z_d = F_{21}X_i + F_{22}X_d \qquad (7.102b)$$

in the linear case, generalizing from Eqs. (7.95) and (7.96). In addition,

$$X_d(t) = \int_{t_0}^{t} \dot{X}_d(t)\, dt + X_d(t_0). \qquad (7.103)$$

If the model contains only 1-port storage elements or if causality does not appear mixed on individual storage elements, then Eqs. (7.101a, b) are not cross-coupled and $F_{12} = F_{21} = 0$ in Eqs. (7.102a, b).

To illustrate the identification process for obtaining the key vectors in a system we shall study a bond-graph model of a lever mechanism. In Figure 7.27a, a model of the lever mechanism of Figure 5.15a is shown in which a damper in parallel with the spring is included. This introduces the 0-junction, bonds 7 and 8, and the R element. The causally augmented bond graph is given in part b of Figure 7.27, and it shows mixed causality.

To start with, we identify bonds internal to the junction structure as (5, 6, 7). Clearly then, bonds (1, 2, 3, 4, 8) are external. Next we classify the external bonds as follows: source bonds are (4), dissipation bonds are (8), integral storage bonds are (2, 3), and derivative storage bonds are (1). Finally, based on the input/output nature of the bond variables, we arrive at the following vector definitions:

Source field $\qquad\qquad U = [e_4] \qquad$ and $\qquad V = [f_4]$

Dissipation field $\qquad D_i = [e_8] \qquad$ and $\qquad D_o = [f_8]$

Integral storage field $\qquad \dot{X}_i = \begin{bmatrix} e_2 \\ f_3 \end{bmatrix} \qquad$ and $\qquad Z_i = \begin{bmatrix} f_2 \\ e_3 \end{bmatrix}$

Derivative storage field $\qquad \dot{X}_d = [e_1] \qquad$ and $\qquad Z_d = [f_1]$

We also note

$$X = \begin{bmatrix} X_i \\ X_d \end{bmatrix} = \begin{bmatrix} p_2 \\ q_3 \\ \overline{p_1} \end{bmatrix} \qquad \text{and} \qquad \dot{X} = \begin{bmatrix} \dot{p}_2 \\ \dot{q}_3 \\ \overline{\dot{p}_1} \end{bmatrix}.$$

At this point, we have shown how it is possible to break up any bond graph into a collection of fields held together by a junction structure. The storage and dissipation fields possess the properties discussed in earlier sections. In practice, the fields arising in this way may contain explicit fields, or they may be mere collections of 1-port —Is, —Cs, or —Rs.

For writing state equations by hand, the methods of Chapter 5 are convenient, but for automatic computer formulation a more formal use of the field equations is convenient. This merely reflects the fact that graphical representations are extremely useful for communicating to sighted human beings, while the use of ordered arrays is suitable for communication with blind computers.

REFERENCES

1. S. H. Crandall and N. C. Dahl, eds., *An Introduction to the Mechanics of Solids*, N.Y.: McGraw-Hill, 1959, p. 378.
2. S. H. Crandall, D. C. Karnopp, E. F. Kurtz, and D. C. Pridmore-Brown, *Dynamics of Mechanical and Electro-mechanical Systems*, N.Y.: McGraw-Hill, 1968, pp. 220, 294, 296.

3. I. Prigogine, *Introduction to the Thermodynamics of Irreversible Processes*, Springfield, Ill.: C. C. Thomas, 1955.

4. J. Meixner, "Thermodynamics of Electric Networks and the Onsager-Casimir Reciprocal Relations," *J. Math. Phys.* **4**, 1963, p. 154.

5. G. Kron, *Tensor Analysis of Networks*, N.Y.: Wiley, 1939.

6. R. C. Rosenberg, "State Space Formulation for Bond Graph Models of Multiport Systems," *Trans. ASME, Journal of Dynamic Systems, Measurement and Control*, **93**, Ser. G, n. 1 (March 1971), pp. 35–40.

7. D. C. Karnopp, "Power-Conserving Transformations: Physical Interpretations and Applications Using Bond Graphs," *J. Franklin Institute*, **288**, (3) (Sept. 1969), pp. 175–201.

8. R. C. Rosenberg, "Modeling and Simulation of Large-scale Linear Multiport Systems," *Automatica*, January 1973.

9. R. C. Rosenberg, *A User's Guide to ENPORT-4*, Wiley, 1975.

PROBLEMS

7-1

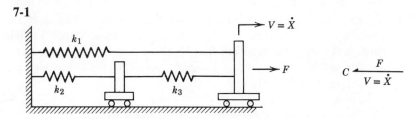

Three linear springs are attached to massless carts, as shown. Make a bond graph for this system, and manipulate the relations for the implicit C-field into the relation for the explicit 1-port C-field shown. Assume that when $X = 0$, all the springs are relaxed.

7-2

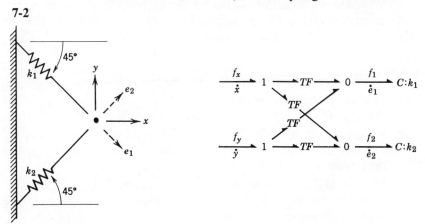

Two linear springs are pin-joined together as shown. Consider *small* motions, x and y, and small spring extensions, e_1, e_2. Show that the implicit field shown represents the system, and find the transformer moduli. Convert to an explicit field form at the x, y ports.

7-3

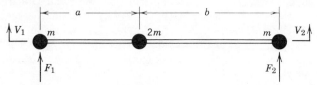

Three concentrated mass points are mounted on a massless rigid bar. Consider plane vertical motion such that the bar moves through only small angles. By expressing the motion of each mass in terms of $V_1 = \dot{X}_1$ and $V_2 = \dot{X}_2$, show that this system may be represented as a 2-port, implicit I-field. Find the constitutive relations for the I-field as seen at the F_1, F_2 ports.

7-4 Three equal pipes carrying an incompressible fluid join at a Tee junction. If each pipe has fluid inertia, $\rho l/A$, show an implicit I-field representation. Demonstrate that all three 1-port inertial elements cannot simultaneously have integral causality. Find a 3-port, explicit I-field representation.

7-5

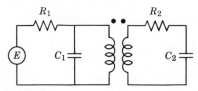

Write the equations of motion for the system shown, assuming the constitutive laws for the transformer are given by Eq. (7.63).

7-6

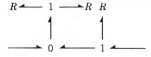

For the R-field shown, show causalities that would result in (a) Onsager forms, (b) Casimir forms.

7-7 Construct an implicit R-field that is expressible in neither an Onsager nor a Casimir form, and write out its constitutive laws as was done in Eq. 7.84.

7-8 Prove that if a resistance matrix were antisymmetric (with zeros on the main diagonal), then the net power dissipated would always be zero, assuming an inward sign convention on all external bonds.

7-9   Two bond-graph representations are shown in (a) and (b) below.

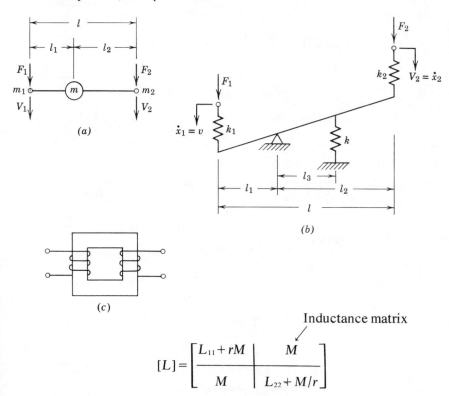

(a) (b)

In (a) assume $q_3 = m_1 q_1 + m_2 q_2$ and $e_3 = C_3^{-1}(q_3)$, a nonlinear capacitor. If the representation of (b) is to be valid, and the 2-port C-field is energy conservative, then

$$\frac{\partial e_1}{\partial q_2} = \frac{\partial e_2}{\partial q_1}.$$

Prove that this expression is valid for the system of (a). Extend the proof to the case in which the TFs in (a) become MTFs, q_3 is a general function of q_1 and q_2, $q_3 = q_3(q_1, q_2)$.

7-10   Study of some 2-port fields.

(1) Draw (a), (b), and (c) as 2-port fields, and, for (a) and (b), show the fields in terms of 0, 1, TF, I, C elements.

(2) Show that the mass matrix for (a) is

$$\begin{bmatrix} m_1 + (ml_2^2/l^2) & m(l_1l_2/l^2) \\ \hline m(l_1l_2/l^2) & m_2 + (ml_1^2/l^2) \end{bmatrix}.$$

Show that the compliance matrix for (b) is

$$\begin{bmatrix} 1/k_1 + (1/k)(l_1^2/l_3^2) & -(1/k)(l_1l_2/l_3^2) \\ \hline -(1/k)(l_1l_2/l_3^2) & 1/k_2 + (1/k)(l_2^2/l_3^2) \end{bmatrix}$$

Compare these results with the inductance matrix for case (c).

(3) Show that

for (a):	for (b):	for (c)
if $m_1 \ll m$,	if $k_1 \gg k$,	if $L_{11} \ll M$,
$m_2 \ll m$,	$k_2 \gg k$,	$L_{22} \ll M$,
then	then	then
$(l_1/l_2)p_1 = p_2$,	$(l_2/l_1)x_1 = x_2$,	$\lambda_1 = r\lambda_2$,
$v_1 = (l_1/l_2)v_2 + (l^2/ml_2^2)p_2$.	$F_1 = (l_2/l_1)F_2 + (kl_3^2/l_1^2)x_1$.	$ri_1 = i_2 + \lambda_1/M$.

(4) Show that for (a) if $m \to \infty$, for (b) if $k \to 0$, for (c) if $M \to \infty$, all systems become $-TF-$.

(5) Show that saturation of the iron in (c) is analogous to the limiting of motion by stops in (b) and, thus, d-c voltages will not pass in (c) and d-c velocities will not pass in (b).

7-11 Consider the bond graphs below:

(a) If the graphs are not in the standard form for classification of the bonds, insert extra 0- or 1-junctions to bring them into the standard form.

(b) For each case, list external bonds, internal bonds, storage-field bonds, dissipative-field bonds, and source-field bonds.

(c) Add causality to the system, and list the key vectors shown in Figures 7.25 and 7.26.

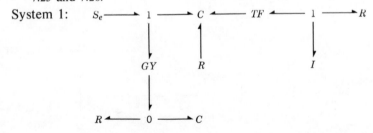

System 1:

System 2:

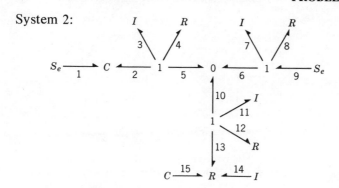

7-12 Consider a linear, multiport C-field with stiffness matrix, K. The constitutive relations are then

$$e = Kq \qquad \text{(i)}$$

where e and q are n-vectors and K is an $n \times n$ matrix. The stored energy, $\mathbf{E}$, is then

$$\mathbf{E} = \tfrac{1}{2}q'Kq. \qquad \text{(ii)}$$

The text argues that if energy is conserved, then K in Eq. (i) must be symmetrical, that is, $K = K'$.

(a) Convince yourself that any square matrix may be decomposed into a symmetrical part and an unsymmetrical part. For example,

$$K = K_s + K_a; \qquad K_s = \frac{(K + K')}{2}, \qquad K_a = \frac{(K - K')}{2};$$

$$K_s = K_s'; \qquad K_a = -K_a'.$$

(b) Using a 2×2 example with

$$K = \begin{bmatrix} k_{11} & k_{12} \\ k_{21} & k_{22} \end{bmatrix}, \qquad k_{12} \neq k_{21},$$

show that $\mathbf{E}(q)$ depends only on the symmetrical part of K, that is, that

$$\tfrac{1}{2}q'Kq = \tfrac{1}{2}q'K_s q,$$

and that, hence, $e = K_s q$ if the constitutive laws are derived by differentiation of the stored-energy function.

7-13

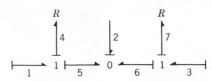

The linear R-field is shown in a mixed-causal form, that is, the input and output port variables are related by a K matrix thus:

$$\begin{bmatrix} e_1 \\ f_2 \\ e_3 \end{bmatrix} = [K] \begin{bmatrix} f_1 \\ e_2 \\ f_3 \end{bmatrix}.$$

(a) What can you predict about K before writing out the R-field equations?

(b) Suppose the R-field were forced to accept resistance or conductance causality. What characteristics would the resistance and conductance matrices have?

(c) Using causality, prove that the resistance matrix does not exist.

(d) Write out the K matrix and the conductance matrix to verify your predictions in (a) and (b). Using the equations for the mixed-causal form, show that the equations cannot be manipulated into the resistance-causality form.

8

TRANSDUCERS, AMPLIFIERS, AND INSTRUMENTS

This chapter deals primarily with models of devices that link two subsystems in two distinct energy domains. In some cases, the efficiency of power transduction is important. Motors, generators, pumps, and transmissions, for example, usually are designed so that they can transduce energy without much loss, at least when operating normally. Instruments and amplifiers, on the other hand, are designed to operate at low power efficiency. An ideal instrument would extract information from a system without power absorption and could communicate at finite power to another system. An amplifier accepts an input signal at near-zero power level and influences another system in response to the input signal at finite power.

The high-efficiency transducers are usually passive; that is, they contain no sources of power. The instruments and amplifiers are often active in the sense that they need a power supply in order to satisfy the first law of thermodynamics. Most practical transducers are themselves rather complex systems that can be modeled in detail by bond graphs. On the other hand, designers of large systems containing transducers as components cannot afford the luxury of modeling transducers in detail and must use good approximate models. In this chapter, we will show the main features of several types of transducers and demonstrate a philosophy of modeling in which nonideal effects may be progressively added to an idealized basic model. We concentrate first on passive transducers and then briefly study instruments and amplifiers.

Passive transducers which contain loss elements can, on the average, only transmit less power than they receive. Some transducers do, however, have the capability of storing energy so that they can temporarily deliver excess power. We call the two types of passive transducers *power transducers* and *energy–storing transducers*.

8.1   POWER TRANSDUCERS

Ideal power transducers were introduced in Chapter 4 as transformers and gyrators. Devices such as hydraulic rams, positive displacement pumps and motors, permanent magnet d-c motors and generators, and the like behave roughly as power-conserving elements. Real devices, of course, do exhibit power losses and also contain energy-storage mechanisms and associated inertial and capacitance effects. Although clever design of transducers can result in rather small values for these parasitic effects, these effects still provide an ultimate limit to the performance of any real device. When a very accurate model of a real transducer is

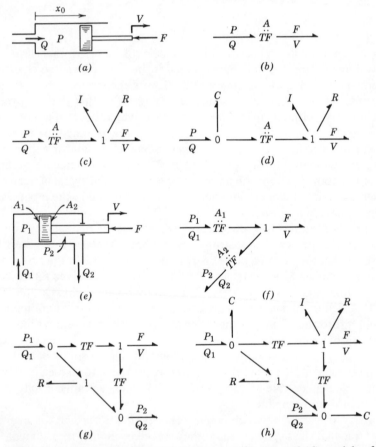

Figure 8.1.   Hydraulic rams. (*a*) Basic ram; (*b*), (*c*) and (*d*) models of basic ram; (*e*) hydraulic cylinder; (*f*), (*g*), and (*h*) models of hydraulic cylinder.

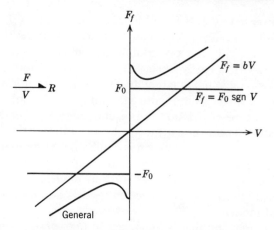

Figure 8.2. Several mechanical friction laws.

desired, it will almost invariably be necessary to replace linear elements of the model with more accurate nonlinear ones. The process of starting with a highly idealized model, adding parasitic elements, and replacing linear elements by nonlinear ones to achieve increasingly accurate transducer models will be illustrated for some typical power transducers.

In Figure 8.1, two hydraulic ram configurations are shown schematically, and a series of bond-graph models are also shown. The model of Figure 8.1b for the basic ram has been used in Chapter 4. This transformer model instantaneously and without loss of power transduces the hydraulic power, PQ, into mechanical power, FV. In some cases, this model may be quite adequate for a system analysis. On the other hand, in real pistons the mass effects and the frictional losses associated with piston rings, packing or tight fits to prevent leakage can be appreciable. The bond graph of Figure 8.1c essentially provides for a loss in force, F, due to the force required to accelerate the piston and the frictional loss.

The mass effect of the piston is quite straightforward, and, in many cases, when the piston rod is stiff and connects directly with a load mass, the piston mass and the load mass may simply be lumped together. But the friction force is more complex. A friction force may be small, and therefore negligible for some purposes, but the next simplest representation, the linear friction force, is virtually never an accurate representation of mechanical friction. In Figure 8.2, several possible friction force laws are shown. The linear law is often used in order to study systems using the analytically convenient linear methods, but usually the friction coefficient is left as a variable so that the model can be "tuned" to reproduce experimental results. Real friction usually includes a component of dry friction, represented by $F_f = F_0 \, \text{sgn} \, V$, where sgn V is $+1$ if $V > 0$ and -1 if $V < 0$. This law is not as convenient to work with as the linear law, particularly near $V = 0$. A more complicated phenomenon is sometimes called "stiction." It is easily observed that it takes more force to start a block lying on a table moving than to keep it moving at low speeds. The general friction law shown in Figure 8.2 attempts to model this phenomenon. When this type of friction is present, there is a tendency for the system to chatter. This is fine if the system is a violin string and bow, but is distinctly unpleasant if the system is chalk on a blackboard, a machine tool and workpiece, or a computer-simulation program with numerical stability problems. The experienced system modeler treads a fine line between realistic but intractable friction models and useful but possibly oversimplified models.

Another phenomenon that is important in high-performance hydraulic systems involves the compliance of the working fluid. Since hydraulic fluids are not usually very compliant, small density changes accompany large pressure changes, and a linear-constitutive relation usually suffices. One may define bulk modulus, β, as the coefficient relating pressure, P, and the change in volume, ΔV, of a mass of fluid that occupies volume, V_0, when $P = 0$.

$$P = \beta \frac{\Delta V}{V_0}. \tag{8.1}$$

Actually, β, which is expressed in pressure units, varies somewhat with mean pressure, temperature, amount of air in the working fluid, and so on, so a more useful form for Eq. (8.1) is

$$P = P_0 + \beta \frac{V}{V_0} = P_0 + \frac{\beta}{V_0} \int^t Q \, dt \tag{8.2}$$

where P_0 is an equilibrium pressure, V_0 is a nominal volume, and $Q = \dot{V}$ is the volume flow rate into the nominal volume. Then β is more obviously the slope of a nonlinear constitutive law relating pressure and the compressed volume of the fluid. The bond graph of Fig. 8.1d shows a $-C$ element that models this compliance effect. Clearly, there is an oscillatory phenomenon that occurs as energy is exchanged between the mass of the piston and the compliance of the hydraulic fluid. Note that for the linear theory of Eq. (8.2), one must pick a nominal volume, $V_0 = Ax_0$, where x_0 is an average position of the piston (see Figure 8.1a). Thus, the hydraulic fluid stiffness is high when x_0 is small and gets lower as x_0 is increased.

Figure 8.1e shows a 3-port transducer in which, because of the area of the piston rod, the areas that transduce the two pressures, P_1 and P_2, into force components are not equal. Also, the flows, Q_1 and Q_2, are not equal. The bond graph of Figure 8.1f represents this system in most highly idealized form. In Figure 8.1g the hydraulic leakage resistance across the piston has been included, and in Figure 8.1h mechanical inertia and friction and hydraulic compliance effects have also been modeled. As one may see from this example, even the simplest transducer models can become rather complex when many nonideal effects are accounted for.

Another useful example of a power transducer is the d-c motor or generator. The basic ideal transducer is shown in Figure 8.3a. The power

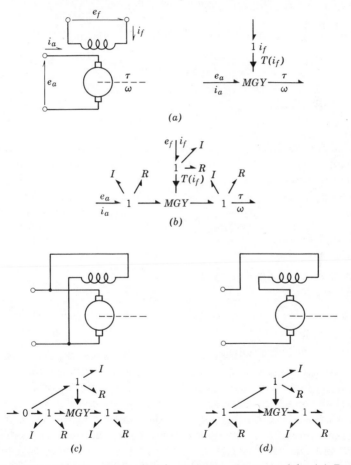

Figure 8.3. Direct-current electric motor-generator models. (a) Basic ideal transducer; (b) self-inductances, resistances, inertia, friction, and windage effects added; (c) shunt motor; (d) series motor.

in the armature circuit, $e_a i_a$, is transduced to shaft power, $\tau\omega$, when the device operates as a motor and the power flow reverses during generator action. The field port establishes an electric field, which provides the coupling between electric variables and mechanical variables for the individual conductors on the rotor. For a permanent magnet motor, this field is constant, but for the separately excited motor shown in Figure 8.3a, the field is a function of the field current, i_f. The commutator of the motor essentially maintains the field due to the armature current in a direction perpendicular to the field generated by i_f. For this reason, although i_f influences the transduction, there is virtually no back effect on e_f due to e_a or i_a. Thus, the field port influence is shown acting on an activated bond.

The equations corresponding to Figure 8.3a are as follows:

$$e_a = T(i_f)\omega,$$
$$T(i_f)i_a = \tau, \tag{8.3}$$
$$e_f = 0.$$

Note that the transduction coefficient, T, which is a gyrator parameter, can assume two different values if electrical and mechanical powers are measured in different units. Since that is the case, it may be worthwhile to rewrite Eq. 8.3 as

$$e_a = T_{em}\omega,$$
$$T_{me}i_a = \tau, \tag{8.3a}$$

with T_{em} expressed in volts per radian per second and T_{me} in foot-pounds per ampere, for example. Power conservation is, then,

$$e_a i_a = \frac{T_{em}}{T_{me}} \tau\omega, \tag{8.4}$$

with T_{em}/T_{me} relating volt-amperes to foot-pounds per second. One must not assume, however, that because T_{me} and T_{em} differ numerically, the device is not power conserving. In the metric system, one volt-ampere is identical to one newton-meter per second so that T_{em} is equal to T_{me}.

The parameter, T, involves the strength of the field as well as the number and effective lengths of the armature conductors that interact with the field. (See the discussion of the modulated gyrator in Chapter 7 for a discussion of this type of interaction with simplified geometry.) In some cases, it is reasonable to assume that the field is proportional to i_f.

Then

$$T(i_f) \cong Ai_f, \tag{8.5}$$

where A is a constant. Thus, if i_f is constant, torque is proportional to i_a, and if i_a is constant, torque is proportional to i_f. In such special cases, the device behaves linearly, even though it is basically nonlinear in a multiplicative sense. In real devices, saturation of the magnetic material limits the field so that the relation between T and i_f does not remain linear for large currents. Also, magnetic hysteresis effects, when present, mean that the field is not a single-valued function of i_f, but rather depends on the previous history of magnetization. (The permanent magnet motor is an extreme example of this—even though there is no i_f, a field exists, and hence T = constant.)

In most cases, the ideal transducer must be supplemented with loss and energy-storage elements to model practically important effects. In Figure 8.3b a common motor model is shown. The self-inductance and resistance of the field and armature coils are included, as well as the moment of inertia of the rotor and a mechanical resistor that models bearing losses and windage from the rotor and any built-in cooling fans. The electrical I and R elements are typically assumed to be linear unless a very accurate model is required. The mechanical resistor is rarely linear in reality, but sometimes an effective linear resistor will give good results if the mechanical friction is small enough that its detailed nature is not critical in determining system performance.

Two common 2-port motors can be constructed from the separately excited motor model. These are the shunt-wound and series-wound motors of Figure 8.3c and d. Note that, in Figure 8.3d, i_a and i_f are identical so that one could simplify the bond graph by combining the field and armature inductances and resistances.

Before leaving the subject of power transducers, let us consider two examples in which the ideal transducer is fairly complicated even when loss and parasitic-energy-storage effects are neglected. In Figure 8.4a, an a-c generator employing a permanent magnet is shown in highly schematic form. As the square coil rotates in the uniform B-field, there is an interaction relating the electric power variables, e and i, and the mechanical variables, τ and ω.

One of the constitutive laws of the device may be found by using the basic definition of flux linkage, λ. The amount of flux linked by the current path is B times the projected area of the coil times the number of turns in the coil, n.

$$\lambda = Bl_2l_1 (\sin \theta)n \tag{8.6}$$

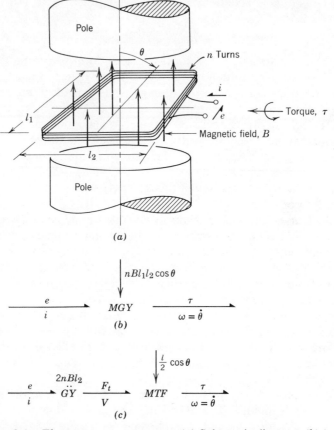

Figure 8.4. Elementary a-c generator. (a) Schematic diagram; (b) bond graph using MGY; (c) bond graph using GY and MTF.

Using the fact that $\dot{\lambda} = e$, one may differentiate Eq. (8.6) to find a relation between e and $\omega = \dot{\theta}$.

$$e = (nBl_1l_2 \cos \theta)\omega \qquad (8.7)$$

Then, noting that power must be conserved according to $ei = \tau\omega$, the remaining constitutive law is

$$(nBl_1l_2 \cos \theta)i = \tau, \qquad (8.8)$$

where the variables are expressed in metric units so that ei and $\tau\omega$ are both in the same units of power.

The constitutive equations of the device, Eqs. (8.7) and (8.8), are embodied in the bond graph of Figure 8.4b. Note that the modulating function involves θ, the integral of the local flow, ω.

Another way to derive the laws for this device involves computing the force, F, on the lengths of wire that cut the flux lines. Note that the lengths of wire associated with l_2 cut flux lines, while the lengths associated with l_1 do not. If F is the force perpendicular to the direction of B and V is the velocity in this direction, then for a typical conductor of length, l_2, the basic equations as discussed in Chapter 7 are

$$e_1 = Bl_2V, \tag{8.9}$$

$$Bl_2i_1 = F. \tag{8.10}$$

where e_1 and i_1 are the voltage and current associated with a single length of conductor. There are, however, $2n$ such lengths of conductor so that the terminal voltage is $2ne_1$:

$$e = 2nBl_2V. \tag{8.11}$$

Also, each length of conductor has the same current, and each produces a force that ultimately will add to produce the torque, τ. Calling the total torque producing force, F_t, we have

$$2nBl_2i = F_t. \tag{8.12}$$

Now, the cutting velocity, V, is

$$V = \left(\frac{l_1}{2}\cos\theta\right)\omega, \tag{8.13}$$

and the relation between F_t and τ is

$$\left(\frac{l_1}{2}\cos\theta\right)F_t = \tau. \tag{8.14}$$

Thus, the entire device can be represented by a gyrator with modulus, $2nBl_2$, and a modulated transformer with modulus, $(l_1\cos\theta)/2$, as shown in Figure 8.4c. Clearly, the two representations of Figure 8.4b and c or Eqs. (8.7) and (8.8) and (8.11), (8.12), (8.13), and (8.14) are equivalent. Loss and energy-storing elements could be added to this power-conserving model.

There are many analogies between rotary electromechanical devices and hydromechanical devices. Just as a d-c motor with multiple windings and a commutator functions as a gyrator, a pump with several pistons and a porting arrangement functions essentially as a transformer. The field port of a d-c motor allows modulation of the gyrator parameter, and a stroke control on a pump, if it exists, allows modulation of the transformer ratio. The a-c machines are often physically simpler, but functionally more complex. The a-c generator of Figure 8.4 is similar to the single piston and crank arrangement of Figure 8.5, which could form part of a pump if suitable valving were added. Although the relation between the piston force and velocity and the pressure and flow of the hydraulic fluid is readily represented by a transformer, the relation between the torque, τ, and the angular speed, ω, and the other power variables is more complex.

One way to find a relation between ω and the piston speed, V, is to start with a relation between x and θ (see Figure 8.5a). Working out the geometry, one finds

$$x = a \cos \theta + (b^2 - a^2 \sin^2 \theta)^{1/2} \tag{8.15}$$

If $\dot{x} = V$, and $\dot{\theta} = \omega$, then the result of differentiating Eq. (8.15) is

$$V = (-a \sin \theta - (b^2 - a^2 \sin^2 \theta)^{-1/2} a^2 \sin \theta \cos \theta)\omega. \tag{8.16}$$

Since $FV = \tau\omega$ if F is the force on the piston, the remaining constitutive

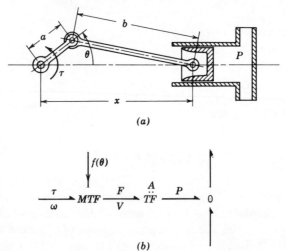

(a)

(b)

Figure 8.5. Crank and piston transducer. (a) Schematic diagram; (b) bond graph.

relation must be

$$(-a \sin \theta - (b^2 - a^2 \sin^2 \theta)^{-1/2} a^2 \sin \theta \cos \theta)F = \tau. \qquad (8.17)$$

Recognizing the complicated function of θ as a transformer modulus, and calling the function $f(\theta)$ for convenience, one may represent the device by the bond graph of Figure 8.5b. Note the similarity between this bond graph and that of Figure 8.4c. It should be evident from these examples that considerations of power conservation can greatly aid in modeling power transducers.

8.2 ENERGY-STORING TRANSDUCERS

The transducers of the previous section were modeled, in ideal form, by junction-structure elements. They were energy conservative, but beyond that, power in one domain was instantaneously transduced into another domain. In this section, we study transducers, which are also ideally energy conservative but in which energy storage plays an indispensable role. Thus, for these energy-storing transducers, energy from one domain may be stored and released in another domain at a later time.

The transducer models in this section are based on C-fields, I-fields, and mixed IC-fields, which were discussed in Chapter 7. Here, we will merely discuss some example systems and use the results of Chapter 7 without much discussion. As in the previous section, models of real transducers can be assembled from ideal models supplemented with loss and dynamic elements to account for effects that are present in real devices, but not accounted for in the ideal transducer.

A typical energy-storing transducer of practical interest is the condenser microphone or electrostatic loudspeaker. The sketch of Figure 8.6a shows roughly how these devices may be constructed. A capacitor is formed by mounting a moveable plate near a rigidly mounted plate and providing electrical connections as one would in a conventional parallel-plate capacitor. In practice, the moveable plate might be a thin diaphragm or tightly stretched membrane on which a thin layer of conducting material is fixed. Such distributed parameter "plates" could actually move in complicated ways, but, for simplicity, we will consider a system with only one mechanical coordinate, X, as shown in Figure 8.6b. (One may think of X as the displacement of the first normal mode shape of a diaphragm for those cases in which the contribution of higher modes may be neglected.) The force, F, which may be due to acoustical pressure in practical cases, is defined such that the mechanical power is $F\dot{X} = FV$ where V is the (generalized) velocity of the moving plate. The electrical power is, of course, ei.

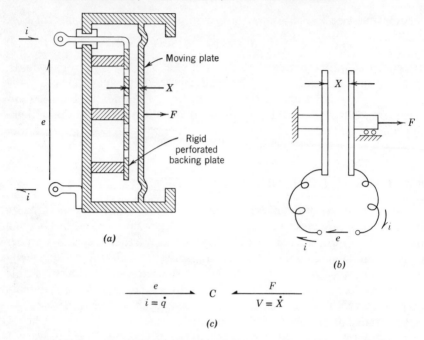

Figure 8.6. The moveable plate capacitor. (*a*) Sketch of microphone or speaker; (*b*) schematic diagram; (*c*) bond-graph representation.

When X is fixed, the device is an ordinary electrical capacitor, and when the charge, q, is fixed, then F depends on X as in a mechanical spring. In general, we may hypothesize that both e and F depend on q and X:

$$e = e(q, X), \tag{8.18}$$

$$F = F(q, X). \tag{8.19}$$

But the constitutive relations above are certainly not arbitrary since the device can at best conserve energy if we temporarily ignore loss effects. The stored energy, $\mathbf{E}$, is readily computed:

$$\mathbf{E}(t) = \mathbf{E}_0 + \int_0^t (ei + FV) \, dt = \mathbf{E}_0 + \int_{0,0}^{q,X} e(q, X) \, dq + F(q, X) \, dX = \mathbf{E}(q, X), \tag{8.20}$$

where $\mathbf{E}_0$, which will generally be assumed to vanish, represents an initial energy at $t = 0$ or when $q = X = 0$.

A close inspection of Eq. (8.20) shows that the constitutive laws of Eqs. (8.18) and (8.19) can be recovered from $E(q, X)$,

$$e = \frac{\partial E}{\partial q}, \qquad F = \frac{\partial E}{\partial X}, \tag{8.21}$$

and the required relation between the two constitutive laws is

$$\frac{\partial e}{\partial X} = \frac{\partial^2 E}{\partial X \partial q} = \frac{\partial F}{\partial q}. \tag{8.22}$$

This result is the integrability or Maxwell reciprocity conditions which the laws of the device must satisfy in order that energy be conserved. These considerations mirror those discussed in Chapter 7 for general C-fields, so the bond-graph representation is simply that shown in Figure 8.6c. The constitutive laws for the C-field are Eqs. (8.18) and (8.19), which must obey Eq. (8.22).

A useful approximation to the constitutive laws may be found by assuming that the device is *electrically linear*, that is, that a capacitance, C, may be defined for every X and that

$$e = \frac{q}{C(X)} \tag{8.23}$$

is the form of Eq. (8.18). Noting that from physical reasoning $F = 0$ when $q = 0$, we may evaluate E using Eq. (8.20) by letting $q = 0$, taking X to any particular value, and then charging the capacitor with $X = $ constant (or $dX \equiv 0$). Then the integral in Eq. (8.20) is just

$$E(q, X) = \int_0^q \frac{q}{C(X)} dq = \frac{q^2}{2C(X)}. \tag{8.24}$$

The force law corresponding to Eq. (8.23) is then

$$F = \frac{\partial E}{\partial X} = \frac{q^2}{2} \frac{d[C(X)]^{-1}}{dX}, \tag{8.25}$$

so that F may be determined from measurements of the variation of C with X. (For an ideal parallel-plate capacitance as sketched in Figure 8.6b, the law would be $C(X) = \varepsilon A/X$, where ε is the dielectric constant of the medium between the plates and A is the area of the plates.) The use of E to find the force law is far more convenient than a direct calculation or an experimental measurement.

Note that even when one assumes that this device is electrically linear, the force law is decidedly nonlinear. In use, the device is normally subjected to both a high polarizing voltage and a fluctuating signal voltage. The plates tend to move together and to short the electrical circuit. This is prevented by the mechanical spring of the diaphragm, which supplies a force in the direction of F (Figure 8.6b). In addition, all real diaphragms have mass and exhibit some energy-loss mechanisms when in motion. Such effects are readily modeled by adding elements to the C-field of Figure 8.6c. This device is discussed in more detail in Reference [1], Chapter 6.

The electrical solenoid shown in Figure 8.7 can serve as a prototype of a mixed IC-field transducer. The device consists simply of a coil of wire in which a soft-iron slug can freely slide. When current flows in the coil, the slug is pulled into the coil. In analyzing this device, it is clear that from the electrical port, the coil will certainly exhibit the characteristic self-inductance effects of any coil, although the position of the coil, X, will presumably affect the electrical behavior. From the point of view of the mechanical port, it seems clear that the force on the slug, F, will depend on X, that is, the device will possess some of the properties of a mechanical spring, although the electrical variables will also affect F. If we assume that the current in the coil depends on the flux linkage, λ, as well as X, and that F also depends on λ and X,

$$i = i(\lambda, X), \tag{8.26}$$

$$F = F(\lambda, X), \tag{8.27}$$

then the stored energy, $\mathbf{E}$, is

$$\mathbf{E} = \mathbf{E}_0 + \int_0^t (ie + FV)\, dt = \mathbf{E}_0 + \int_{0,0}^{\lambda, X} i\, d\lambda + F\, dX, \tag{8.28}$$

where $\dot{\lambda} = e$, and $\dot{X} = V$, and we will assume that $\mathbf{E}_0 = 0$.

By direct examination of Eq. (8.28) one may see that

$$i = \frac{\partial \mathbf{E}}{\partial \lambda}, \qquad F = \frac{\partial \mathbf{E}}{\partial X}, \tag{8.29}$$

and

$$\frac{\partial i}{\partial X} = \frac{\partial^2 \mathbf{E}}{\partial X\, \partial \lambda} = \frac{\partial F}{\partial \lambda}, \tag{8.30}$$

which is the integrability condition or Maxwell reciprocal condition constraining Eqs. (8.26) and (8.27). The bond graph of Figure 8.7b

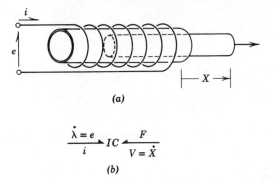

(a)

$$\dot{\lambda} = e \atop i \quad IC \quad {F \atop V = \dot{X}}$$

(b)

Figure 8.7. The solenoid. (a) Sketch of device; (b) bond-graph representation.

represents this ideal transducer. For mnemonic purposes the electrical port is shown impinging on the I and the mechanical port on the C.

Further insight into the device may be gained by assuming that the device is electrically linear. We assume that an inductance, $L(X)$, exists that relates i and λ for any position of the slug. Thus Eq. (8.26) becomes

$$i = \frac{\lambda}{L(X)}. \tag{8.31}$$

Also, on physical grounds, when i and λ vanish, F must also vanish, so that in evaluating **E** in Eq. (8.28), we may establish the slug at some particular position, X, with $\lambda = 0$, without doing any work on the device, and then hold X fixed while λ is brought to its final value. During the change in λ, $dX = 0$, so only electrical energy is stored. The energy is then

$$\mathbf{E}(\lambda, X) = \int_0^\lambda \frac{\lambda}{L(X)} \, d\lambda = \frac{\lambda^2}{2L(X)}. \tag{8.32}$$

The constitutive law corresponding to Eq. (8.27) can now be found using Eq. (8.29).

$$f = \frac{\lambda^2}{2} \frac{d[L(X)]^{-1}}{dX} = -\frac{\lambda^2}{2} \frac{L'}{L^2}, \tag{8.33}$$

where L' is dL/dX. The general forms for $L(X)$ and $L'(X)$ are sketched in Figure 8.8. From these sketches and Eq. (8.33), one can see that the slug will experience a force tending to center it, but even when the simple law of Eq. (8.31) is assumed, the corresponding force law is rather complex

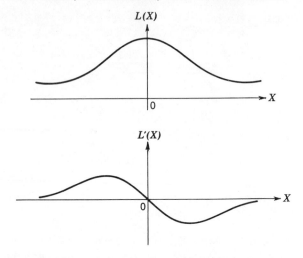

Figure 8.8.   Inductance functions for the solenoid.

and inherently nonlinear in both the electrical and the mechanical variables.

The IC-field concept is particularly useful for electromagnetic devices, which often involve interacting magnetic fields associated with moving parts. Alternate-current motors and generators, for example, typically involve coils that rotate with respect to each other. The electrical ports of these devices are inertial and the rotary mechanical port is capacitive when the device is described as an IC-field. Examples of such devices appear in the problems and in Reference [1], Chapter 6.

8.3   AMPLIFIERS AND INSTRUMENTS

The central idea behind the words "amplifier" and "instrument" is a low-power or a one-way interaction without back effect. The description of ideal amplifiers and instruments is functional rather than physical, and thus the physical, bilateral, power interactions of bond graphs must be degenerated into signal interactions by the use of activated bonds in order to represent these devices. One assumes frequently that an ideal amplifier supplies an output power variable such as a voltage or current at finite power in response to an input signal at essentially zero power. Similarly, an instrument is supposed to extract information about some variable without affecting the system in which the variable appears and to transmit the information, often at finite power levels.

Clearly, ideal instruments and amplifiers violate even more physical laws than the ideal transducers discussed previously. For example, the usual functional descriptions violate even the first law of thermodynamics since finite output power is somehow produced from zero input power. In reality, of course, most amplifiers have a readily identifiable power supply that is built into the functional relationships. The bond graphs of Figure 8.9 show the power supply and show that power is not created from a vacuum.

On the other hand, the active bond that indicates a signal flow with no associated power can only be approximate. At the microscopic level, the Heisenberg uncertainty principle is essentially a statement that signal interactions without back effect are impossible, but at a macroscopic level we know that amplifiers with extremely high power gain can be built and that instruments that have virtually no effect on the observed system are available in many cases. But no absolute statements about the appropriateness of an active bond representation can be made.

Probably every engineering student has had the experience of thinking of some real instrument as an ideal signal transducer and amplifier only to find that, in some cases, the attachment of the instrument to some system greatly distorted the behavior of the system to be measured. An oscilloscope, for example, has a high but finite input impedance. Therefore, it cannot be expected to measure voltages in a system having impedances of the same order of magnitude as the oscilloscope input impedance without a large effect on those voltages due to the current flowing to the instrument. Similarly, the final stage of a hydraulic amplifier of the size, say, of a ship steering engine requires sizeable power levels at its input.

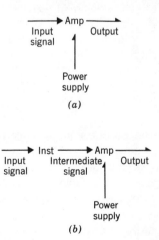

Figure 8.9.   Amplifiers and instruments. (*a*) Basic bond-graph representation of an amplifier; (*b*) instrumentation system with signal level transducer and associated amplifier.

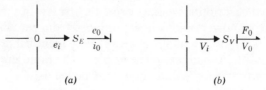

(a)   (b)

Figure 8.10. Amplifier models using controlled sources. (a) Voltage-controlled voltage source; (b) vibration shaker shown as a velocity source.

The input can be considered to be an active bond only under restricted circumstances, for example, if the next-to-last stage of the amplifier is sized such that the load of the input to the last stage does not cause a significant effect on the response of the next-to-last stage.

In low-power applications, power efficiency is generally not very important, and amplifiers are designed to provide a drastic decoupling of input and output back effects. In high-power applications on the other hand, amplifiers must usually be treated more physically since it is not feasible to build such high-power gain components that the input to an amplifier stage can reasonably be considered an active signal. Thus, it is a modeling decision whether or not to represent an amplifier or instrument with an active bond, and such a representation must be justified for each system of which the real device is to be a part.

With these caveats in mind, let us consider some useful bond-graph representations of ideal elements incorporating active bonds. The simplest type of amplifier model consists of a signal-controlled effort or flow source. In Figure 8.10a, the ubiquitous electrical voltage amplifier is shown in bond-graph form. The output voltage, e_0, is assumed to be a static or dynamic function of the input voltage, e_i. In the static, linear case, there is a voltage gain, G, and

$$e_0 = Ge_i(t). \tag{8.34}$$

In the dynamic case, the output voltage may be related to the input voltage by a differential equation, or in the common linear case by means of a transfer function. In the latter case, one may find internal state variables for the amplifier that yield the desired transfer function. Generally, these state variables are nonphysical and merely serve to provide state equations equivalent to the specified frequency domain representation of the amplifier. Such state equations may be used in time-domain analysis of the complete system.

The input voltage, e_i, in Figure 8.10a is shown as coming from a 0-junction as a signal on an active bond. In writing the current sum relation for all the bonds incident on the 0-junction, one assumes that there is no current associated with the active bond. Thus, the ideal amplifier does not affect the 0-junction from which it obtains its input voltage. At the output, a finite current, i_0, may exist, but e_0 is not affected by i_0. Thus, the power gain is infinite in this ideal case.

In well-designed systems, amplifiers may indeed function as controlled sources, but in some cases, the controlled-source assumption is made even when a more detailed and physical model would be preferable. In Figure 8.10b, for example, a vibration shaker is modeled as a controlled-velocity source in which a desired velocity time history is generated as V_i and a complex servomechanism system is supposed to enforce $V_0(t)$ at a port of a test system in response to $V_i(t)$. In many cases, the reaction force does affect the servo so that V_0 does not faithfully track V_i. In such a case, the simple model of Figure 8.10b is clearly inadequate, but the simplicity of controlled-source models often tempts system engineers to use them, particularly when the dynamics of the real devices have not been well explored or documented.

Many physical devices function essentially as amplifiers or transducers, but are inadequately modeled as simple controlled sources. For example, a gasoline engine clearly amplifies the power of a human or automatic controller. The torque-speed curves of an engine, as sketched in Figure 8.11a are drastically modified by the position of the throttle valve, which has been indicated by the angle, θ. Also, the power required to move the throttle valve is often extremely small, not only in comparison with the engine power, but also in comparison with spring forces, inertial forces, and pivot-friction forces in the throttle linkage. Thus, it may be reasonable to consider that the torque associated with movement of the throttle linkage is negligible, but we cannot merely assume that θ controls a source since the torque, τ, is always a function of the angular velocity of the output shaft, ω. Since torque is related to speed, one may describe the engine as a resistor, albeit an unusual one in which power is normally supplied rather than dissipated. (The power supply in the gasoline has been absorbed into the torque-speed curves.) Thus, the amplifier representation of the engine can be shown as a controlled resistance as in Figure 8.11b.

The active bond indicating that there is no torque on the throttle linkage corresponding to $\dot\theta$ is equivalent to a signal, so one may use block diagram notation to show that $\dot\theta$ is integrated into the signal, θ, that controls the resistor representing the engine. In general, when amplifiers and instruments are part of a system, the use of block diagrams for showing dynamic relations for active-bond signals in a bond graph can be very useful.

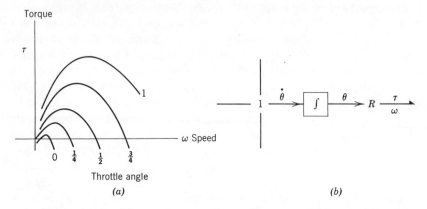

Figure 8.11. Static model of gasoline engine as amplifier. (*a*) Torque-speed curves as a function of throttle linkage angle, θ; (*b*) amplifier model of engine using controlled resistance.

As another example of the use of a controlled resistance, consider the strain-gage instrument shown in Figure 8.12. The idea behind a resistance strain gage is that the electrical resistance of the gage is a function of the mechanical strain of the member to which the gage is attached. In many applications, the member functions as an elastic element that relates strain to stress and, ultimately, to force in the structure. Thus, when the strain gage is put into the bridge circuit of Figure 8.12*a*, and the bridge is supplied with the voltage, e_i, then the voltage, e_0, reacts to force, F, in the structure. A bond-graph representation of the basic instrument can be constructed out of a force-controlled resistance and the bond graph version of a bridge circuit, as shown in Figure 8.12*b* and *c*. Of course, one could continue to simplify this instrumentation system. For example, if e_i were supplied by a constant source, then one could reduce the system to a force-modulated resistance at the output port with the voltage supply built in to the resistance relation. Furthermore, if an amplifier were connected to the output port, it could be arranged to supply an output voltage as a function of the force. Thus, the entire system might simplify to a force-controlled voltage source.

Although virtually any transducer might be arranged to act as an instrument under certain conditions and, in conjunction with a power supply, could function as an amplifier, variable-resistance elements are particularly important. In this category fall most electronic devices such as transistors, vacuum tubes, and the like, and the important valve-controlled hydromechanical devices. Perhaps a final example will help indicate how such devices may be modeled in bond-graph terms.

Valves are devices in which the position of a mechanical part influences the hydraulic resistance. This change in resistance can be converted to a change in pressure drop across the valve and, thus, one may deduce the position of the moveable member; that is, the valve can be made into a position-indicating instrument. Also, by moving the valve, large amounts of fluid power may be controlled, and the valve may be made into an amplifier. It is this latter application that we now consider.

The so-called four-way valve shown in Figure 8.13 is common in hydraulic power systems. Actually, four resistances in the valve are modulated by the valve spool position, z, simultaneously. When the valve is connected to a pressure supply and a load (typically a hydraulic ram or

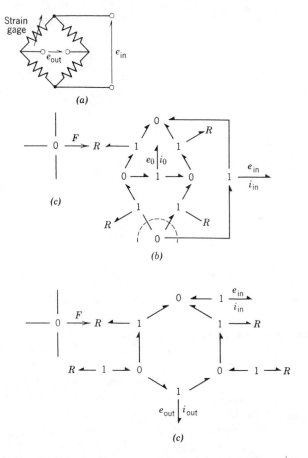

Figure 8.12. Strain-gage instrument. (a) Circuit diagram; (b) bond graph; (c) simplified bond graph after choice of ground voltage.

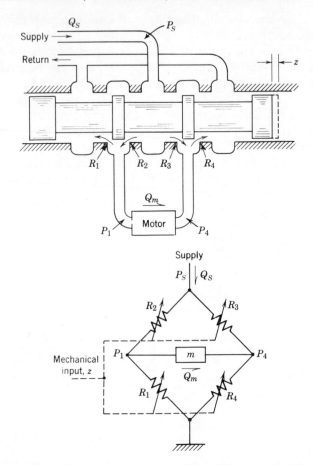

Figure 8.13.   The general four-way valve and its equivalent circuit.

a positive displacement rotary hydraulic motor), a large flow of hydraulic power, $P_m Q_m$, is controlled by a small amount of power associated with the movement of the spool, $\dot{z}$. As shown in Figure 8.13, the hydraulic circuit is a bridge.

Such systems involving hydraulic valves are complicated by the intrinsic nonlinearity of hydraulic resistors. These systems have been extensively studied, however, and rather simple models may often suffice. From Chapter 7 of Reference [2], for example, we find the curves of Figure 8.14, which give the relation between motor flow, Q_m, and motor pressure, P_m ($P_m = P_1 - P_4$ in Figure 8.13), for various values of valve spool position, z. Clearly, this amplifier is represented as a displacement-modulated resistance, and the curves of Figure 8.14 are the constitutive

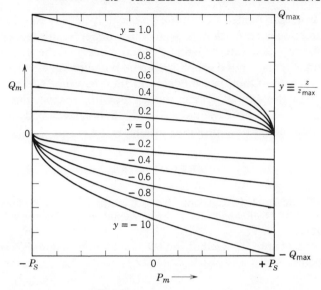

Figure 8.14. Pressure-flow characteristics for a four-way valve.

laws of the device. (Various characteristic curves for different valve geometries are given in Reference [2].) The bond graphs of Figure 8.15 show how the amplifier may be represented. In Figure 8.15a, the supply pressure is merely absorbed into the constitutive laws of the resistor. In Figure 8.15b, the valve is shown in 2-port resistance form. In this case, the effect on the amplifier of operating with various supply pressures can be

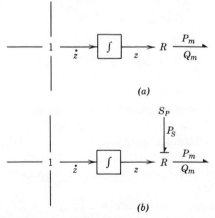

Figure 8.15. Bond-graph models of the four-way hydraulic valve. (a) Displacement-modulated resistance with P_s built into constitutive law; (b) representation showing pressure supply.

studied. In either model a complex physical system is represented in a simple functional form that is valid only under restricted conditions. Naturally, more accurate physically based models of the device may be made using standard bond-graph methods, but such models will be more complex than the functional model, and they may not add much to the usefulness of the overall system model. For this reason, the functional model may at least serve the purposes of a first system analysis. Later, a more refined model may prove desirable.

REFERENCES

1. S. H. Crandall, D. C. Karnopp, E. F. Kurtz, and D. C. Pridmore-Brown, *Dynamics of Mechanical and Electromechanical Systems*, N.Y.: McGraw-Hill, 1968.
2. J. F. Blackburn, G. Reethof, and J. L. Shearer, *Fluid Power Control*, Cambridge, Mass.: The M.I.T. Press, 1960.

PROBLEMS

8-1

The diagram shows a positioning system using a separately excited d-c motor. On the diagram physical-system variables and parameters are identified, where

θ_0 = output position angle;

e_{in} = input voltage;

e_a = output voltage of linear amplifier;

i_a = motor armature current;

i_f = motor field current, assumed constant;

K_a = gain of linear amplifier, assumed to have no significant time constants;

R_a = resistance of armature winding;

L_a = inductance of armature winding;

J = inertial load;

β = viscous-damping constant;

K_T = torque constant of motor;

K_v = back-emf constant of motor.

The differential equations that govern the dynamics of the system are

$$J\ddot{\theta} + \beta\dot{\theta} = K_T i_a,$$

$$L_a \dot{i}_a + R_a i_a = V_a - K_v \dot{\theta}_0.$$

(a) Construct a bond graph for the system.

(b) Write state space equations, and verify that your equations are equivalent to those listed above.

(c) Compare the two methods for analyzing this system. For example, is the system third or second order? Are K_T and K_v related in any way?

8-2 Consider the seismometer sketched below:

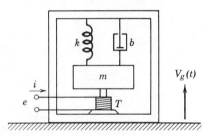

The input is ground motion, $V_g(t)$, and an electrical transducer using a permanent magnet moving in a coil reacts to the relative motion between the case and the seismic mass, m.

(a) Construct a bond graph for the device, leaving the electrical port as a free bond and neglecting coil resistance and inductance.

(b) Assume that the device is connected to a voltage amplifier so that $i \cong 0$. Find the transfer function between $V_g(t)$ and e.

(c) It is sometimes preferable to use current rather than voltage as a signal for reasons connected with noise pickup. Suppose that a current amplifier is connected to the seismometer terminals so that $e \cong 0$. Show that when the coil resistance is neglected, one can still find a transfer function between V_g and i even though the system state equations degenerate.

(d) Reconsider the case of (c) when the coil resistance, R_c, is not neglected.

8-3 Consider the solenoid shown in Figure 8.7, but include the mass, m, of the moving element and a coulomb friction force, F_f, with the constitutive law

$$F_f = F_0 \text{ sgn } V = F_0 V/|V|,$$

where F_0 is the magnitude of the friction force. Show a bond graph for this transducer, and write state equations for it assuming electrical linearity and a voltage source input.

8-4 The float shown is cylindrical and is immersed in water in a cylindrical container.

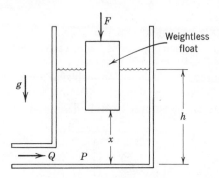

The cross-sectional area of the float is not negligible with respect to the tank area. The pressure, P, at the tank inlet is γh, where γ is the weight density of water and h is the height of the water. The volume of water, V, is the time integral of the flow rate, Q. If we assume that

$$P = P(V, x),$$

$$F = F(V, x),$$

what type of bond-graph element is this device? Are the P and F functions related in any way? Can you sketch the P and F functions?

8-5 An electrostatic loudspeaker system is shown in which a high charging voltage, E, and a signal voltage, $e(t)$, are applied through a current-limiting resistor, R. The moving plate or membrane may be assumed to have effective mass, m, and a mechanical spring of constant, k. (The mechanical spring represents the combined effect of membrane tension and the air-spring formed by the sealed cabinet.) The effect of the acoustic loading is shown by the bond-graph fragment in which A is the effective area of the speaker and P represents the acoustic overpressure. The I and R elements may be adjusted at any single frequency to match the impedance of the air.

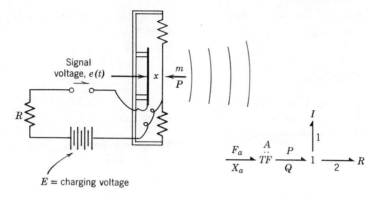

E = charging voltage

Assuming the basic transducer is an electrically linear device, with capacitance $C(x) = \varepsilon A/x$, where ε is the dielectric constant for the air in the cabinet, construct a bond graph for the system. Let the mechanical spring be relaxed at $x = x_0$ when $E = 0$. Write the state equations for the system, letting the acoustic inertia and resistance be called I_1 and R_2, respectively.

8-6

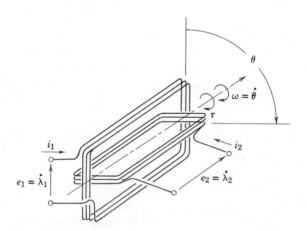

The basic transduction mechanism for electrical alternators and motors may be understood by studying the ideal energy-storing transducer shown above, consisting of a fixed and a moving coil. Assume the two currents and the torque, τ, are related to two flux-linkage variables, λ_1, λ_2, and the angular position, θ.

$$i_1 = i_1(\lambda_1, \lambda_2, \theta); \qquad i_2 = i_2(\lambda_1, \lambda_2, \theta); \qquad \tau = \tau(\lambda_1, \lambda_2, \theta).$$

For the *electrically linear* case, it is conventional to specify self- and mutual-inductance parameters; for example,

$$\begin{bmatrix} L_1 & L_0 \cos\theta \\ L_0 \cos\theta & L_2 \end{bmatrix} \begin{bmatrix} i_1 \\ i_2 \end{bmatrix} = \begin{bmatrix} \lambda_1 \\ \lambda_2 \end{bmatrix}; \qquad \begin{array}{c} L_0, L_1, L_2 \text{ are constants,} \\ L_1 L_2 \geq L_0^2. \end{array}$$

(a) What kind of bond graph element describes this device?

(b) Derive the torque relation from the inductance matrix above by computing the stored energy at fixed θ when λ_1 and λ_2 are brought from zero to final values and then differentiating the energy function with respect to θ (see Reference [1], pp. 322–323).

8-7

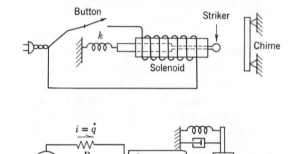

A model of a doorbell chime is shown that features a solenoid, return spring, frictional force, and a striker and chime. The schematic diagram attempts to depict the physical effects to be modeled for the case in which the button-switch is closed.

Make a bond graph for the system, including all physical effects. Let the striker-chime interaction be modeled by a nonlinear spring with zero force until the striker contacts the chime.

Write the equations of motion for this device assuming electrical linearity for the solenoid. Leave your results in functional form, that is, let $L(x)$ be the inductance and $F(x)$ be the striker spring force. Sketch the shape of $L(x)$ and $F(x)$.

8-8

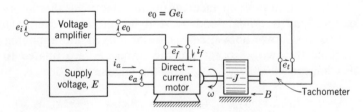

A speed control system uses a d-c motor with the following parameters:

L_f = field inductance;
R_f = field resistance;
L_a = armature inductance;
R_a = armature resistance;
J = total moment of inertia;
B = rotary dashpot coefficient;
$T(i_f)$ = transduction coefficient (see Figure 8.3).

The control voltage, e_i, drives an amplifier that functions as a voltage-controlled voltage source:

$$\xrightarrow{\;e_i\;} S_e \xrightarrow{\;e_0\;}, \qquad e_0 = Ge_i(t).$$

The tachometer is a permanent magnet device that functions as an instrument, so that its mechanical bond may be activated:

$$1 \xrightarrow[\omega]{} \overset{K_t}{\ddot{G}Y} \xrightarrow[i_t]{e_t}, \qquad e_t = K_T\omega.$$

Write a bond graph for the system, write state equations, and find the transfer function between e_i and ω when i_a = constant.

8-9

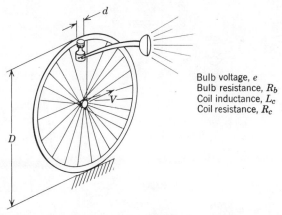

Bulb voltage, e
Bulb resistance, R_b
Coil inductance, L_c
Coil resistance, R_c

An a-c bicycle generator is shown in which a small wheel of diameter, d, bears upon the bicycle wheel of diameter, D. Assume the generator functions exactly as the one shown in Figure 8.4, although, in reality, it is more common for a magnet to rotate inside a fixed coil. Set up a bond-graph representation that models the coil self-inductance and resistance and the bulb resistance and that would suffice to predict how the bulb voltage would vary with forward speed, V.

Explain the self-regulation feature of the $L–R$ circuit that allows the bulb voltage to rise less than proportionately to V due to frequency response effects.

8-10

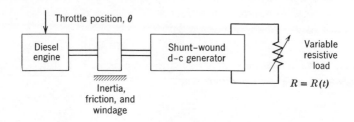

It is desired to study the response of the generating system shown to changes in throttle setting, $\theta(t)$, and load resistance, $R(t)$. Use the bond-graph model of Figure 8.11 to represent the engine, and let the load resistance be controlled by an active bond, $— R \prec$.

Show a bond graph for the system, and write the equations of motion in functional form. See Figure 8.3.

8-11

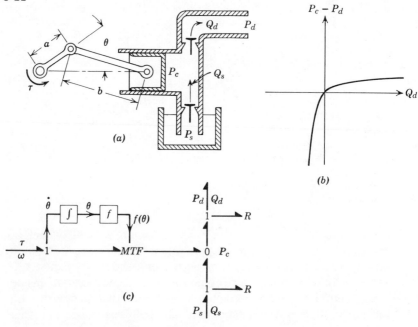

A simple pump is sketched in part *a* above. The check valves are represented by nonlinear resistors with characteristics such as that shown in part *b*.

(a) Verify the bond graph shown in part *c*, and compute the modulation function, $f(\theta)$.

(b) Suppose the speed, $\omega = \theta$, were essentially constant. Sketch how the discharge flow and the suction flow vary as functions of time.

(c) You should be able to see that this pump acts like an electrical half-wave rectifier circuit. Invent a pump that acts like a full-wave rectifier circuit, and show a bond graph for your pump.

8-12 Consider the high-performance speed-control system shown below. The idea is to drive a hydraulic pump with a shunt-wound d-c motor and to control speed, ω_2, by stroking the bypass valve in a hydrostatic transmission. The valve stroke is $x(t)$, which may be changed manually or, ultimately, automatically by a servo-control system. The system should have fast response since, if the valve is slammed shut, the hydraulic motor pump will be suddenly directly coupled so that energy stored in the rotating inertia, J_1, would be available to accelerate J_2 very rapidly.

(a) Draw a bond graph for the system including all the effects shown on the sketch. Augment the bond graph, and list the state variables required.

(b) Either write out the state equations or construct a block diagram corresponding to the bond graph. Identify the input variables to your dynamic equations.

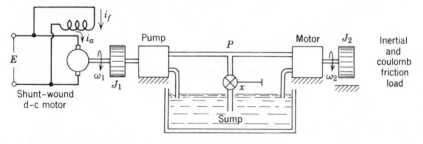

The basic motor characteristic is

$$T(i_f)i_a = \tau,$$

$$e_a = T(i_f)\omega_1.$$

The pressure-flow law for a valve depends on stroke, x, that is,

$$P = A(x)Q\,|Q| = A(x)Q^2\,\mathrm{sgn}\,Q.$$

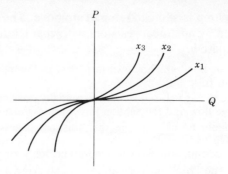

The model should include armature and field inductance and resistance. The constitutive laws for a pump and a hydraulic motor are

(a) For a pump:

$$\tau_p = \alpha_p P_p, \qquad \alpha_p \omega_p = Q_p$$

(b) For a motor:

$$\tau_m = \alpha_m P_m, \qquad \alpha_m \omega_m = Q_m$$

Neglect leakage in the pump and motor and compressibility in the oil lines.

8-13   In the diagram, a conventional hydropneumatic suspension system such as those found on certain automobiles has been made active

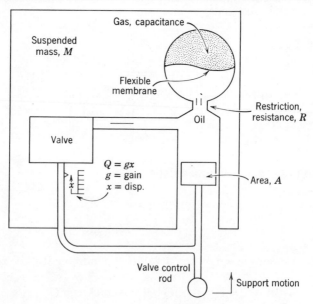

by the addition of a valve. A simple model of the valve is indicated in which the flow of oil, Q, is simply proportional to the relative displacement of the suspension, x. We assume a very high-pressure source of oil (not shown) so that the valve can act as a source of flow modulated by x,

$$\xrightarrow{\ \ }_{x} \ S_Q \longmapsto,$$

independent of the pressure in the suspension.

With this system, the "static deflection" of the suspension is always zero for any gain, g, but as g is increased, it is not so clear that the system will remain stable.

(a) Find a bond graph for the system.
(b) Decide on appropriate system inputs, and write a set of state space equations for the system.
(c) Assuming linearized characterizations for all elements of the system, set up an expression that will yield system eigenvalues for any particular numerical values of parameters.

8-14 A servomechanism has been constructed by connecting a spool valve and hydraulic ram with a feedback linkage.

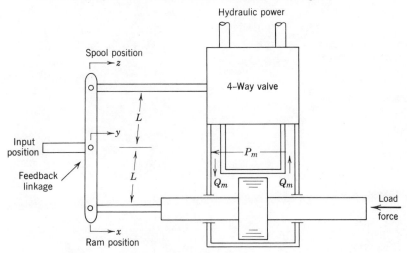

Using the text discussion and Figures 8.13, 8.14, and 8.15, you are to construct a bond graph for this device. Assume the following:

(a) The force required to move the valve is $F_r = F_0 \, \text{sgn} \, z$, where $F_0 =$ constant.
(b) The load force is inertial and resistive and oil compressibility is included.

    (c) The linkage is light and frictionless and moves only through very small angles.

    (d) The working area of the ram is A, and the supply pressure is P_s.

Using your bond graph, answer the following questions:

    (1) What is the force required to move the input, and what is the maximum load force possible?

    (2) With zero load force, what is the maximum velocity of the ram?

    (3) Write the equations of motion, leaving the valve constitutive laws in a general functional form, for example, $Q_m = Q_m(z, P_m)$.

9

OTHER APPLICATIONS OF MULTIPORT MODELING

9.1 INTRODUCTION

In preceding chapters, bond-graph models for systems containing mechanical, electrical, hydraulic, and heat transfer elements were developed. Here we indicate how bond-graph methods may be extended to wider classes of physical systems. First, we show how to deal with the geometric nonlinearities that arise when mechanical systems undergo finite angular motion. The type of nonlinear junction structure that results has no parallel in many other types of systems. We then go on to study magnetic systems, thermodynamic systems, and fluid-dynamic systems. In the last two sections, a deeper insight into the restrictions inherent in the previous treatment of heat transfer and hydraulic systems is provided.

The class of systems that can be treated with bond-graph methods is by no means exhausted by the systems described here. In References [1–4], for example, systems involving chemical reactions, diffusion, and electrochemical interactions are discussed using bond graphs. Such systems are particularly important in the life sciences, and the bond-graph approach provides insight into the nature of physiological processes. Indeed, bond graphs may even prove useful for nonenergetic systems such as social or economic systems, but we resist the temptation to speculate on such matters in this book.

9.2 MULTIPORT MODELS IN MECHANICS

This section is addressed to the following problem: Given a schematic representation of a mechanical system plus accompanying verbal and mathematical descriptions of the parts, find a convenient way to model the system so as to predict its dynamic behavior.

317

The importance of the problem hardly needs emphasizing. It has been worked on in essentially modern form at least since the time of Newton, and has attracted the attention of many notable scientists, including Hamilton and Lagrange. The development since the time of Newton has relied, not upon pictorial or graphical representation, but upon analytically oriented notation in the form of operators and equations of various types. Since we now know how to formulate mechanics problems in a variety of ways, it is fair to say that the basic problem has been solved [5, 6].

The principal purpose of this chapter is to present and develop the use of certain bond-graph forms as standard models in mechanics, thereby bringing the study of this very important class of generally nonlinear problems into the multiport-systems pattern [7]. We shall study mechanical systems involving the large-scale motion of particles and rigid bodies in both conservative and nonconservative force fields. A variety of techniques is used, including the selection by the modeler of key variables for formulation, the determination of certain required transformations by simple analytic means, and the combination of all the parts into a unified representation by a bond graph.

The development begins with the simplest type of system involving straightforward coupling between inertial and compliance elements, extends the form to include static coupling among inertial elements and among compliance elements, and arrives at a general symmetric bond-graph model.

9.2.1 The Basic Modeling Procedure

Let us start by considering the result we would like to achieve. For a given problem in mechanics involving particles, rigid bodies, and force fields, we wish to obtain a set of first-order differential equations in terms of variables that yield physical insight into system behavior. We anticipate that the equations will be coupled and, if possible, explicit (i.e., one derivative in each equation), having the form of Eqs. (9.1–9.4).

$$C\text{-field} \qquad\qquad \mathbf{f}_C = \boldsymbol{\phi}_C(\mathbf{q}_C) \qquad\qquad (9.1)$$

$$I\text{-field} \qquad\qquad \mathbf{v}_I = \boldsymbol{\phi}_I(\mathbf{p}_I)^* \qquad\qquad (9.2)$$

$$\text{Junction structure} \qquad \dot{\mathbf{q}}_C = (\mathbf{T}_{CI}(\mathbf{q}_C))\mathbf{v}_I \qquad\qquad (9.3)$$

$$\dot{\mathbf{p}}_I = (-\mathbf{T}_{CI}^t(\mathbf{q}_C))\mathbf{f}_C \qquad\qquad (9.4)$$

---

* For rigid bodies, the translational and angular velocities in $\mathbf{v}_I$ must properly define the kinetic coenergy. A safe choice is to use the center of mass velocity and the angular velocity in an inertial frame. For fixed-axis notation, only the angular velocity is needed. See Reference [8] for the correct means of computing kinetic coenergy.

where $\mathbf{f}_C$ is the set of forces defining the potential co-energy; $\mathbf{q}_C$ is the set of displacements defining the potential energy; $\mathbf{v}_I$ is the set of (inertial) velocities defining the kinetic co-energy; $\mathbf{p}_I$ is the set of momenta defining the kinetic energy; and $\dot{\mathbf{q}}_C$ and $\dot{\mathbf{p}}_I$ are the time derivatives of $\mathbf{q}_C$ and $\mathbf{p}_I$, respectively.

Inspection of the equations indicates that Eq. (9.1) arises from the compliance elements' constitutive laws directly. The set of relations may be linear or nonlinear and coupled or decoupled according to the nature of the C-field. Equation (9.2) is obtainable directly from the inertia elements in the system. The vector, $\mathbf{v}_I$, is defined with respect to an inertial frame and represents translational velocities of the centers of mass of rigid bodies and angular velocities with respect to nonrotating coordinate systems.

Typically, the most difficult aspect of mechanics is the generally nonlinear coupling of C-fields and I-fields introduced by the geometry of large-scale motions. Such coupling is represented in Eqs. (9.3) and (9.4) by the transformation array $\mathbf{T}_{CI}(\mathbf{q}_C)$ and its transpose. Each element of $\mathbf{T}_{CI}$ is a scalar function of all the elements of the $\mathbf{q}_C$ vector, in principle. Usually one of the most difficult problems in mechanics is constructing the transformation, $\mathbf{T}_{CI}(\mathbf{q}_C)$. However, multiport techniques for doing this are straightforward.

Before we discuss an example, let us observe that Eqs. (9.1–9.4) may be combined to eliminate $\mathbf{f}_C$ and $\mathbf{v}_I$, giving

$$\dot{\mathbf{q}}_C = (\mathbf{T}_{CI}(\mathbf{q}_C))\boldsymbol{\phi}_I(\mathbf{p}_I), \tag{9.5}$$

$$\dot{\mathbf{p}}_I = (-\mathbf{T}_{CI}^t(\mathbf{q}_C))\boldsymbol{\phi}_C(\mathbf{q}_C). \tag{9.6}$$

These are the equations of a nonlinear system, conservative if $\boldsymbol{\phi}_I$ and $\boldsymbol{\phi}_C$ represent conservative fields, expressed in terms of state variables, $\mathbf{q}_C$ and $\mathbf{p}_I$. In some cases, they can be modified easily to include nonconservative force effects (e.g., dissipation), as well as geometric constraints not already incorporated in the transformation structure.

Bond-graph models representing the type of system discussed thus far are shown in Figure 9.1. The compliances are represented by the coupled C-field in part a and the "M" C-elements in part b. The inertias are represented by the coupled I-field in part a and the "P" I-elements in part b. The transformation coupling is represented by the M-port $\times P$-port MTF element, and the 1-junctions are introduced for the sake of clarity, not necessity. Notice that the moduli of the MTF depend on the $\mathbf{q}_C$ vector.

As an illustration of the type of formulation in which we are interested, consider the nonlinear oscillator of Figure 9.2a. The mass is free to slide

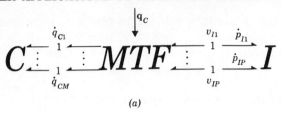

(a)

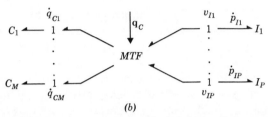

(b)

Figure 9.1.   Symbolic bond-graph model of basic nonlinear conservative system. (a) Coupled C- and I-fields; (b) 1-port C- and I-fields.

on the rod and is constrained by a spring. A second spring constrains the rod as it rotates in the X–Y plane. Key geometric variables are identified in part b as r, θ, x, y, v_x, and v_y. Constitutive relations for the linear springs and inertias, in the form of Eqs. (9.1) and (9.2), are given by Eqs. (9.7) and (9.8), respectively:

$$F = k_1(r - R), \tag{9.7a}$$

$$\tau = k_2\theta, \tag{9.7b}$$

$$v_x = m^{-1}p_x, \tag{9.8a}$$

$$v_y = m^{-1}p_y, \tag{9.8b}$$

where R is the free length of the rod spring, F is the rod spring force, τ is the torque of the torsional spring on the rod about an axis through the origin normal to the X–Y plane, and k_1 and k_2 are spring constants. Both the C-field and the I-field are linear and decoupled.

If we identify r and θ as elements of the $\mathbf{q}_C$ vector, and v_x and v_y as elements of the $\mathbf{v}_I$ vector, the $\mathbf{T}_{CI}(\mathbf{q}_C)$ array may be found from Eqs. (9.9a) and (9.9b),

$$\dot{r} = (\sin\theta)v_x + (\cos\theta)v_y, \tag{9.9a}$$

$$\dot{\theta} = \left(\frac{\cos\theta}{r}\right)v_x + \left(-\frac{\sin\theta}{r}\right)v_y, \tag{9.9b}$$

as

$$\mathbf{T}_{CI}(\mathbf{q}_C) = \begin{bmatrix} \sin\theta & \cos\theta \\ \dfrac{\cos\theta}{r} & -\dfrac{\sin\theta}{r} \end{bmatrix}. \tag{9.10}$$

It remains for us to calculate $\mathbf{p}_I$ in terms of the spring forces. By appropriate resolution of the force, F, and torque effect, τ, in the directions of X and Y (i.e., p_x and p_y), we obtain Eqs. (9.11a, b), namely,

$$\dot{p}_x = (-\sin\theta)F + \left(-\frac{\cos\theta}{r}\right)\tau, \tag{9.11a}$$

$$\dot{p}_y = (-\cos\theta)F + \left(\frac{\sin\theta}{r}\right)\tau + mg. \tag{9.11b}$$

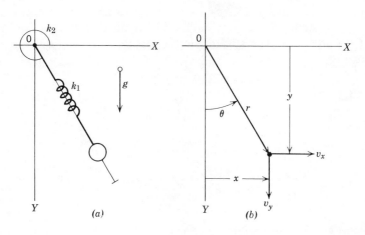

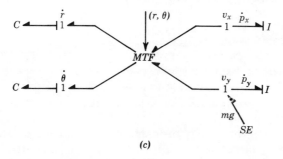

Figure 9.2. A nonlinear mechanical oscillator. (*a*) Schematic diagram; (*b*) key geometric quantities defined; (*c*) bond-graph model.

Careful inspection of Eqs. (9.11) shows that indeed the transformation array, $\mathbf{T}_{CI}(\mathbf{q}_C)$, is imbedded as the negative transpose, namely,

$$-\mathbf{T}^t_{CI}(\mathbf{q}_C) = \begin{bmatrix} -\sin\theta & -\dfrac{\cos\theta}{r} \\ -\cos\theta & \dfrac{\sin\theta}{r} \end{bmatrix}. \tag{9.12}$$

The influence of gravity is merely added in the appropriate way (i.e., to directly influence p_y) and is shown in Figure 9.2c as an effort source, *SE*.

At this point, the thoughtful reader might well ask "If we have obtained $\mathbf{T}_{CI}(\mathbf{q}_C)$ as in Eq. (9.10) once, is it really necessary to obtain it by separate development again, as in Eq. (9.12)?" The answer is "No," and a systematic procedure that takes note of this result is the next topic. First we should combine Eqs. (9.9) and (9.11), eliminating F, τ, v_x, and v_y by Eqs. (9.7) and (9.8). The system state equations are

$$\dot{r} = (\sin\theta)m^{-1}p_x + (\cos\theta)m^{-1}p_y, \tag{9.13a}$$

$$\dot{\theta} = \left(\frac{\cos\theta}{r}\right)m^{-1}p_x + \left(-\frac{\sin\theta}{r}\right)m^{-1}p_y, \tag{9.13b}$$

$$\dot{p}_x = (-\sin\theta)k_1(r-R) + \left(-\frac{\cos\theta}{r}\right)k_2\theta, \tag{9.13c}$$

$$\dot{p}_y = (-\cos\theta)k_1(r-R) + \left(\frac{\sin\theta}{r}\right)k_2\theta + mg. \tag{9.13d}$$

A bond-graph representation of the complete system is given in Figure 9.2c. The fields are shown explicitly, and the junction-structure transformation is implied by the two-by-two *MTF*.

Definitions of Key Geometric Quantities

As the first step in specifying a procedure for constructing bond-graph models in mechanics, we must identify key variables. The approach of this chapter is geometric, meaning that we shall use displacement and velocity quantities to organize the system, with forces and momenta playing a secondary role.

Two key vectors have already been described. They are $\mathbf{q}_C$, the vector of displacements that define the potential energy, and $\mathbf{v}_I$, the vector of velocities that define the kinetic co-energy. The $\mathbf{q}_C$ vector is directly associated with the *C*-field and the $\mathbf{v}_I$ vector with the *I*-field.

If it were always possible to relate $\mathbf{v}_I$ to $\dot{\mathbf{q}}_C$ in terms of $\mathbf{q}_C$ as nicely as was done in the previous example, then mechanics problems would not be the bete noire they typically are. It is often the case that there are more velocities and displacements in the problem than the coupling constraints permit to be independent. To treat this situation it will be useful to identify another vector, $\mathbf{q}_k$, the kinematic displacement vector or generalized coordinate vector in the language of Lagrangian mechanics. Basically, the elements of $\mathbf{q}_k$ are necessary and sufficient to fix the configuration of the system at any instant. Therefore the $\mathbf{q}_C$ vector can be found in terms of $\mathbf{q}_k$, and the necessary velocity relations for $\dot{\mathbf{q}}_C$ can be suitably evaluated, as we show next. Furthermore, if the vector, $\dot{\mathbf{q}}_k$, represents a necessary and sufficient set of velocities to fix all the motions at any instant, then the system is holonomic.* We shall consider only holonomic systems here, but a slight extension to the development would permit treatment of nonholonomic systems. If $\mathbf{q}_k$ describes all the motions, then the vector, $\mathbf{v}_I$, must be expressible in terms of $\dot{\mathbf{q}}_k$, given $\mathbf{q}_k$.

Therefore, the modeling process begins with the identification of three key geometric vectors: $\mathbf{q}_k$, the vector of kinematic displacements, or generalized coordinates; $\mathbf{q}_C$, the vector of C-field displacements; and $\mathbf{v}_I$, the vector of I-field velocities.

In Figure 9.2b the kinematic displacement vector could have been chosen to contain (r, θ) or (x, y), or even other pairs, such as (θ, x). Notice that if

$$\mathbf{q}_k = \begin{bmatrix} r \\ \theta \end{bmatrix},$$

then

$$\mathbf{q}_k = \mathbf{q}_C = \begin{bmatrix} r \\ \theta \end{bmatrix},$$

that is, $\mathbf{q}_k$ would be identical to $\mathbf{q}_C$. That would be convenient for many reasons. On the other hand, if

$$\mathbf{q}_k = \begin{bmatrix} x \\ y \end{bmatrix},$$

then

$$\dot{\mathbf{q}}_k = \begin{bmatrix} \dot{x} \\ \dot{y} \end{bmatrix} = \begin{bmatrix} v_x \\ v_y \end{bmatrix} = \mathbf{v}_I,$$

that is, $\dot{\mathbf{q}}_k$ would be identical to $\mathbf{v}_I$. That would be convenient for many reasons, too. To a significant extent, the wise choice of $\mathbf{q}_k$ to balance the

* See, for example, Reference [8].

"needs" of $\mathbf{q}_C$ and $\mathbf{v}_I$ can ease the formulation of equations for a complicated system. In this regard, experience is the best teacher. It is an important aspect of the method being presented that a range of choices and some of their implications are made available in clear fashion.

As an example of a problem involving some significant choices, consider the spring-pendulum system shown in Figure 9.3a. The pivot point of the pendulum is constrained to slide on the Y axis. The pendulum swings in the X–Y plane. A number of geometric quantities are shown in Figure 9.3b.

The configuration of the system can be specified at any instant by the position of the pivot (y_k) and the angle of the pendulum (α). Other choices are possible, such as y_m and α. However, let us use

$$\mathbf{q}_k = \begin{bmatrix} y_k \\ \alpha \end{bmatrix}. \tag{9.14}$$

The potential energy of the system resides in the spring, and y_s is an obvious way to specify it. Hence,

$$\mathbf{q}_C = [y_s]. \tag{9.15}$$

Finally, the kinetic co-energy is associated with the particle (the rod being assumed massless), so

$$\mathbf{v}_I = \begin{bmatrix} v_x \\ v_y \end{bmatrix}. \tag{9.16}$$

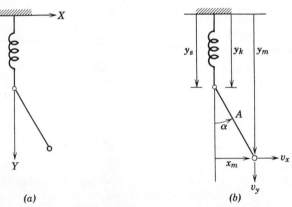

(a)   (b)

Figure 9.3.   A mass-spring oscillator example. (a) Schematic diagram; (b) key geometric quantities.

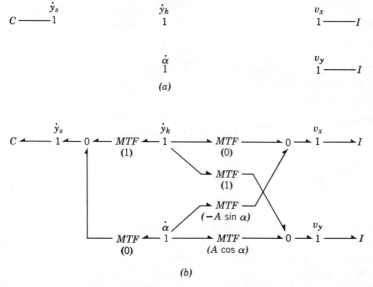

Figure 9.4. Explicit bond-graph model for the mass-spring oscillator. (a) Key geometric variable sets; (b) insertion of the transformations, T_{Ck} and T_{Ik}.

A bond graph identifying the variables and the C- and I-fields is shown in Figure 9.4a. It will become standard practice to write a column of 1-junctions for $\dot{\mathbf{q}}_C$, a column for $\dot{\mathbf{q}}_k$, and a column for $\mathbf{v}_I$. The next steps are to develop transformation relations among the vectors and to represent them in the bond graph.

Calculating Velocity Transformations

The basic idea in calculating velocity transformations is first to write displacement relations and then to differentiate them [9]. The relations between $\mathbf{q}_k$ and $\mathbf{q}_C$ may be written as

$$\mathbf{q}_C = \boldsymbol{\phi}_{Ck}(\mathbf{q}_k), \tag{9.17}$$

where $\boldsymbol{\phi}_{Ck}$ is a set of relations defining each element of $\mathbf{q}_C$ in terms of the elements of $\mathbf{q}_k$. By differentiating each relation in Eq. (9.17) with respect to time we get

$$\dot{\mathbf{q}}_C = \frac{\partial \boldsymbol{\phi}_{Ck}(\mathbf{q}_k)}{\partial \mathbf{q}_k} \dot{\mathbf{q}}_k = \mathbf{T}_{Ck}(\mathbf{q}_k)\dot{\mathbf{q}}_k. \tag{9.18}$$

The ith relation is

$$\dot{q}_{Ci} = \sum_{j=1}^{N} \frac{\partial \phi_{Cki}}{\partial q_{kj}} \dot{q}_{kj}, \tag{9.19}$$

where there are N displacements in $\mathbf{q}_k$.

For example, in the pendulum problem of Figure 9.3, $\mathbf{q}_C$ is related to $\mathbf{q}_k$ as follows, where the vectors are given by Eqs. (9.14) and (9.15):

$$y_s = 1 y_k + 0 \alpha, \tag{9.20}$$

then,

$$\dot{y}_s = 1 \dot{y}_k + 0 \dot{\alpha}, \tag{9.21}$$

and

$$\mathbf{T}_{Ck}(\mathbf{q}_k) = [1 \quad 0]. \tag{9.22}$$

Since the displacements are linearly related, $\mathbf{T}_{Ck}$ does not depend on $\mathbf{q}_k$ explicitly.

As the next step, we express the inertia-velocity vector, $\mathbf{v}_I$, in terms of $\mathbf{q}_k$ and $\dot{\mathbf{q}}_k$, thereby obtaining $\mathbf{T}_{Ik}(\mathbf{q}_k)$. Generally it is easiest to do this by first creating an inertia-displacement vector, $\mathbf{q}_I$, chosen so that

$$\dot{\mathbf{q}}_I = \mathbf{v}_I. \tag{9.23}$$

Then, we may write

$$\mathbf{q}_I = \boldsymbol{\phi}_{Ik}(\mathbf{q}_k), \tag{9.24}$$

from which we can calculate $\mathbf{v}_I$ as

$$\mathbf{v}_I = \dot{\mathbf{q}}_I = \left(\frac{\partial \boldsymbol{\phi}_{Ik}}{\partial \mathbf{q}_k}\right) \dot{\mathbf{q}}_k = (\mathbf{T}_{Ik}(\mathbf{q}_k)) \dot{\mathbf{q}}_k. \tag{9.25}$$

The ith relation is

$$v_{Ii} = \sum_{j=1}^{N} \left(\frac{\partial \phi_{Iki}}{\partial q_{kj}}\right) \dot{q}_{j}, \tag{9.26}$$

where there are N elements in $\mathbf{q}_k$.

In the pendulum example, we may write

$$\mathbf{q}_I = \begin{bmatrix} x \\ y \end{bmatrix} \quad \text{and} \quad \dot{\mathbf{q}}_I = \mathbf{v}_I = \begin{bmatrix} \dot{x} \\ \dot{y} \end{bmatrix} = \begin{bmatrix} v_x \\ v_y \end{bmatrix},$$

so $\mathbf{q}_I$ is related to $\mathbf{q}_k$ by

$$x = A \sin \alpha, \tag{9.27a}$$

$$y = y_k + A \cos \alpha. \tag{9.27b}$$

From Eqs. (9.27) we obtain T_{Ik} by differentiating, namely,

$$v_x = \dot{x} = (0)\dot{y}_k + (A \cos \alpha)\dot{\alpha}, \tag{9.28a}$$

$$v_y = \dot{y} = (1)\dot{y}_k + (-A \sin \alpha)\dot{\alpha}, \tag{9.28b}$$

and

$$\mathbf{T}_{Ik}(\mathbf{q}_k) = \begin{bmatrix} 0 & A \cos \alpha \\ 1 & -A \sin \alpha \end{bmatrix}. \tag{9.29}$$

In this case, the transformation depends on one element of $\mathbf{q}_k$, that is, α.

Establishing the Junction Structure

Since the basic pattern of both $\mathbf{T}_{Ck}$ and $\mathbf{T}_{Ik}$ transformations is that the elements of $\dot{\mathbf{q}}_k$ are multiplied by constants or functions of $\mathbf{q}_k$ and added together, we may represent these transformations by 0-junctions and 2-port MTFs. The MTFs are used to multiply by the appropriate function, and the 0-junctions are used to add up velocity terms.

Referring to the partial bond graph of Figure 9.4a, $\mathbf{T}_{Ck}$ and $\mathbf{T}_{Ik}$ of Eqs. (9.22) and (9.29) are introduced in the graph of Figure 9.4b between the appropriate sets of 1-junctions. To check the graph, one could write an expression for each of the velocities—$\dot{y}_s$, v_x, and v_y—in terms of $\dot{y}_k$ and $\dot{\alpha}$. Clearly, the role of the 0-junctions is to add velocity terms, and the role of the MTFs is to multiply. We now have a basic bond-graph model of the spring-pendulum system of Figure 9.3a.

Now consider that we wish to add two more dynamic effects—dissipation in the pivot joint and a gravity force. The dissipation torque arises due to relative motion, $\dot{\alpha}$, directly and is shown by the R element appended to that junction in Figure 9.5. The gravity force acts on v_y and is appended directly to the v_y junction as a constant source of force of

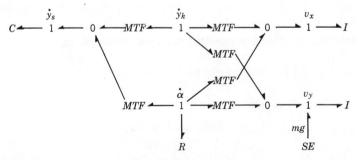

Figure 9.5. The mass-spring oscillator of Figure 9.4 with dissipation and gravity effects added.

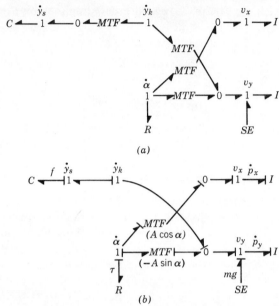

Figure 9.6. Simplification of the oscillator bond graph. (*a*) MTFs with zero modulus removed; (*b*) further simplification of the graph.

magnitude, mg. We observe that any or all of the fields C, I, and R can be linear or nonlinear. Typically, I will be linear, but C and R might not be. The junction structure is often nonlinear due to the geometric coupling represented by MTF moduli. In practice, the difficulty of generating state equations depends somewhat upon the nature of the nonlinearities [10].

The bond-graph model of Figure 9.5 has been reproduced in Figure 9.6a with MTFs having zero moduli eliminated. In part b, additional simplifications have been made, and causality has been added.

In selecting formulation variables for expressing the state equations, one would do well to include the q_k variables (y_k and α), because they are intimately involved in the system structure. Since y_s is expressible in terms of y_k, this represents no difficulty. Our other choice could be the momenta, p_x and p_y, or the velocities, v_x and v_y, if an entirely geometric formulation is sought.

The C-, I-, and R-field relations are

$$f = k(y_s - Y) \qquad \text{(spring)}, \qquad (9.30a)$$

$$v_x = m^{-1}p_x \qquad \text{(mass)}, \qquad (9.30b)$$

$$v_y = m^{-1}p_y \qquad \text{(mass)}, \qquad (9.30c)$$

$$\tau = R\dot{\alpha} \qquad \text{(dissipation)}, \qquad (9.30d)$$

respectively, where Y is the free length of the spring in Eq. (9.30a), m is the mass in Eqs. (9.30b, c), and R of Eq. (9.30d) is the viscous dissipation parameter in the joint.

The connection or junction-structure relations are, in initial form,

$$\dot{p}_x = (-\tan \alpha)f - \left(\frac{R}{A \cos \alpha}\right)\dot{\alpha}, \tag{9.31a}$$

$$\dot{p}_y = -f + mg, \tag{9.31b}$$

$$\dot{y}_k = (\tan \alpha)v_x + v_y, \tag{9.31c}$$

$$\dot{\alpha} = \left(\frac{1}{A \cos \alpha}\right)v_x. \tag{9.31d}$$

If we use Eqs. (9.30) to eliminate f, τ, v_x, and v_y from Eqs. (9.31), we get a momentum-displacement set of state equations,

$$\dot{p}_x = (-\tan \alpha)k(y_k - Y) - \frac{Rm^{-1}}{(A \cos \alpha)^2} p_x, \tag{9.32a}$$

$$\dot{p}_y = -k(y_k - Y) + mg, \tag{9.32b}$$

$$\dot{y}_k = (\tan \alpha)m^{-1}p_x + m^{-1}p_y, \tag{9.32c}$$

$$\dot{\alpha} = \frac{1}{A \cos \alpha} m^{-1}p_x, \tag{9.32d}$$

and if we eliminate p_x and p_y in terms of v_x and v_y, we get a velocity-displacement set of equations,

$$\dot{v}_x = (-\tan \alpha)\frac{k}{m}(y_k - Y) - \frac{R}{m(A \cos \alpha)^2} v_x, \tag{9.33a}$$

$$\dot{v}_y = -\frac{k}{m}(y_k - Y) + g, \tag{9.33b}$$

$$\dot{y}_k = (\tan \alpha)v_x + v_y, \tag{9.33c}$$

$$\dot{\alpha} = \left(\frac{1}{A \cos \alpha}\right)v_x. \tag{9.33d}$$

Let us check Eqs. (9.33) for the case when $\alpha = 0$ and $v_x = 0$ (i.e., the pendulum hangs straight down and is not moving in the X direction). Then

Eqs. (9.33) become

$$\dot{v}_x = 0, \tag{9.34a}$$

$$\dot{v}_y = -\frac{k}{m}(y_k - Y) + g, \tag{9.34b}$$

$$\dot{y}_k = v_y, \tag{9.34c}$$

$$\dot{\alpha} = 0. \tag{9.34d}$$

In other words, the system behaves like a simple mass-spring oscillator subject to gravity.

Summary of the Procedure

The procedure for obtaining a bond-graph model may be summarized in the following series of steps:

1. Identify the key vectors, $\mathbf{q}_k$, $\mathbf{q}_C$ and $\mathbf{v}_I$ (or $\mathbf{q}_I$). Write the 1-junctions corresponding to $\dot{\mathbf{q}}_k$, $\dot{\mathbf{q}}_C$, and $\mathbf{v}_I$.

2. Obtain the displacement transformation relating $\mathbf{q}_C$ to $\mathbf{q}_k$. Differentiate with respect to time to obtain a velocity transformation relating $\dot{\mathbf{q}}_C$ to $\dot{\mathbf{q}}_k$ in terms of $\mathbf{q}_k$. Then

$$\dot{\mathbf{q}}_C = \mathbf{T}_{Ck}(\mathbf{q}_k)\dot{\mathbf{q}}_k.$$

   Write the results into the bond graph using *MTF* and 0-junction elements.

3. Obtain the velocity transformation relating $\mathbf{v}_I$ to $\dot{\mathbf{q}}_k$. Do this directly, or by relating $\mathbf{q}_I$ to $\mathbf{q}_k$ and differentiating with respect to time (being sure that $\dot{\mathbf{q}}_I = \mathbf{v}_I$). Then

$$\mathbf{v}_I = \mathbf{T}_{Ik}(\mathbf{q}_k)\dot{\mathbf{q}}_k.$$

   Write the results into the bond graph using *MTF* and 0-junction elements.

4. Append the C-field and I-field elements to the $\dot{\mathbf{q}}_C$ and $\mathbf{v}_I$ junctions as indicated.

5. Append dissipation effects, force sources, and geometric constraints not yet included, as appropriate. If necessary construct additional velocities, using the transformation method.

6. Simplify the bond graph by removing *MTF*s with zero modulus, making into direct bonds *MTF*s with unit modulus, and combining 2-port junctions, when power directions permit.

9.2.2 Systems with Constraints

In mechanics there are often more elements in the inertia-field velocity vector, $\mathbf{v}_I$, than there are generalized velocities (elements in $\mathbf{v}_k$). When this is the case, not all of the components of $\mathbf{v}_I$ are independent. Similarly, it is not uncommon to have more compliance-field displacement variables in $\mathbf{q}_C$ than the system configuration vector, $\mathbf{q}_k$, allows, due to geometric constraints. The procedure outlined in Section 9.2.1 is applicable, but some interpretation of the resulting model is in order.

In this section we shall consider systems having velocity constraints and systems having displacement constraints of a static nature.

Systems with Statically-Coupled Inertias

We anticipate that a system will have statically-coupled inertias when the size of $\mathbf{v}_k$ (i.e., N) is smaller than the size of $\mathbf{v}_I$ (i.e., P). As an example of such a system, consider the crank-slider mechanism depicted in Figure 9.7a. Significant geometric variables are identified in part b; we may choose the vectors $\mathbf{q}_C$, $\mathbf{q}_k$, and $\mathbf{v}_I$ as follows:

$$\mathbf{q}_C = [d], \qquad \mathbf{q}_k = [\theta_1], \qquad \text{and} \qquad \mathbf{v}_I = \begin{bmatrix} \omega_A \\ v_{BX} \\ v_{BY} \\ \omega_B \end{bmatrix},$$

where d is the spring length, θ_1 is the angle of rotation of bar A about the fixed pivot, ω_A is the angular velocity of bar A, v_{BX} is the x component of the center-of-mass velocity of bar B, v_{BY} is the y component of the center-of-mass velocity of bar B, and ω_B is the angular velocity of bar B. We note in passing that consistent choice of center-of-mass velocities for rigid bodies leads to a highly systematic approach to constructing bond-graph models, and is recommended as good practice. This is based on the fact that the net force on a rigid body is the rate of change of the momentum and that the momentum is the mass times the velocity of the center of mass. Also, the torque about the center of mass is the rate of change of angular momentum, and the angular momentum is the inertia tensor times the angular velocity vector. With increasing experience the analyst can choose other velocities based on insight into the problem that may yield simpler bond-graph models in particular cases. For example, in

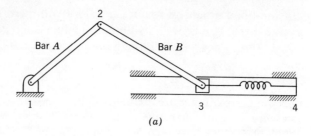

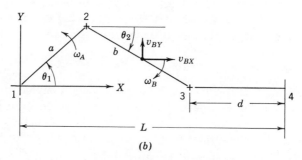

Figure 9.7. Crank-slider mechanism. (*a*) Schematic mechanism; (*b*) geometry of the mechanism.

Figure 9.7*a* bar A could have been represented by three velocities, namely, the angular velocity, ω_A, and x and y components of the center-of-mass velocity. Then the fixed pivot point would enter into the model explicitly as a geometric constraint. The simpler method is to represent the bar as fixed at the pivot point since the torque about the fixed pivot is equal to the rate of change of angular momentum about the pivot.

The next step in the procedure is to represent the velocities derived from the choices of $\mathbf{q}_k$, $\mathbf{q}_C$, and $\mathbf{v}_I$. This is done in Figure 9.8*a*, where the columns are arranged in correspondence with the vectors. Careful inspection of that figure shows that $\dot{\theta}_2$ also appears. This is because it will be easier to write the displacement relations using both θ_1 and θ_2 (see also Figure 9.7*b*), and then to constrain θ_1 and θ_2 subsequently. In other words, θ_2 is introduced on a temporary basis for algebraic reasons. The modified $\mathbf{q}_k$ vector is then

$$\mathbf{q}'_k = \begin{bmatrix} \theta_1 \\ \theta_2 \end{bmatrix}.$$

Now we relate $\mathbf{q}_C$ to $\mathbf{q}_k$ as follows:

$$d = L - a \cos \theta_1 - b \cos \theta_2, \qquad (9.35)$$

and

$$\dot{d} = (a \sin \theta_1)\dot{\theta}_1 + (b \sin \theta_2)\dot{\theta}_2. \qquad (9.36)$$

Therefore,

$$\mathbf{T}_{Ck}(\theta_1, \theta_2) = [a \sin \theta_1 \quad b \sin \theta_2],$$

which we write in Figure 9.8b as $[m_{11} \quad m_{12}]$ for convenience.

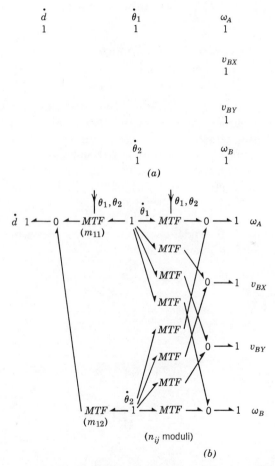

Figure 9.8. Junction structure representing system coupling for the crank-slider mechanism. (a) Key geometric variables identified; (b) velocity transformations inserted.

The inertia-velocity components may be written as

$$\omega_A = (1)\dot{\theta}_1, \tag{9.37a}$$

$$v_{BX} = (-a \sin \theta_1)\dot{\theta}_1 - \left(\frac{b}{2} \sin \theta_2\right)\dot{\theta}_2, \tag{9.37b}$$

$$v_{BY} = (a \cos \theta_1)\dot{\theta}_1 - \left(\frac{b}{2} \cos \theta_2\right)\dot{\theta}_2, \tag{9.37c}$$

$$\omega_B = (1)\dot{\theta}_2. \tag{9.37d}$$

From Eqs. (9.37) the $\mathbf{T}_{Ik}$ transformation array may be identified as

$$\mathbf{T}_{Ik} = \begin{bmatrix} 1 & 0 \\ -a \sin \theta_1 & -(b/2) \sin \theta_2 \\ a \cos \theta_1 & (b/2) \cos \theta_2 \\ 0 & 1 \end{bmatrix} = \begin{bmatrix} n_{11} & n_{12} \\ n_{21} & n_{22} \\ n_{31} & n_{32} \\ n_{41} & n_{42} \end{bmatrix},$$

where n_{ij} are written for convenience. The bond-graph model representing the transformation is shown in Figure 9.8b, with the n_{ij} moduli implied.

When the spring effect and the inertias are added to the model, as in Figure 9.9a, we obtain a virtually complete bond graph. Note that J_A is the moment of inertia of bar A about the pivot point, and J_B is the moment of inertia about the center of mass of bar B. In Figure 9.9 the explicit transformation structure has been compacted for convenience. It is implied by the m_{ij}, n_{ij} moduli of the $\mathbf{T}_{Ck}$ and $\mathbf{T}_{Ik}$ arrays, respectively.

Now let us impose the slider constraint on the angles θ_1 and θ_2, retaining θ_1 as the independent quantity. Since

$$a \sin \theta_1 = b \sin \theta_2,$$

or

$$\sin \theta_2 = \frac{a}{b} \sin \theta_1, \tag{9.38}$$

then

$$(\cos \theta_2)\dot{\theta}_2 = \left(\frac{a}{b} \cos \theta_1\right)\dot{\theta}_1 \tag{9.39}$$

and

$$\cos \theta_2 = \left[1 - \left(\frac{a}{b}\right)^2 \sin^2 \theta_1\right]^{1/2}, \tag{9.40}$$

so

$$\dot{\theta}_2 = \frac{\left(\dfrac{a}{b} \cos \theta_1\right)\dot{\theta}_1}{\left[1 - \left(\dfrac{a}{b}\right)^2 \sin^2 \theta_1\right]^{1/2}}, \tag{9.41}$$

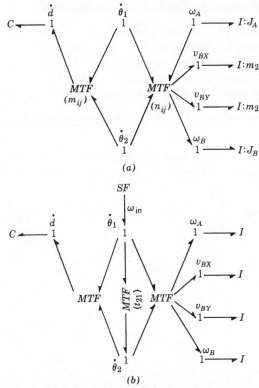

Figure 9.9. Bond-graph model for the crank-slider mechanism. (a) C- and I-fields adjoined to model; (b) model with slider constraint and velocity input.

or

$$\dot{\theta}_2 = [t_{21}(\theta_1)]\dot{\theta}_1. \tag{9.42}$$

This constraint on velocities, expressed in terms of θ_1, may be inserted into the model as shown in Figure 9.9b. Now only θ_1 and $\dot{\theta}_1$ are required to specify the configuration and motion of the mechanism.

If we also assume that the angular velocity of the arm A is prescribed, a velocity source (S_f) may be added directly to the 1-junction for $\dot{\theta}_1$, as is shown in Figure 9.9b. The effect is that all other velocities are statically determined by the input velocity, ω_{in}, which is consistent with the usual definition of a kinematic mechanism [11]. In such a case, the bond graph indicates that all inertial elements must have derivative causality. The forces in the mechanism could be found using the bond graph in terms of the time history of the input angular velocity.

Systems with Statically Constrained Compliances

The procedure for constructing bond-graph models of systems containing statically coupled compliances is essentially the same as we have developed already. One effect we anticipate is that the geometric transformation on the C-field side of the graph may become more complicated.

As an illustration, consider the mass particle suspended by three springs in Figure 9.10a. The fixed ends of the springs are at $(X_1, 0)$, $(0, Y_2)$, and (X_3, Y_3), respectively. Some key geometric quantities are shown in part b and include l_1, l_2, and l_3 (the spring lengths), x and y (the mass position), and v_x and v_y (the mass motion). Note that the common connection point and the mass are at the same location. A set of vectors is

$$\mathbf{q}_C = \begin{vmatrix} l_1 \\ l_2 \\ l_3 \end{vmatrix}, \qquad \mathbf{q}_k = \begin{vmatrix} x \\ y \end{vmatrix}, \qquad \text{and} \qquad \mathbf{v}_I = \begin{vmatrix} v_x \\ v_y \end{vmatrix}.$$

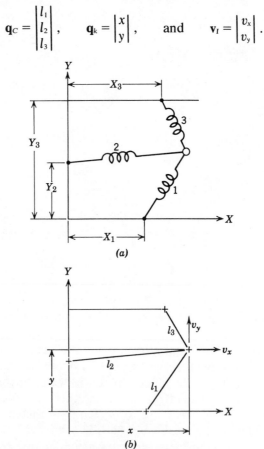

(a)

(b)

Figure 9.10. Mass-spring oscillator with statically coupled springs. (a) Mass-spring oscillator schematic; (b) geometric variables.

Since

$$\mathbf{v}_I = \begin{vmatrix} v_x \\ v_y \end{vmatrix} = \begin{vmatrix} \dot{x} \\ \dot{y} \end{vmatrix} = \dot{\mathbf{q}}_i, \qquad (9.43)$$

then

$$\mathbf{T}_{Ik} = \begin{bmatrix} 1 & 0 \\ 0 & 1 \end{bmatrix}. \qquad (9.44)$$

That is, the inertia velocities and the generalized motions are in one-to-one correspondence.

On the other hand, the $\mathbf{q}_C$ elements are

$$l_1 = [(x - X_1)^2 + (y^2)]^{1/2}, \qquad (9.45a)$$

$$l_2 = [(x)^2 + (y - Y_2)^2]^{1/2}, \qquad (9.45b)$$

and

$$l_3 = [(x - X_3)^2 + (Y_3 - y)^2]^{1/2}, \qquad (9.45c)$$

giving $\mathbf{q}_C$ in terms of $\mathbf{q}_k$. By differentiating Eqs. (9.45) we obtain a set of velocity relations, namely,

$$\dot{l}_1 = (m_{11}(x, y))\dot{x} + (m_{12}(x, y))\dot{y}, \qquad (9.46a)$$

$$\dot{l}_2 = (m_{21}(x, y))\dot{x} + (m_{22}(x, y))\dot{y}, \qquad (9.46b)$$

and

$$\dot{l}_3 = (m_{31}(x, y))\dot{x} + (m_{32}(x, y))\dot{y}, \qquad (9.46c)$$

where $m_{ij} = \partial l_i / \partial q_{kj}$. As is typical in this type of problem, all the hard work is done in obtaining the geometric velocity transformations. For example,

$$m_{22}(x, y) = \frac{\partial l_2}{\partial y} = \left(\frac{-(y - Y_2)}{[(x)^2 + (y - Y_2)^2]^{1/2}} \right). \qquad (9.47)$$

A bond-graph model representing the system is shown in Figure 9.11. Because of the procedure employed in setting up the model, the static constraint among l_1, l_2, and l_3 is handled implicitly by the proper use of x and y. In Figure 9.11, the n_{ij} moduli are implied since they are both unity. Clearly, the resulting graph could be simplified before equations were written.

We may anticipate some of the features of writing state equations by studying the bond graph closely. The three 0-junctions represent the spring forces. The two 1-junctions labeled $\dot{x}$ and $\dot{y}$ show that the respective inertias have a net force on them composed of three components, one from each spring. The actual component values are given by

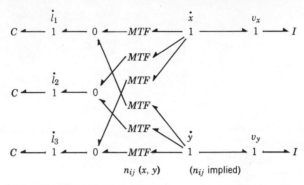

Figure 9.11. Bond graph model for mass-spring oscillator with coupled springs.

the spring force multiplied by an appropriate transformer modulus, already derived on the basis of geometry.

9.2.3. A Symmetric Multiport Model for Mechanics

Having developed several problem situations in a similar fashion, we may summarize the bond-graph model for mechanics in a symmetric structure about the column of kinematic or generalized velocity 1-junctions, as shown in Figure 9.12. In that figure the central column represents the N elements of the $\mathbf{q}_k$ vector, defining the system motion. The column of 1-junctions to the left represents the M elements of the $\dot{\mathbf{q}}_C$

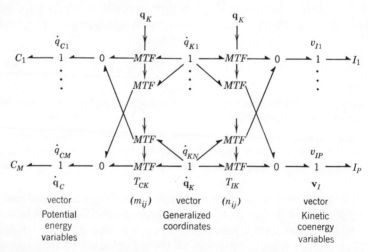

Figure 9.12. A symmetric bond-graph model for multiport mechanics.

vector, which relate to the compliance field directly. The right-hand column of 1-junctions represents the P elements of the $\mathbf{v}_I$ vector, defining the state of the inertia field directly.

The couplings between the $\dot{\mathbf{q}}_k$ and $\dot{\mathbf{q}}_C$ vectors are expressed by the modulated $M \times N$ transformation in terms of $\mathbf{q}_k$, with m_{ij} being the moduli. Similarly, the couplings between the $\dot{\mathbf{q}}_k$ and $\mathbf{v}_I$ vectors are expressed by the modulated $P \times N$ transformation in terms of $\mathbf{q}_k$, with n_{ij} being the moduli. This graph is obtained by geometric arguments, although it contains all required force transformations as well.

Once such a model has been established, it is relatively simple to introduce dissipation effects, force sources and constraints, and additional geometric constraints. If the required motions do not already exist in the model (as they most likely do), then they may be constructed as required by using the $\mathbf{q}_k$, $\dot{\mathbf{q}}_k$ vectors in a transformation structure.

The use of bond-graph models in mechanics as developed in this chapter is recommended for several reasons. One is that the structural characteristics of complicated mechanical systems can be displayed in a concise, uniform style that lends itself to communication between people, and between a person and a computer. A second reason is that the physical interpretation of various key force and velocity terms and relations is aided considerably by the bond-graph picture. Until a given graph is simplified, every variable is displayed in unambiguous fashion somewhere in the graph. A third reason is that there are usually a number of alternative generalized variable sets that can be used. Once a bond graph has been obtained for a given alternative, causal inspection of the graph will reveal the relative utility of that alternative in obtaining state equations *before* equations are actually written. With practice this can become a significant aid to the mechanics analyst. This chapter has only hinted at these points, most of which require experience to appreciate.

9.3 MAGNETIC CIRCUITS AND DEVICES

Many useful electrical and electromechanical devices contain magnetic circuits. Multiport models for some of these devices have already been studied in previous chapters, but here the magnetic flux paths will be modeled in detail. A detailed model is required, of course, if one is to design a motor, solenoid, transformer, or the like, although a system analyst may be content with an overall multiport model that adequately predicts only the external port behavior.

The toroidal coil shown in Figure 9.13 can serve as an ideal configuration for defining magnetic variables and establishing bond-graph representations for magnetic circuits. If the toroid is made of soft iron, we expect

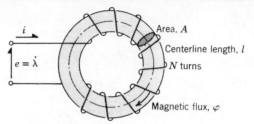

Figure 9.13. Toroidal coil.

the device to behave at the single electrical port as an inductance with a linear relation between current, i, and flux linkage, λ, at least for moderate values of i. This model shows no processes inside the coil, however.

Inside the iron core, magnetic *flux lines* are set up whenever current flows through the coil. Let the total flux be called ϕ, measured in the mks system in *webers*. (In this unit system one weber is equivalent to one volt-second. There are several units for magnetic variables corresponding to the several metric and English systems in use, but for simplicity, only the units appropriate to the mks system will be presented here.) The magnetic field may be described by the *magnetic flux density vector*, **B** with the units of *webers per square meter*. The length of **B** corresponds to the amount of flux passing through an area element perpendicular to the flux lines, and the direction of **B** is along the flux lines. The **B** vector points in the direction the north pole of a compass needle would point if it were free to line itself up with the magnetic field. The **B** vector is also often called the *magnetic induction*.

In Figure 9.13, if we can assume that each of the N turns of conductor around the toroidal core links all the flux, ϕ, in the toroid, then the flux linkage, λ, is related to ϕ by

$$\lambda = N\phi. \qquad (9.48)$$

In practice, when many layers of wire are wound on a core, some turns of wire do not link all of the flux lines. In this case, sometimes described by saying that some flux "leaks" out of the coil, Eq. (9.48) can be still valid except that N is a nondimensional number representing an *effective number of turns*, rather than an actual number of turns. In what follows, we will refer to N as a number of turns without necessarily including the adjective "effective."

By differentiating Eq. (9.48) with respect to time, the relation between the port voltage, e, and the magnetic variable, ϕ, is found.

$$e = \dot{\lambda} = N\dot{\phi}, \qquad (9.49)$$

which is just Faraday's law applied to the coil.

The driving force which tends to set up ϕ in the core is the *magneto-motive force*, M, which is proportional both to the number of turns of conductor and to the current flowing in the coil,

$$M = Ni. \tag{9.50}$$

The magnetomotive force, which is analogous to the electromotive force, has the units of current, amperes (although it is conventionally given in ampere-turns, despite the fact that the number of turns is really dimensionless). In the ideal, dissipationless case, M and ϕ for the coil are related by means of a nonlinear or linear relation. This relation is sketched in Figure 9.14.

In Figure 9.14a, a typical relation for a ferromagnetic material core is sketched. A *soft ferromagnetic* material is one that is easily magnetized and demagnetized by means of a current carrying coil, that is, a material for which M and ϕ are related by a single-valued curve so that $\phi = 0$ when $M = 0$. Hard ferromagnetic materials, used for permanent magnets, show a hysteresis loop when M is cycled, and ϕ can remain high even when M is zero. For now, we will concentrate on soft ferromagnetic materials. Such materials exhibit a saturation effect as shown. The increase in ϕ as M is increased slows down and, practically, there is a limiting value of flux that the core can attain when M is very large.

In order to characterize a core material, it is convenient to exhibit a curve independent of any particular core configuration. For this purpose the so-called B-H curve is frequently used. In Figure 9.14b, the magnitude of the **B** vector, B, is shown as ϕ divided by the core cross-sectional area, A. This way of computing B would be exact if the flux

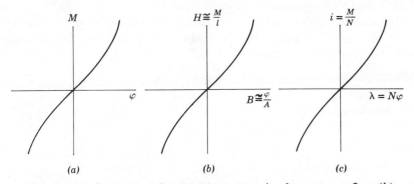

Figure 9.14. Core properties. (a) Magnetomotive force versus flux; (b) magnetizing force versus magnetic flux density; (c) current versus flux linkage.

were uniformly distributed throughout the core. The *magnetizing force*, H, is the magnetomotive force per unit length with units of ampere-turns per meter. For the toroidal coil, H is approximately M divided by the centerline length of the coil, l. When only a single B-H curve is given, it is implied that the material is *isotropic*, that is, that the relation between B and H is the same no matter how the field is oriented in the material. For certain crystal structures, this is not the case, and several different B-H curves for different orientations of **B** can be found.

In Figure 9.14c, the port characteristics of the inductor are sketched by modifying the M-ϕ or B-H curve using N. We see that the saturation effect corresponds to a nonlinear region of the $-I$ element representing the inductance. Since most B-H curves exhibit saturation, electrical chokes must be nonlinear, even though in normal operation of these devices, a linear inductance model is often sufficiently accurate.

It is of interest to compute the inductance for the coil. Assuming that the curves of Figure 9.14 can be approximately represented by a straight line near the origin, L may be found as follows:

$$L = \frac{\lambda}{i} = \frac{N\phi}{M/N} = \frac{N^2 AB}{lH}. \tag{9.51}$$

The initial slope of the B-H curve is given the symbol, μ, and called the *permeability*:

$$\mu = \frac{B}{H}, \tag{9.52}$$

so that $L = N^2 A\mu/l$. (Free space itself has a permeability, μ_0, and sometimes μ is expressed as a relative permeability, that is, the value of B/H is given by $\mu\mu_0$, rather than just μ.)

For the entire coil, one may express the slope of the M-ϕ curve in the linear region in two ways. The *permeance*, **P**, is given by the following formula:

$$\mathbf{P} = \frac{\phi}{M} = \frac{\mu A}{l}. \tag{9.53}$$

The *reluctance*, **R**, is given by

$$\mathbf{R} = \frac{1}{\mathbf{P}} = \frac{M}{\phi} = \frac{l}{\mu A}. \tag{9.54}$$

As a mnemonic device, one may say that a long small-area flux path in a material of small permeability is reluctant to permit the establishment of much flux.

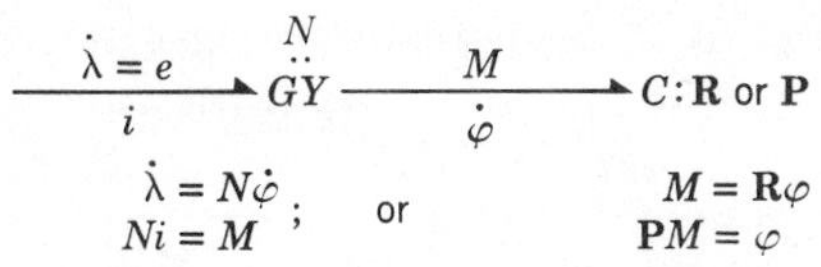

$$\xrightarrow[i]{\dot\lambda = e} \overset{N}{GY} \xrightarrow[\dot\varphi]{M} C:\mathbf{R} \text{ or } \mathbf{P}$$

$$\begin{matrix} \dot\lambda = N\dot\varphi \\ Ni = M \end{matrix}; \quad \text{or} \quad \begin{matrix} M = \mathbf{R}\varphi \\ \mathbf{P}M = \varphi \end{matrix}$$

Figure 9.15.   Bond graph for coil of Figure 9.13.

In order to proceed further with the analysis of magnetic circuits and devices, it is useful to begin classifying variables such as M and ϕ and parameters such as $\mathbf{P}$ and $\mathbf{R}$. Historically, the reluctance $\mathbf{R}$ was sometimes thought of as analogous to electrical resistance, which implies that flux was thought of as analogous to current and the magnetomotive force was analogous to the electromotive force. Although this analogy can be used, it is not satisfying as a basis for a bond graph since an electrical resistor dissipates power while a coil exhibiting reluctance stores energy. In fact, the reluctance and the permeance are linear parameters of a $-C$ or $-I$ element. From a bond-graph point of view it seems reasonable that M, the magnetomotive force, should be an effort quantity, but the flow should not be flux as in the more traditional analogy, but rather the time rate of change of flux, $\dot\phi$.

The deficiencies of the reluctance–resistance analogy are apparent when one attempts to study a dynamic system containing electrical and mechanical elements as well as a magnetic circuit (see Reference [12]). It was only after the gyrator had been accepted as a useful network element, however, that the analogy most natural for bond graphs could be proposed [13].

Suppose we return to the basic equations, Eqs. (9.49) and (9.50), and identify $\dot\phi$ as a flow variable and M as an effort variable. Then N is clearly a gyrator parameter. In addition, since ϕ is a displacement variable, then Eq. (9.53) shows that $\mathbf{P}$ is a capacitance parameter and $\mathbf{R}$, the reluctance, is the inverse capacitance or stiffness parameter. All the relations are elegantly summed up in the bond graph of Figure 9.15. Of course, the $-G-C$ combination does behave like $-I$ at the external port, as it must. The gyrator is necessary if we wish to consider both electromotive force and magnetomotive force as effort variables.

The variable classification which will be used subsequently is listed in Table 9.1 for convenience. Note that the capacitance parameter only is to be applied in the linear case. The magnetic circuit generally involves nonlinear C elements with characteristics as sketched in Figure 9.14.

Using the effort-flow identifications of Table 9.1, it is possible to begin a study of lumped-parameter elements for magnetic circuits. In Figure 9.16a, a coil around a length of core is shown contributing an increase in

TABLE 9.1. Electrical and magnetic variables and parameters

General	Electric	Magnetic
Effort	Electromotive force, e	Magnetomotive force, M
Flow	Current, i	Flux rate, $\dot{\phi}$
Displacement	Charge, q	Flux, ϕ
Capacitance parameter	Capacitance, C	Permeance, $\mathbf{P}$
Stiffness parameter	$1/C$	Reluctance, $\mathbf{R} = 1/\mathbf{P}$

magnetomotive force (MMF). In this model, the length of core has no reluctance. In Figure 9.16b, an MMF drop is shown related to the flux, ϕ. When two flux paths are joined as shown in Figure 9.16c, the result can be modeled with a 0-junction since the three flux rates sum to zero and there is a single MMF. Finally, in Figure 9.16d, an air gap is modeled essentially in the same way as a length of core material. The air gap has high reluctance for its length or low permeability (essentially the permeability of free space, μ_0) and does not exhibit saturation as iron cores do. In computing reluctance or permeance parameters for $-C$ elements of flux

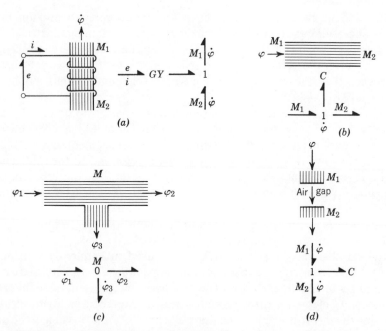

Figure 9.16. Bond graphs for magnetic circuit elements. (a) Driver coil; (b) piece of ferromagnetic core; (c) core junction; (d) air gap.

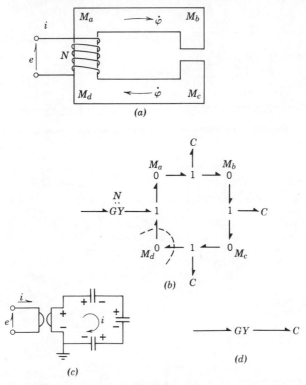

Figure 9.17. Bond-graph model of magnetic circuit with air gap. (*a*) Sketch of device; (*b*) bond graph; (*c*) analogous electric circuit; (*d*) simplified bond graph.

paths, we use Eqs. 9.53 and 9.54, in which areas and lengths for the flux path elements must be estimated judiciously, with consideration given to the actual paths of the flux lines.

Using the elements of Figure 9.16, a bond graph for a simple magnetic circuit such as the one shown in Figure 9.17 may be constructed. The *MMF* can be treated just as voltage in an electric circuit. In Figure 9.17*b*, the *MMF* variables have been each assigned to a 0-junction, and drops in *MMF* across —*C* elements and a rise due to the coil have been indicated using 1-junctions. Since only differences in *MMF* are significant ultimately, M_d is chosen as a zero *MMF* point, and the bond graph can be simplified. It may be helpful to consider the previously studied case of the analogous electric circuit shown in Figure 9.17*c*.

After the ground node has been eliminated and the 2-port 0-junctions removed, it is found that three —*C* elements are joined to a single

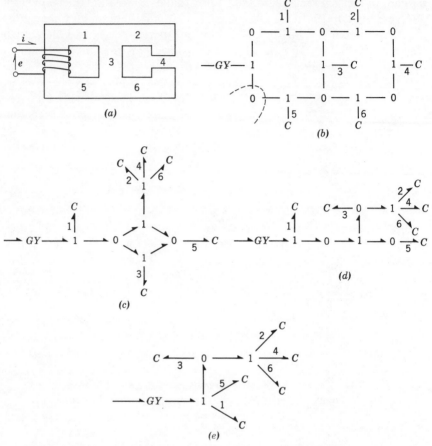

Figure 9.18. Circuit with extra flux path to model leakage flux. (*a*) Sketch; (*b*) basic bond graph; (*c*), (*d*), and (*e*) simplified bond graphs.

1-junction. This means that all *C*s have the same flow variable and the effort drops add. In magnetic terms, this means that for the linear case, a single *C* element with a reluctance equal to the sum of the three reluctances of the original *C*s can be substituted as shown in Figure 9.17*d*. In the nonlinear case, an equivalent —*C* relation can be found by simply adding *MMF* drops for the three original *C*s as the common flux is varied over the range of interest.

A more complicated magnetic circuit appears in Figure 9.18. The flux path associated with ϕ_3 could be physically present, or it could be a lumped representation of the path taken by leakage flux that bypasses the air gap. In fact, there always is some flux that escapes from the main path and spreads into the surrounding space. Although the real leakage paths are distributed in space, it is often sufficient to include a single equivalent leakage path and leakage flux. Since the leakage-path reluctance is usually dominated by the high reluctance of the air portion, it is common to assume that the leakage path can be represented by a linear, high-reluctance —C. The basic bond graph of Figure 9.18b is simplified by the choice of a zero *MMF* point, and the bond graph of Figure 9.18c results. The loop may be eliminated using the bond-graph identity for the 0-1-0-1 ring since the signs are proper for the reduction. The bond graphs of Figure 9.18d and e result. In the last form, it is clear that $C2$, $C4$, and $C6$ could be combined, as could $C1$ and $C5$. If these Cs were linear, their reluctances would be summed. Ultimately, a single equivalent C could be found for the entire collection of Cs, 0s and 1s but then, of course, none of the internal *MMF*s or fluxes could be found.

When no loss mechanisms are included in magnetic circuits, then they appear internally as networks of Cs, and it is often convenient to reduce the system by finding equivalences among the subfields. At a more detailed level, however, there are energy losses associated with eddy currents in the core and other effects. Laminations in the core of a magnetic circuit help in reducing the loss associated with eddy currents, but accurate models require the insertion of loss elements. Analysis of a laminated core can show that an R-C transmission line model for the core may be used for the linear case [13]. A study of such detailed models would take too long for present purposes, but, in a practical sense, a simpler approach often suffices. It is often sufficient to append a magnetic resistance to the 1-junction of Figure 9.16b. This implies an extra *MMF* drop associated with a length of core that is a function of $\dot{\phi}$ in addition to the drop in *MMF* that is associated with ϕ. As might be expected, it is not easy to predict the magnitude of the resistance in the absence of experimental evidence. On the other hand, particularly for a limited frequency range, it is often possible to adjust a linear resistance relation to provide a quite accurate representation of losses in the core.

Before exhibiting magnetic circuits containing loss elements, it is important to show a method of handling magnetomechanical transducers. (The solenoid studied in Chapter 7 was treated in multiport fashion without analyzing the magnetic circuit in detail.) The basic idea behind many transducers is that mechanical motion can alter the relation between flux and *MMF* in a flux path. This type of transducer is often called a "variable-reluctance" transducer, although this name is best used only for the magnetically linear case. The bond-graph model will be valid even when the magnetic relations are not linear.

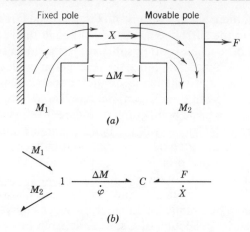

Figure 9.19. Magnetomechanical transducer. (*a*) Sketch of typical device; (*b*) bond-graph representation.

Consider the device of Figure 9.19, in which a force, F, is associated with a moveable pole piece that communicates with a fixed pole through an air gap of variable length, X. A flux, ϕ, passes through the gap and is associated with an MMF drop, ΔM. The device can be represented by a C-field, as shown in Figure 9.19*b* and is analogous to the moveable plate capacitor of Chapter 8.

Since this particular device uses an air gap, it is reasonable to assume that the device is linear, but that the reluctance, $\mathbf{R}$, is a function of X, that is,

$$\mathbf{R}(X)\phi = \Delta M. \tag{9.55}$$

Using arguments similar to those in Chapter 7, we may write the energy, E, as

$$E(X, \phi) = \tfrac{1}{2}\mathbf{R}(X)\phi^2. \tag{9.56}$$

Then,

$$\Delta M = \frac{\partial E}{\partial \phi} = \mathbf{R}(X)\phi$$

and

$$F = \frac{\partial E}{\partial X} = \frac{1}{2}\frac{d\mathbf{R}}{dX}\,\phi^2. \tag{9.57}$$

From Eq. 9.54, we might expect

$$\mathbf{R}(X) \cong \frac{X}{\mu_0 A}.$$

So

$$F \cong \frac{\phi^2}{2\mu_0 A},$$

indicating that F would be roughly constant for small values of X if ϕ were held constant. In most cases, however, ϕ is more nearly inversely proportional to **R** (when an MMF is held constant somewhere in the circuit), so F tends to vary as X^{-2} varies for small X. For very large values of X, the circuit model itself breaks down, and the flux does not follow the paths assumed. Any movement of core pieces will be associated with changes in stored energy and, hence, with forces or torques. The bond graph of Figure 9.19b may be used even when the magnetic relations are nonlinear due to saturation, but the computation of the energy is more complicated.

As an example of the utility of the bond-graph models for magnetic circuits developed above, consider the relay of Figure 9.20. Bond graphs are an aid in the study of such systems since three energy domains are involved. From the schematic diagram of the device, Figure 9.20a, one may begin assembling a bond-graph model by indicating MMF values and then representing MMF drops with 1-junctions and C and R elements. At this point, some judgment is required since it is not clear how many flux paths should be represented, nor is it clear where the eddy current losses will be most severe. When the bond graph is simplified after choosing M_0 as the zero MMF, as in Figure 9.20c, it becomes clear that several C and R elements can be combined.

Discounting the loss elements, the transduction from electrical energy to mechanical energy is accomplished by a C-field addressed at one port through a gyrator or, in other words, by an IC-field transducer. This would have been predicted by the methods used in Chapter 8. However, the present approach using the detailed model of the magnetic circuit allows a designer to study the flux paths in detail and to model internal losses. From a designer's point of view, a single relay is a rather complex dynamic system. Conventional descriptions of models of such systems using equations are not particularly easy to follow. For an example of a real device modeled at about the level of complexity shown in Figure 9.20, see Reference [14]. The reader who takes the trouble to read this paper carefully may become convinced that the bond-graph representation yields a compact and insightful means of displaying the physical assumptions used in creating a multiple-energy-domain model.

Although this brief introduction to magnetic-circuit bond graphs by no means exhausts the subject, the basic ideas have been set forth, and it is

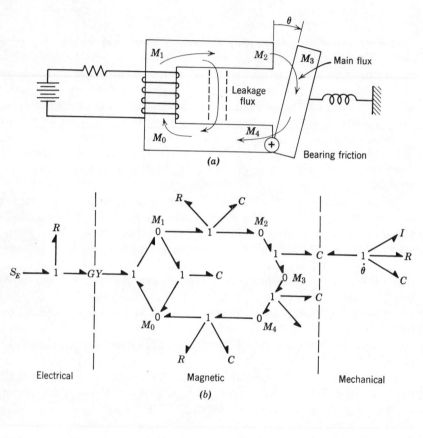

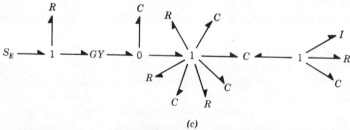

Figure 9.20. Relay. (*a*) Schematic diagram; (*b*) basic bond graph; (*c*) simplified bond graph.

hoped that the interested reader can extend bond-graph techniques to more complicated situations in magnetic systems.

9.4 THERMODYNAMIC SYSTEMS

Thermodynamics is essentially the universal science of physical processes. All of the models of physical systems that have been made so far may be studied from the point of view of thermodynamics and, indeed, when C- and I-fields were constrained to conserve energy or R-fields were arranged to dissipate power, thermodynamic arguments were used. On the other hand, since heat flow and temperature dependence has not yet been discussed, most of our dynamic models can be characterized as "isothermal models;" that is, we treated our multiports as if they were immersed in an infinite reservoir that maintains constant temperature even when finite energy is lost or supplied. In fact, we know that all elements change their constitutive laws with temperature to greater or lesser extent, and the transfer of heat and change in temperature is the special province of thermodynamics.

Much of engineering thermodynamics is not very dynamic at all. It deals with changes between equilibrium states, but often is not concerned with the process of change itself. In many cases, useful information may be extracted from a knowledge of the constraints that end-point equilibrium states of a system must satisfy. But if we are really interested in the path a system will follow in state space, then we must study what has come to be called "irreversible thermodynamics," which involves nonequilibrium systems.

In recent years, there has been an upsurge of interest in making dynamic models for thermodynamic systems in which a variety of electrical, chemical, mechanical, fluid-dynamic, and heat-transfer effects are important. Here we can only give an introductory account in which the model systems discussed previously are shown to have thermodynamic implications. We do this by the device of considering systems of such physical extents and time scales that local pseudoequilibrium conditions exist. In this way, we may modify some familiar relations from thermostatics into a slowly varying form of thermodynamics.

Another restriction for the present is to consider only closed systems, that is, systems in which no mass passes into or out of the system boundaries. This restriction to a Lagrangian point of view will be lifted in the section on fluid mechanics, but the control-volume or Eulerian point of view, which is often used for certain types of systems, introduces a set of modeling difficulties that are best left for later discussion.

Consider first, then, a fixed amount of a pure substance which is in at least pseudoequilibrium so that all parts of the substance are at essentially identical conditions of pressure, temperature, density, and so forth.

We consider below a unit mass of the substance. *Extensive* quantities such as volume, internal energy, and entropy for a unit mass will be designated by the lower-case letters: v, u, s. When a mass of the substance, m, is involved, the corresponding variables will be designated by upper-case letters: $V = mv$, $U = mu$, and $S = ms$. The substance is supposed to be of such small extent and the disturbances of such a slow time scale that wave motion, turbulence, and so forth are negligible. Then in the absence of motion and electromagnetic or surface-tension forces, we may assume that such a pure substance has only two independent properties. All other properties are related to any two independent properties by equations of state for the substance.

A useful, and logical, way to describe the constitutive relations for a pure substance begins with the statement that the internal energy per unit mass, u, of the substance is a function of the volume, v, and entropy, s, per unit mass:

$$u = u(s, v). \tag{9.58}$$

The well-known Gibbs equation relates changes in u to changes in v and s,

$$du = T\,ds - p\,dv, \tag{9.59}$$

where T is the thermodynamic temperature and p is the pressure. (In all that follows, T will indicate thermodynamic or absolute temperature, a positive quantity.)

The relation of Eq. (9.59) involves energy or work. If the substance changes state slowly enough, we may treat it as if it were a lumped parameter multiport with the relation,

$$\frac{du}{dt} = T\frac{ds}{dt} - p\frac{dv}{dt}. \tag{9.60}$$

The term, $T\,ds/dt$, is associated with a flow of heat or, sometimes, with dissipative work such as might be done by a paddle wheel slowly stirring a fluid substance. The term, $p\dot{v}$, represents the sort of reversible power with which we have been dealing frequently in dynamic models.

Since

$$du = \frac{\partial u}{\partial s}\,ds + \frac{\partial u}{\partial v}\,dv,$$

it is clear that

$$T = \frac{\partial u}{\partial s} \quad \text{and} \quad -p = \frac{\partial u}{\partial v}. \tag{9.61}$$

Also,

$$\left.\frac{\partial T}{\partial v}\right|_s = \frac{\partial^2 u}{\partial v\, \partial s} = \left.\frac{\partial(-p)}{\partial s}\right|_v, \tag{9.62}$$

which we may recognize as a reciprocal relation similar to those found for other energy-storing fields. In thermodynamics, Eq. (9.62) is called a *Maxwell reciprocal relation*.

The constitutive laws of the pure substance may be represented by a *C*-field, thus,

$$\underset{\dot{s}}{\overset{T}{\longrightarrow}} C \underset{\dot{v}}{\overset{P}{\longrightarrow}},$$

if we are willing to call T an effort, $\dot{s}$ a flow, and s a displacement. Integral causality,

$$\left|\underset{\dot{s}\cdot}{\overset{T}{\longrightarrow}} C \underset{\dot{v}}{\overset{P}{\longrightarrow}}\right|,$$

implies

$$T = T(s, v), \qquad p = p(s, v), \tag{9.63}$$

which are also implied by Eqs. (9.58) and (9.59). The only unusual feature of the pure-substance *C*-field is the sign convention on the $p\dot{v}$ port that is negative in the traditional form of the Gibbs equation, Eq. (9.59).

The internal energy is associated with all integral causality and with constitutive laws in the form of Eqs. (9.63). For mixed and all derivative causality, the constitutive laws are switched around. In thermodynamics this switching around of independent and dependent variables is often accomplished using the Legendre transformations of u. The enthalpy, h, Helmholtz free energy, f, and Gibbs free energy, ϕ, are all Legendre transformations of the internal energy, u, and correspond to different causal patterns.

The enthalpy is

$$h(s, p) \equiv u + pv, \tag{9.64}$$

and its derivatives are

$$\frac{\partial h}{\partial s} = T(s, p), \tag{9.65}$$

$$\frac{\partial h}{\partial p} = v(s, p) \tag{9.66}$$

Using $h(s, p)$, another Maxwell relation can be found:

$$\left.\frac{\partial T}{\partial p}\right|_s = \frac{\partial^2 h}{\partial s\, \partial p} = \left.\frac{\partial v}{\partial s}\right|_p. \tag{9.67}$$

The causal pattern for the C-field corresponding to h is shown below:

$$h = h(s, p) \Leftrightarrow \left|\frac{T}{s}\right.\!\!\!\!\rightarrow C \left|\frac{p}{v}\right.\!\!\!\!\rightarrow .$$

The Helmholtz free energy is defined by the transformation,

$$f(T, v) \equiv u - Ts. \tag{9.68}$$

Its derivatives are

$$\frac{\partial f}{\partial T} = -s(T, v), \tag{9.69}$$

$$\frac{\partial f}{\partial v} = -p(T, v). \tag{9.70}$$

And yet another Maxwell relation is

$$\left.\frac{\partial(-s)}{\partial v}\right|_T = \frac{\partial^2 f}{\partial T\, \partial v} = \left.\frac{\partial(-p)}{\partial T}\right|_v. \tag{9.71}$$

The causal pattern corresponding to f is

$$f = f(T, v) \Leftrightarrow \frac{T}{s}\!\!\rightarrow\!\!\left| C \right.\frac{p}{v}\!\!\rightarrow\!\!| .$$

Finally, the Gibbs free energy is defined by a double-Legendre transformation on u:

$$\phi(T, p) = u + pv - Ts. \tag{9.72}$$

The derivatives of ϕ are

$$\frac{\partial \phi}{\partial T} = -s(T, p), \tag{9.73}$$

$$\frac{\partial \phi}{\partial p} = v(T, p), \tag{9.74}$$

and the Maxwell relation is

$$\left.\frac{\partial(-s)}{\partial p}\right|_T = \frac{\partial^2 \phi}{\partial T \partial p} = \left.\frac{\partial v}{\partial T}\right|_p. \tag{9.75}$$

The Gibbs free energy corresponds to all derivative causality for the C-field.

$$\phi = \phi(T, p) \Leftrightarrow \begin{matrix} T \\ \overline{} \\ \dot{s} \end{matrix} \!\!\dashv\! C \!\!\vdash\!\! \begin{matrix} p \\ \overline{} \\ \dot{v} \end{matrix}.$$

The energy function, u, and the co-energy functions, h, f, and ϕ, which are Legendre transformations of u, have been expressed in terms of their own natural variables. When this is done, the constitutive functions for the substance may be found by differentiation, and such constitutive laws automatically insure that the C-field will be conservative. On the other hand, one could write out constitutive laws in any form without the use of a state functions such as u, h, f, or ϕ; the only difficulty is that arbitrary constitutive laws would not, in general, allow the existence of an internal energy function. In a cycle, a substance with arbitrary constitutive laws might allow net energy production and, thus, the construction of a perpetual motion machine of the first kind. In a sense, deriving constitutive laws from u, h, f, or ϕ is safer than not using state functions since conservation of energy will be built into the relations.

A classical example of constitutive laws not stated in terms of state function derivatives is the perfect gas. The equation of state for a perfect gas is

$$pv = RT, \tag{9.76}$$

in which R is a constant. It is often stated, in addition, that u is a function of temperature only. This means that if the constitutive laws for the gas were used to express $u(s, v)$ in terms of, say, T and v, u would be a function only of T and not of v. Let us derive this fact using the state functions.

If Eq. (9.76) is solved for p in terms of T and v, the constitutive law in the form of Eq. (9.70) is evaluated.

$$\left.\frac{\partial f}{\partial v}\right|_T = -p(T, v) = \frac{-RT}{v}. \tag{9.77}$$

Using this result, the Helmholtz free energy can be found by integration:

$$f(T, v) = -RT \ln v + \psi(T), \tag{9.78}$$

where $\psi(T)$ is some function of temperature alone. Then, using Eq. (9.69),

$$-s(T, v) = \frac{\partial f}{\partial T} = -R \ln v + \frac{d\psi(T)}{dT}. \tag{9.79}$$

Finally, substituting Eq. (9.79) into Eq. (9.68) and solving for u,

$$u(T, v) = f(T, v) + Ts = -RT \ln v + \psi(T) + RT \ln v - T\frac{d\psi(T)}{dT}$$
$$= \psi(T) - T\frac{d\psi(T)}{dt}, \tag{9.80}$$

which demonstrates that u is only a function of T.

Although u is a function of T, we need u in terms of s and v in order to evaluate the complete equations of state. There are two complete equations of state because the gas is a 2-port C-field. We need more information than just Eq. (9.76). (We could solve for the function, ψ, in Eq. (9.78), for example.) It is more common to assume that the two so-called specific heats are constant. The specific heat at constant pressure, c_p, is defined as

$$c_p = \frac{\partial h}{\partial T}\bigg|_p, \tag{9.81}$$

and the specific heat at constant volume, c_v, is

$$c_v = \frac{\partial u}{\partial T}\bigg|_v. \tag{9.82}$$

By substituting Eq. (9.76) into Eq. (9.64) and differentiating as in Eq. (9.81), we find that

$$c_p = c_v + R. \tag{9.83}$$

If one now assumes that c_v is constant, then Eq. (9.82) may be integrated to yield

$$u = c_v(T - T_0), \tag{9.84}$$

and, noting that Eqs. (9.64) and (9.76) imply that h is also a function of T alone, Eq. (9.81) may be integrated similarly to yield

$$h = c_p(T - T_0), \tag{9.85}$$

in which the subscript 0 stands for a state in which it is assumed that $u = h = s = 0$.

Rearranging the basic equation, Eq. (9.59), and using Eq. (9.84), one can find s.

$$ds = \frac{du}{T} + p\frac{dv}{T} = c_v\frac{dT}{T} + R\frac{dv}{v},$$

or

$$s = c_v \ln\frac{T}{T_0} + R \ln\frac{v}{v_0},$$

which may be solved for $T(s, v)$,

$$T = T_0 e^{s/c_v}\left(\frac{v}{v_0}\right)^{-R/c_v}. \tag{9.86}$$

Further manipulations of Eq. (9.59) yield

$$ds = c_p\frac{dv}{v} + c_v\frac{dp}{p},$$

$$s = c_p \ln\frac{v}{v_0} + c_v \ln\frac{p}{p_0},$$

which yields $p(s, v)$,

$$p = p_0 e^{s/c_v}\left(\frac{v}{v_0}\right)^{-c_p/c_v}. \tag{9.87}$$

Equations (9.86) and (9.87) are the complete relations for the C-field in the form of Eqs. (9.61). It is desirable, however, to show that these equations do indeed derive from $u(s, v)$. Using Eqs. (9.84) and (9.86), we have

$$u(s, v) = c_v T_0\left[e^{s/c_v}\left(\frac{v}{v_0}\right)^{-R/c_v} - 1\right], \tag{9.88}$$

from which we find

$$T = \frac{\partial u}{\partial s} = T_0 e^{s/c_v}\left(\frac{v}{v_0}\right)^{-R/c_v},$$

which agrees with Eq. (9.86), and

$$-p = \frac{\partial u}{\partial v} = c_v \frac{T_0}{v_0} e^{s/c_v}\left(\frac{-R}{c_v}\right)\left(\frac{v}{v_0}\right)^{(-R/c_v)-1} = \frac{-RT}{v_0} e^{s/c_v}\left(\frac{v}{v_0}\right)^{-(R+c_v)/c_v}, \tag{9.89}$$

which does agree with Eq. (9.87) upon using Eqs. (9.76) and (9.83).

The reader probably feels that all of the above manipulations of the perfect gas laws has only produced some complicated laws in Eqs. (9.86) and (9.87) in place of the simple law, $pv = RT$. But it is important to remember that (1) $pv = RT$ is not a complete characterization of the C-field of the gas, and (2) $pv = RT$ is in a form that only allows differential causality on the $T\dot{S}$ port. Equations (9.86) and (9.87) are complete and in integral-causality form at both C-field ports and, hence, are useful in making a dynamic model of a quantity of gas as a thermomechanical transducer.

9.4.1 Representation of Heat Transfer

The $T\dot{S}$ port for the pure-substance C-field may be used to model all types of power flow that are associated with entropy increase or, in other words, irreversible effects. Changes in entropy of a pure substance can be accomplished by a variety of dissipative effects such as stirring with a paddle wheel, heating with an electrical resistor immersed in a fluid, or the like. In each case, one may identify the power dissipated with $T\dot{S}$ in order to find the change in state of the substance. An important use of $T\dot{S}$ ports is in modeling the effects of heat flow. Using $\dot{Q}$ to stand for the rate of transfer of heat in power units, we may often identify $\dot{Q}$ with $T\dot{S}$.

Consider first the simple case of conduction heat transfer shown in Figure 9.21a. The idea is that two reservoirs of thermal energy at absolute temperatures, T_1 and T_2, are allowed to communicate through a thermal resistance, but in no other way. Generally, a heat flow, $\dot{Q}$, will be set up between the two reservoirs. It is common experience that heat flows from higher toward lower temperature, and, indeed, this observation is behind the second law of thermodynamics. In Figure 9.21b, possible relations between $\dot{Q}$ and $T_1 - T_2$ are sketched for the thermal resistance. All we need to assume is that $\dot{Q}$ is related to $T_1 - T_2$ in such a way that any values of $\dot{Q}$ and $T_1 - T_2$ would plot in the first and third quadrants of the $\dot{Q}$ versus $T_1 - T_2$ plane. (Although it is common to assume that $\dot{Q}$ is a function of $T_1 - T_2$ as sketched, the argument is true if $\dot{Q}$ is any function of T_1 and T_2 such that $\dot{Q}$ is positive when $T_1 - T_2$ is positive, negative when $T_1 - T_2$ is negative, and zero when $T_1 - T_2$ is zero.)

If we now write

$$T_1\dot{S}_1 = \dot{Q} = T_2\dot{S}_2, \tag{9.90}$$

which implies that the heat flow out of one body is instantaneously equal to the heat flow into the other body, then the net entropy flow rate, $\dot{S}_2 - \dot{S}_1$, may be found:

$$\dot{S}_2 - \dot{S}_1 = \frac{\dot{Q}}{T_2} - \frac{\dot{Q}}{T_1} = \dot{Q}\left(\frac{1}{T_2} - \frac{1}{T_1}\right) = \dot{Q}\,\frac{T_1 - T_2}{T_1 T_2}. \tag{9.91}$$

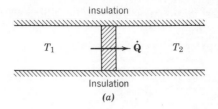

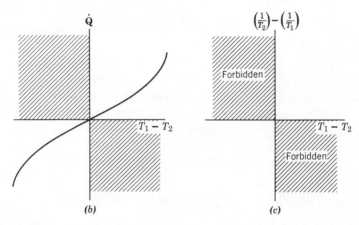

Figure 9.21. Conduction heat transfer. (*a*) Thermal resistance; (*b*) allowed relations between Q and $T_1 - T_2$; (*c*) forbidden regions for $(1/T_2) - (1/T_1)$.

In Figure 9.21*c*, it is shown that $1/T_2 - 1/T_1$ is positive, negative, or zero according to whether $T_1 - T_2$ is positive, negative, or zero. This is true because the absolute temperatures are inherently positive quantities. Thus, the net entropy production rate, which is the product of $\dot{Q}$ and $1/T_2 - 1/T_1$, is positive for any finite value of $T_1 - T_2$ and only vanishes when $\dot{Q}$ vanishes. The thermal resistor may be represented by the 2-port field shown below:

$$\overset{T_1}{\underset{\dot{S}_1}{\longrightarrow}} R \overset{T_2}{\underset{\dot{S}_2}{\longrightarrow}} .$$

Although the 2-port resistive field is power conservative [Eq. (9.90)], it is neither a transformer nor a gyrator. It has the peculiar property that, with the sign convention shown above,

$$\dot{S}_2 - \dot{S}_1 \geq 0,$$

which implies that the entropy flow leaving the field is greater than the entropy flow entering the field no matter which direction the heat is flowing. When the thermal resistor is part of a system, it will tend to increase the entropy of the system whenever any heat flows through the resistor.

A possible constitutive law for the resistor is

$$\dot{Q} = H(T_1 - T_2),$$

or

$$\dot{S}_1 = \frac{H(T_1 - T_2)}{T_1},$$

$$\dot{S}_2 = \frac{H(T_1 - T_2)}{T_2},$$

(9.92)

where the heat-transfer coefficient, H, is assumed to be constant or a slowly varying function of the average temperature, $(T_1 + T_2)/2$. As a check, the net entropy production rate is

$$\dot{S}_2 - \dot{S}_1 = \frac{H(T_1 - T_2)}{T_2} - \frac{H(T_1 - T_2)}{T_1} = \frac{H(T_1 - T_2)^2}{T_1 T_2} > 0.$$

(9.93)

Note that Eq. (9.92) is written in the causal form,

$$\frac{T_1}{\dot{S}_1} \rightarrow\!\!\mid R \mid\!\frac{T_2}{\dot{S}_2} \rightarrow\ .$$

Given any two temperatures, $T_1 > 0$, $T_2 > 0$, it is possible to solve for $\dot{Q}$, $\dot{S}_1$, and $\dot{S}_2$. The remaining causal forms are not so useful. For example,

$$\mid\!\frac{T_1}{\dot{S}_1} \rightarrow T \frac{T_2}{\dot{S}_2}\!\mid$$

implies that given any $\dot{S}_1$ and $\dot{S}_2$, the resistance should provide T_1 and T_2. But we cannot actually impose arbitrary $\dot{S}_1$ and $\dot{S}_2$ variables since Eq. (9.93) must be obeyed. This causality is, therefore, not useful for dynamic systems. Even mixed causalities such as

$$\frac{T_1}{\dot{S}_1} \rightarrow\!\!\mid R \frac{T_2}{\dot{S}_2}\!\mid$$

are fraught with difficulties. Inverting the last of Eqs. (9.92), we find

$$T_2 = \frac{HT_1}{\dot{S}_2 + H},$$

which seems to imply that T_2 can be negative if $\dot{S}_2 < H$. Actually, $\dot{S}_2$ is limited, given T_1, by the requirement that T_2 must be positive. Thus, the values $\dot{S}_2$ may be assigned are limited by the choice of T_1. In this sense, the mixed causal patterns are less useful than the causality of Eq. (9.92), in which the (positive) temperatures, T_1 and T_2, can be assigned arbitrarily. In what follows, the thermal resistor will be assumed to accept only the causality, $\dashv R\vdash$.

9.4.2 A Simple Example

Figure 9.22 shows a system incorporating two chambers containing compressible fluids. One chamber is of constant volume so that the internal energy of the fluid contained in the chamber varies only because of heat flow through the partition, which is modeled as a thermal resistance. The volume of the other chamber is changed as a piston moves back and forth.

The bond graph for the system shown in Figure 9.22b incorporates two capacitance elements for the fluids and an R-field for the thermal resistance. The 0-junction is used simply to achieve inward sign conventions for both the $T_1\dot{S}$ port of the left-hand C and the resistive field so that the sign conventions assumed previously are reflected in the model. The 0-junction assures that $T_1 = T_2$ but $\dot{S}_1 = -\dot{S}_2$.

In Figure 9.22c, integral causality is shown for the system. The preferred R-field causality is compatible with integral causality on both C

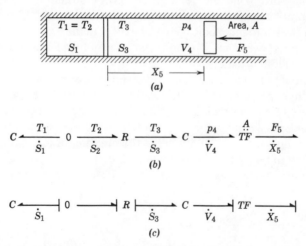

Figure 9.22. Example system. (a) Schematic diagram; (b) bond graph; (c) bond graph with integral causality.

elements. The state variables are S_1 (the entropy of the fixed volume fluid), S_3 (the entropy of the variable volume fluid), and V_4 (the volume of the variable chamber). The velocity, $\dot{X}_5(t)$, plays the role of an input variable.

The constitutive laws for the fluids could be found by starting with the internal energy per unit mass, $u(s, v)$, a function of entropy per unit mass and specific volume, and then converting to the total energy, U, as a function of total entropy, S, and volume, V. This would require a knowledge of the total mass and the initial state of the fluids. The pressure and temperature constitutive relations could then be found by differentiating U with respect to S and V. The constitutive relations for the thermal resistance have been discussed previously.

Clearly, the model for the system represented by the bond graph can only be appropriate for relatively slow changes. We assume, for example, that the fluids are in pseudoequilibrium so that the temperature and pressure are essentially uniform throughout the volumes at every instant. This would not be the case if one needed to worry about acoustic waves or the details of heat transmission in the fluids themselves. Also, although the entire system is power conservative, the entropy of the two C elements will increase if any heat flows. From the point of view of the external port, this irreversibility makes it appear that energy is lost. For example, if X_5 is taken through a cycle starting from an equilibrium state at which $T_2 = T_3$, the fluid temperature in the variable volume will change, heat will flow, and, during the cycle, a net energy loss will be observed at the external port. What the bond graph shows is that energy is not lost, but rather converted into thermal energy that cannot be completely converted back into mechanical energy except in the limited case in which X_5 moves so slowly that T_2 and T_3 are virtually identical and the net entropy production almost vanishes.

In the model of Figure 9.22 no irreversible phenomena other than heat transfer have been included. It would not be hard to include other dissipative effects. For example, if the piston had coulomb friction and if all the mechanical energy lost in friction were converted into thermal energy, a simple model would involve the creation of another equivalent entropy flow to the fluid C-field equal to the dissipated power divided by the fluid temperature. A bond graph for this case is shown in Figure 9.23.

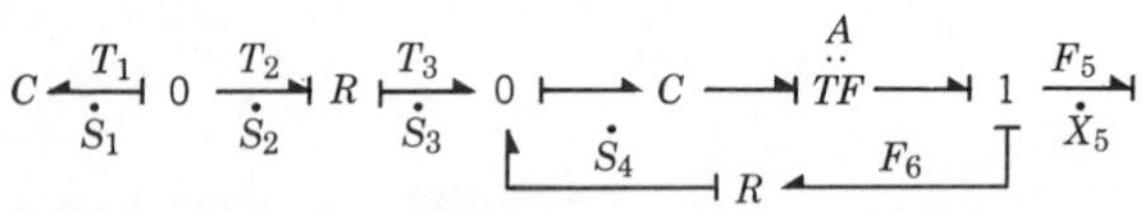

Figure 9.23. System of Figure 9.22 with piston friction.

The coulomb friction resistor may be described by a relation such as

$$F_6 = A \text{ sgn } \dot{X}_5,$$

so that the power dissipated is

$$F_6 \dot{X}_5 = A \dot{X}_5 \text{ sgn } \dot{X}_5 \geq 0.$$

The entropy flow, $\dot{S}_4$, is then just

$$\dot{S}_4 = \frac{1}{T_3} A \dot{X}_5 \text{ sgn } \dot{X}_5 \geq 0.$$

The 2-port R-field version of a mechanical resistor can accept either causality at the mechanical port, but only the causality shown in Figure 9.23 at the thermal port. Actually, the frictional energy may not reach the fluid instantaneously, as assumed in the model: it may heat the piston and wall material first and then warm the fluid. A more complex model with other thermal capacitances and resistances could model such effects.

9.4.3 An Electrothermal Resistor

All dissipation results in thermal effects that can sometimes be neglected or treated separately. For example, electrical resistors heat up as electrical power is dissipated, but in many cases the heating does not change the characteristics of the device very much if a means of cooling the resistor is provided. In such cases, ordinary circuit models assume that the circuit components remain essentially at a constant temperature. Here, we construct a model that contains both electrical and thermal effects.

The resistor is sketched in Figure 9.24a. The body of the resistor is assumed to have a fairly uniform temperature, T, and to be immersed in an atmosphere at temperature, T_0. The resistor is assumed to dissipate electrical power,

$$ei \geq 0. \tag{9.94}$$

As long as Eq. (9.94) is obeyed, we may assume that the relation between e and i for the resistor has some temperature dependence.

The power lost electrically is often spoken of as being converted to heat, so we will assume that

$$ei = \dot{Q} = T\dot{S}, \tag{9.95}$$

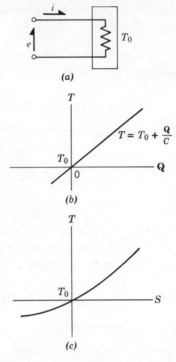

Figure 9.24. The electrothermal resistor. (a) Schematic diagram; (b) definition of thermal capacity; (c) capacitance relationship in terms of entropy.

in which it is a moot point whether $\dot{Q}$ should be regarded as a heat flow in the usual sense. Two causal forms for an R-field representation are useful:

$$\vdash \frac{e}{i} \blacktriangleright R \vert \frac{T}{\dot{S}} \blacktriangleright$$

$$e = e(i, T),$$

$$\dot{S} = \frac{ie(i, T)}{T}, \tag{9.96}$$

and

$$\frac{e}{i} \dashv R \vert \frac{T}{\dot{S}} \blacktriangleright$$

$$i = i(e, T),$$

$$\dot{S} = \frac{ei(e, T)}{T}. \tag{9.97}$$

As in previous examples, there is a preferred causality at the thermal port.

The temperature of the resistor body depends on how much thermal energy has been stored up in the resistor. In what follows, we neglect any work done against the pressure of the atmosphere by expansion of the resistor, that is, the assumption is that no significant power flows at the $P\dot{V}$ port. The temperature, then, will depend on the amount of energy in the form, $T\dot{S}$, that has been absorbed.

It is common to assume that T is a function of Q, but, for consistency, we express T as a function of S. These two points of view are readily reconciled since $\dot{Q}$ and S are related by Eq. (9.95). Suppose, for example, that a thermal capacity, C (approximately constant), has been defined as in Figure 9.24b and in the equation below:

$$T = T_0 + \frac{Q}{C}, \tag{9.98}$$

in which we begin integrating $\dot{Q}$ to find Q at a time when $T = T_0$. Using Eqs. (9.95) and (9.98), S may be found.

$$S = \int_0^S dS = \int_0^Q \frac{dQ}{T_0 + Q/C} = \frac{C \ln (T_0 + Q/C)}{T_0}. \tag{9.99}$$

where S and Q are both assumed to vanish at the initial time. Solving Eq. (9.99) for Q in terms of S and then substituting the result into Eq. (9.98), we find

$$T = T_0 e^{S/C}, \tag{9.100}$$

which is the relationship for the thermal capacitance when S is used instead of C. (In reality, since T is an absolute temperature, $Q \cong T_0 S$ and $T \cong T_0(1 + S/C)$ for modest excursions of T from T_0.)

The bond graph of Figure 9.25 shows the complete model for the resistor. The entire system is conservative, but power is lost from the electrical port, since the power on some heat-flow bonds cannot reverse. Both the electrothermal and the heat-transfer R-fields represent irreversible effects and generate entropy. As long as the temperature dependence

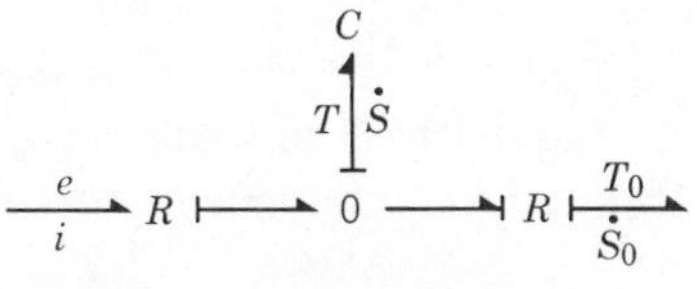

Figure 9.25. Bond graph for electrothermal resistor.

of the electrical resistance is not strong, one may choose not to use the complicated model of Figure 9.25 in favor of a simple 1-port R. However, in principle, the electrical and thermal systems are always coupled bilaterally so that models of this type are required.

The reader will have noticed in this brief introduction to bond-graph models for thermodynamic systems that only closed systems were modeled; that is, no mass crossed the boundary of the system. Much of the science of thermodynamics is concerned with flow processes in which not only power interchanges between the system and the environment, but also mass flows into and out of the system. While such systems can often be modeled with bond graphs, the convection of energy as mass moves through the boundary of a control volume complicates the modeling process. Some flow-process models, in which the distinction between Lagrangian and Eulerian descriptions of fluid motion parallels that between closed and open thermodynamic systems, will be discussed in the section on fluid mechanics.

9.5 FLUID-DYNAMIC SYSTEMS

We have already established effort, flow, displacement, and momentum variables for fluid systems of the closed-circuit hydraulic type, in which the static pressure times the volume flow rate represents most of the transmitted power. We now take a deeper look at fluid systems in general.

Much of fluid-mechanics work is concerned with field problems in which a flow field in two or three dimensions is to be determined. Such problems are usefully described by partial differential equations. Unless an analytic solution to such a problem happens to be known, one must resort to finite-difference or finite-element techniques. Such techniques represent the continuum with a large number of similar lumps. Although bond graphs can be made for the "micro lumps" involved in partial differential equations or their finite approximations, there often is little to be gained by such a representation. The micro lumps are all very similar, and they interact with their fellows only in standardized ways.

In this section, the main concern will be with the gross type of lumping commonly done when a fluid system is a small part of a larger system, and therefore, cannot practically be represented in great detail. Typical examples of such cases occur when hydraulic or pneumatic elements form part of a control system [15]. In what follows, the main concern will be with interior flows and with fluid-dynamic interaction with solid-mechanical elements through the influence of forces, motions, pressures, and volume flow rates.

As a simple example of the sort of approximations often made in the

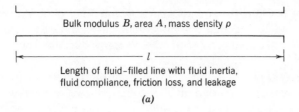

Bulk modulus B, area A, mass density ρ

$\longleftarrow$ ———————— l ———————— $\longrightarrow$

Length of fluid–filled line with fluid inertia,
fluid compliance, friction loss, and leakage

(a)

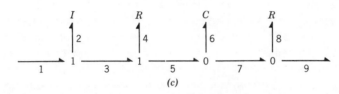

Q_1 Q_3 === Q_5 Q_7 Q_9

p_1 p_3 p_5 p_7 p_9

(b)

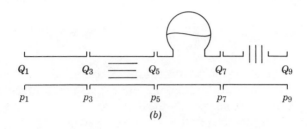

$$
\begin{array}{cccc}
I & R & C & R \\
\end{array}
$$

(c)

Figure 9.26. Lumped representation of fluid-filled lines. (a) Sketch of
line with some parameters; (b) schematic diagram; (c) bond graph.

analysis of fluid systems, consider the line element shown in Figure 9.26.
The model, which in the linear case is analogous to electrical-transmission
line models, attempts to treat a variety of effects clearly present in real
lines. The model is plausible for short segments of line and for linear
elements—yet it is often used for long lengths, l, and for nonlinear
elements. As will be seen, there are some philosophical snags with this
model if it is used without careful thought.

First of all, it is clear that the model is one-dimensional. The volume
flow rate past a cross-sectional area, Q, and the pressure, p, clearly must
be thought of as averaged quantities over a field of velocities and
pressures that vary over the area. Since the velocity profile of the flow
may vary widely during transient conditions, it is clear that the p and Q
variables and the parameters of the $-R$, $-C$, and $-I$ elements in Figure
9.26 cannot be evaluated exactly unless the nature of the flow over the
cross-sectional area is known in advance. Nevertheless, in system-design

studies, one rarely is sure in advance how the flows in various parts of the system will behave, so that one must make at least a preliminary estimate of the system parameters before the dynamics of the flow can even be estimated.

Consider first the problem of estimating the inertia of the fluid in the line. An elementary derivation of the inertia coefficient (see Reference [16], for example) proceeds as follows from Figure 9.26: The mass of the fluid in the line is $\rho A l$, the force tending to accelerate the fluid is $A(p_1 - p_3)$, and the velocity of the fluid is Q_2/A. Thus,

$$\rho A l \frac{d}{dt} \frac{Q_2}{A} = A(p_1 - p_3) \qquad \text{or} \qquad Q_2 = \frac{A}{\rho l} p_{p_2}, \qquad (9.101)$$

where p_{p_2} is the pressure momentum of bond 2 (or the time integral of p_2). This result shows that the inertia coefficient expressed as a "mass" is $\rho l / A$ when p, Q variables are used. That the inertia of small-area tubes is larger than the inertia of larger-area tubes comes as something of a surprise.

There are several problems with the simple derivation given above. First, in keeping with the assumption of one-dimensionality, the fluid in the pipe was treated as if it moved as a rigid body. It is difficult to improve on this assumption until it is known how the velocity profile of the fluid changes in space and time. The types of flows being considered here are often called "quasi-one-dimensional flows," and in the steady state, when a well-developed velocity profile is known at each cross-section, one may identify Q/A as an *average* velocity. In later calculations, the average of the square of the velocity over the cross section will be needed, and this quantity may be taken as $\beta Q^2 / A^2$, where the correction factor, $\beta \geq 1$, can be calculated if the velocity profile is known (see Reference [17]). The β factor is unity only for a uniform velocity distribution that can occur for frictionless, irrotational flow. For transient conditions the velocity distribution is often very hard to estimate, so in the remainder of the section we will assume a uniform distribution even in cases in which this cannot strictly be true. The accuracy of some results may be slightly improved by the introduction of factors such as β, although the estimation of the factors in the absence of experimental data may be difficult.

Second, unless one is willing to assume that the fluid is incompressible, the proper value of density, ρ, is open to question. If $l = dx \rightarrow 0$, then $\rho(x, t)$ might represent an instantaneous value of ρ at position x, but if l is finite, then existence of capacitance in Figure 9.26 is not compatible with using ρ as a constant. If ρ varies, then, of course, the two ends of the slug of fluid do not move with the same velocity. Later on, a more sophisticated look at the problem will be taken. For now, let it simply be noted that for hydraulic systems in which the density changes are small, an average density for ρ is often sufficient, despite the contradiction implied by the use of a compliance together with a constant inertia.

What is really wrong with this elementary derivation is that the slug of fluid is treated as a rigid body, even though it is clear that a control volume comprising a length of the pipe through which mass and momentum flow is being considered. As will be demonstrated, the derivation happens to be essentially correct when the two ends of the pipe have identical cross-sectional areas since the momentum flow terms at the ends then cancel. The elementary derivation does not generalize readily when pipes of varying area are encountered, and it is surprising that few authors of elementary system-dynamics texts even mention the control-volume basis for their fluid-mechanical system dynamics.

The element, $R4$, in Figure 9.26 represents a loss in pressure beyond that required to accelerate the fluid. The relation between p_4 and Q_4 would be easy to specify as a generally nonlinear relationship if one could use the data for flow in pipes that has been determined experimentally. But almost all the data on friction factors is for fully developed steady flow and would not apply exactly to any but the slowest transient conditions. For this reason, the analyst must be prepared to experiment with the friction loss law until the model response matches experimental data sufficiently well.

The next element in Figure 9.26 is intended to model the compliance of the fluid or of the fluid and pipe walls. The flow, Q_6. which is the difference between Q_5 and Q_7, represents a loss in flow between the ends of the pipe. The pressure, p_6 (which equals p_5 and p_7), can be determined from the integral of Q_6 when one can define a bulk modulus, B.

$$p_6(t) = p_6(0) + \frac{B}{V_0} \int_0^t Q_6 \, dt = p_6(0) + \frac{B}{V_0} V_6(t), \qquad (9.102)$$

where V_0 is the volume of the fluid in the pipe at $t = 0$ when the pressure is $p_6(0)$. One might think that simply by replacing the linear relation, Eq. (9.102), by some nonlinear relation between p_6 and V_6 to model compressibility effects of a gas, the same model would serve in the case in which large density changes occur, but the situation is not quite so simple. As is demonstrated below, when density variations are significant, the thermodynamics of the situation must be studied.

Finally, R_8 represents a loss of flow in the pipe section due to leakage. Clearly, when the p s represent absolute pressure rather than gage pressure, R_8 should react to the difference between p_8 and the pressure external to the pipe, rather than just to p_8 itself. This requires that R_8 be attached to a 1-junction that computes the difference between the internal and external pressure and imposes the condition that flow out of the pipe is the same as flow into the external atmosphere.

From the brief introduction to the lumped line elements given above, it is clear that there are some fundamental difficulties in justifying the lumping process. Many difficulties disappear when only a linearized lumped model is desired since it becomes plausible to neglect small changes in quantities that have finite mean values. In order to see more clearly the nature of some of the more commonly used fluid dynamic models, one needs to consider the partial differential equations of fluid mechanics and nonlinear effects.

9.5.1 One-Dimensional Incompressible Flow

In order to illuminate the connection between standard techniques in fluid mechanics and a lumped representation using bond-graph elements, a derivation of Bernoulli's equation is given first. Consider the problem of Figure 9.27a. Let s represent distance along the center line of a curved rigid pipe of length, l. Then $A(s)$ is the pipe cross-sectional area, $x(s)$ and $z(s)$ describe the horizontal and vertical positions of the pipe center line, respectively, $v(s, t)$ represents the (average) fluid velocity at position s and time t, and $p(s, t)$ represents the pressure.

Newton's law yields

$$\rho \frac{Dv}{Dt} = \rho \frac{\partial v}{\partial t} + \rho v \frac{\partial v}{\partial s} = -\frac{\partial p}{\partial s} - \rho g \frac{dz}{ds}. \tag{9.103}$$

Noting that, because of the incompressibility, the volume flow rate, $Q(t)$, is constant with s, v is given by

$$v(s, t) = \frac{Q(t)}{A(s)}. \tag{9.104}$$

Substituting Eq. (9.104) into Eq. (9.103), we have

$$\rho \left(\frac{\dot{Q}(t)}{A(s)} + \frac{Q^2(t)}{A(s)} \frac{\partial 1/A(s)}{\partial s} \right) = -\frac{\partial p}{\partial s} - \rho g \frac{dz}{ds}, \tag{9.105}$$

which may be integrated in s from $s = 0$ to $s = l$.

$$I\dot{Q} + \frac{\rho Q^2}{2} \left(\frac{1}{A_l^2} - \frac{1}{A_0^2} \right) = p_0(t) - p_l(t) - \rho g(z_l - z_0), \tag{9.106}$$

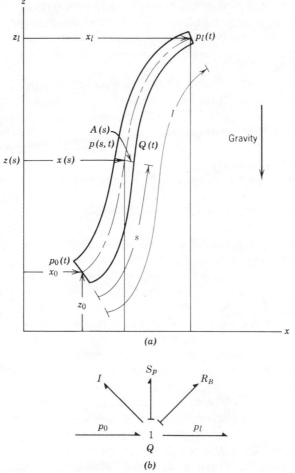

Figure 9.27. Constrained motion of an incompressible fluid. (*a*) Sketch of system; (*b*) bond graph.

where

$$I = \rho \int_0^l \frac{ds}{A(s)}, \tag{9.107}$$

and one could substitute $\beta Q^2/A^2$ for the Q^2/A^2 terms as discussed previously.

It is interesting to note that Bernoulli's equation [essentially Eq. (9.106)] can be represented exactly by the lumped elements shown in Figure 9.27*b*. The linear inertia coefficient defined in Eq. (9.107) reduces to that found by the nonrigorous method of Eq. (9.101) if $A(s)$ is constant, but several other terms appear in Eq. (9.107) that did not appear in Eq. (9.101). The gravity term that is represented by a constant-pressure source in Figure 9.27b is easily understood, but the term involving Q^2 requires explanation.

First of all, if $A(s)$ is constant, then $A_l = A_0$ and the Q^2 term vanishes, thus showing that Eq. (9.101) is essentially correct for the constant-area case. However, in deriving Eq. (9.101), the flow of fluid through the ends of the pipe section was not properly considered. To use Newton's law in its simplest form one must follow the flow of the fluid; that is, one must use a Lagrangian description. However, one really wants to treat the pipe as a control volume through which fluid passes, that is, with an Eulerian description. The Q^2 term in Eq. (9.107) may be thought of as a dynamic pressure-correction term which was fortuitously absent from the case of Eq. (9.101).

Another way to interpret the result of Eq. (9.107) is to note that the power flow, $p(t)Q(t)$, does not represent the total flow of energy past a stationary point of the pipe. Terms of the form, $\rho Q^2/2A^2$, represent dynamic pressure associated with the kinetic energy of the fluid. In the system under study, p and Q contain all the information required to find the dynamic pressure since the velocity is Q/A. The Q^2 term in Eq. (9.107) may be represented as a resistance in Figure 9.27b since it is a relation between Q and a pressure (the dynamic pressure). When the resistance characteristic is plotted as in Figure 9.28, it is clear that if one keeps track only of the pQ power, then the dynamic pressure may be converted back and forth into static pressure as in nozzles and diffusers.

The resistance required to represent Bernoulli's equation is unusual in two ways. First, no matter whether $A_0 > A_l$ or $A_0 < A_l$, there are regimes of operation in which the resistance supplies power. Physically, this means that dynamic pressure is being partially converted to static pressure and, hence, to an apparent power flow in the form, pQ. Second, although the pressure in the resistance is uniquely given for any flow, if one gives the pressure, then there is either no corresponding flow or two of them. Thus, there is a very strong causal preference for this resistance, as shown in Figure 9.27b.

The latter feature is readily explained by the observation that the basis for our derivation was the assumption that the fluid filled the pipe, and, hence, $v = Q/A$. For some pressure conditions at the ends of the pipe, this is not true—a jet can form in the pipe. Also, the equations derived may be valid only for one direction of flow. In studying the flow of water through a nozzle into the atmosphere, for example, it may make no sense to consider reverse flow because the nozzle would fill with air and the

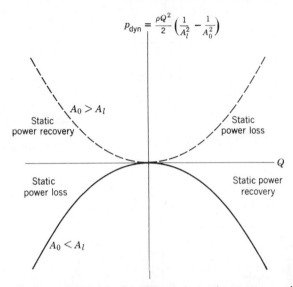

$$p_{dyn} = \frac{\rho Q^2}{2} \left(\frac{1}{A_l^2} - \frac{1}{A_0^2} \right)$$

Figure 9.28. Constitutive law for the dynamic pressure resistor.

equations would no longer be meaningful. Suffice it to say that in any application of Bernoulli's equation including the bond-graph representation shown in Figure 9.27, one must use care that the proper branch of the resistance relation is being used, to avoid nonsensical results. Such difficulties with resistance relations are rare in other types of physical systems.

As an illustration of the utility of the bond-graph representation, consider the classical elementary problem of estimating the time required for a tank to empty through a pipe. The system is shown in Figure 9.29. For simplicity, no friction losses will be considered, but the Bernoulli resistor may be thought of as indicating the loss in kinetic energy of the fluid that leaves the system. The tank has a capacitance of $A_T/\rho g$, where A_T, the tank area, is large compared to the pipe area. As in the classical analysis, we imagine the fluid entering the end of the pipe from a large area, A_1, at pressure, p_1, and essentially zero velocity. The Bernoulli resistor then gives a pressure, $\rho Q^2/2A_2^2$, where Q is related to the pressure momentum by a relation of the form of Eq. (9.101). Using the notation shown in Figure 9.29 and noting from the bond graph that two state variables, $V(t)$, the volume, and $p_p(t)$, the pressure momentum, are required, we find

$$\dot{V} = -\frac{A_2}{\rho L} p_p, \tag{9.108}$$

$$\dot{p}_p = \frac{-\rho}{2A_2^2} \left(\frac{A_2}{\rho L} p_p \right)^2 + \frac{\rho g V}{A_T}, \tag{9.109}$$

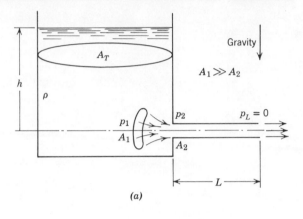

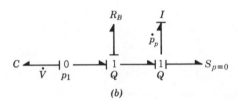

(b)

Figure 9.29. Tank-emptying problem. (a) Sketch of system; (b) bond graph.

or

$$-\frac{L}{A_2}\ddot{V} = \frac{-\rho}{2A_2^2}(\dot{V})^2 + \frac{\rho g}{A_T}V \qquad (9.110)$$

as long as $V \geq 0$. Note that, although no static power flows into the zero-pressure source representing atmosphere, there is a flow, and with a diffuser one could recover power from this flow.

Control volume problems can become quite complex if the control volume itself is not fixed in inertial space. Bond-graph methods can clarify the modeling of such systems, but space limitations prohibit a full discussion here. An example of such a system appears in Reference [21]. We now go on to discuss the implications of dropping the incompressibility assumption, and we find that further difficulties await us and that more approximations are involved in the hydraulic-system bond graphs used previously than first might have been imagined.

9.5.2 Representation of Compressibility Effects

For small changes in density, it is easy to see that compressibility effects are readily modeled by means of linear capacitors, as discussed previously. For large changes in density, however, it is best to begin with a study of the thermodynamics of a pure substance.

As we have seen in the previous section, the internal energy per unit mass of a fluid, u, depends on two independent properties: the density, ρ, and the entropy, s. Changes in these quantities are related by Gibbs equation,

$$du = T\,ds - p\,d\left(\frac{1}{\rho}\right),\tag{9.60}$$

where T is the thermodynamic temperature, p is the pressure, $1/\rho$ is the specific volume, and

$$u = u\left(s, \frac{1}{\rho}\right).\tag{9.58}$$

The characteristics of the fluid are conveniently summed up by Eq. (9.58), in which we have used $1/\rho$ for specific volume since we will use v for velocity in this section. The Gibbs equation for unit mass or a fixed amount of matter can be represented by a C-field as we have seen.

Using the C-field representation, it is straightforward to model systems in which fixed amounts of a fluid are compressed and expanded and heated. When the fluid passes into and out of a control volume, however, the situation is somewhat more complex. In Figure 9.30, for example, fluid is compressed in a fixed volume by allowing fluid to pass slowly in and out of a port. In this case, the total energy contained in the volume, V_0, denoted by U, depends on the mass, m, contained in the volume and the internal energy per unit mass, u. Changes in U occur not only because of the flow work, pQ, but also because of the convection of energy. If one uses the mass flow rate, $\dot{m} = \rho Q$, instead of Q as a flow variable, then

$$dU = u\,dm + \frac{p}{\rho}\,dm\tag{9.111}$$

in the isentropic case. Defining the enthalpy, h, by the relation

$$h = u + \frac{p}{\rho},\tag{9.112}$$

it now appears that the power flow past the port is given by

$$h\dot{m} = u\dot{m} + \frac{p}{\rho}\dot{m} = u\rho Q + \frac{p}{\rho}\rho Q = (\rho u + p)Q,\tag{9.113}$$

which shows that pQ is only the hydrostatic power. As in the previous section, in which it was found necessary to supplement the static pressure

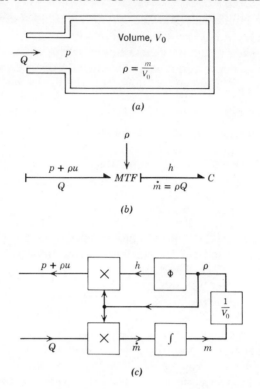

Figure 9.30.   Isentropic compression of a fluid. (a) Fluid flow into a rigid
volume; (b) bond graph; (c) block diagram.

with a dynamic pressure term properly to account for real power flow, so
here it is necessary to supplement p with an extra term, ρu, to account for
convected internal energy. Figure 9.30b shows a bond graph based on
Eqs. (9.112) and (9.113), and a block diagram corresponding to integral
causality is given in Figure 9.30c. Note that the dynamic pressure term
has not been incorporated, so this model is only valid for low flow rates
(as indeed is the C-field representation for the Gibbs equation). Also, note
that the pressure and internal energy depend on density alone in the
isentropic case, so the enthalpy also is determined by conditions inside
the vessel. In more complicated cases, the enthalpy at the port is
determined by conditions inside the vessel for outflow, $\dot{m} < 0$, but is
determined by external conditions for inflow, $\dot{m} > 0$. This kind of causal
switching with changes in direction of flow is known to occur in
heat-exchanger systems in which the flow can reverse, but has not

received much study in terms of bond graphs or any other means of system analysis.

Figure 9.31 shows how the bond graph is modified when isentropic compression occurs partly due to variable volume and partly due to inflow and outflow, as in the power cylinder of a pneumatic servomechanism. The C-field in Figure 9.31b has one $h\dot{m}$ port and one $p\dot{V}$ port. The implication is that the total energy, U, depends on m and V with

$$h = \frac{\partial U}{\partial m} ; \qquad -p = \frac{\partial U}{\partial V} . \tag{9.114}$$

It may be worthwhile to verify that the C-field is really conservative by checking the validity of the Maxwell reciprocal relation,

$$\frac{\partial h}{\partial V} = \frac{\partial (-p)}{\partial m} . \tag{9.115}$$

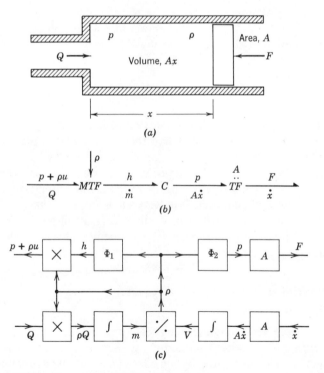

Figure 9.31. Compression in a ram. (a) Sketch of system; (b) bond graph; (c) block diagram.

With the isentropic assumption, the internal energy per unit mass is a function of specific volume only:

$$u = u\left(\frac{1}{\rho}\right); \qquad -p = \frac{du(1/\rho)}{d(1/\rho)}. \tag{9.116}$$

Expressing the specific volume as

$$\frac{1}{\rho} = \frac{V}{m}, \tag{9.117}$$

we have

$$p = p\left(\frac{V}{m}\right), \qquad u = u\left(\frac{V}{m}\right), \qquad h = u + p\frac{V}{m}; \tag{9.118}$$

then

$$\frac{\partial h}{\partial V} = u'\frac{1}{m} + p'\frac{V}{m^2} + p\frac{1}{m} = p'\frac{V}{m^2}, \tag{9.119}$$

where the primes denote the derivatives of the u and p functions and Eq. (9.61) has been used to cancel two terms in Eq. (9.119).

Differentiating p with respect to m yields

$$\frac{\partial p}{\partial m} = \frac{-V}{m^2}p', \tag{9.120}$$

which, upon comparison with Eq. (9.119), validates Eq. (9.115).

The bond graph of Figure 9.31 and the equations it represents are a sophisticated version of models of compressibility effects in fluid servomechanisms in which the volume of fluid concerned is variable. In practice, a linearized compressibility parameter may simply be incorporated into the system equations (see Reference [15], for example), but, as is often the case, it is virtually impossible to proceed from a linearized model of an effect to a more accurate nonlinear model without starting again from the basic physics of the situation.

9.5.3 Inertial and Compressibility Effects in One-Dimensional Flow

The simple model of Figure 9.26, neglecting for the moment the loss elements, is a lumped approximation of the inertial effects and compressibility effects in a length of line. When a large number of such models with parameters appropriate to a length, Δl, are cascaded, and when $\Delta l \to 0$, the state equations for the model form an approximation to the partial differential equation called the *one-dimensional wave equation.* In the sections above, models for the inertia effect in incompressible flow and the compliance effect when no inertia is considered were developed. The question now arises whether one can simply cascade a series of lumps that alternately account for inertia and compressibility effects in the manner of Figure 9.26 and thereby construct a model of a distributed line.

The answer is in the affirmative in several important cases, namely,

1. *The acoustic approximation.* In this case, small deviations in pressure and density are modeled with linear equations. The inertial and compliance coefficients are calculated at the mean pressure and density. See References [8] and [20]. Such a model is useful in studying compressibility effects in oil-hydraulics and water-hammer problems, as well as in acoustics.

2. *Lagrangian descriptions.* When one is willing to follow the motion of a group of particles, then the inertia of the group is constant. The system of equations can be nonlinear only due to nonlinear elastic or dissipative effects as long as Newtonian mechanics is used. For the vibration of bars and strings, for example, a Lagrangian description is natural, since the particles never move very far anyway. For beams and plates, the differential equations are more complex than the wave equation, but the Lagrangian description still allows one to use finite models with *I*s and *C*s representing inertia and compliance effects. The junction structure is more complex than the simple 0–1–0–1 string of Figure 9.26, however. For fluid systems in which the particles move through fixed boundaries, the Lagrangian descriptions are used rather rarely (see Reference [8]).

For the cases above one may easily construct finite element models from normal bond-graph components that, when reduced to differential volumes, mirror the usual derivations of the partial differential equations of motion for the continuum model. Essentially, the inertial, compliance and resistance aspects of the system may be treated separately in the finite elements, even though these aspects refer to a single point when the volume of the element is made to approach zero in the continuum description. As will be demonstrated, the Eulerian description common to fluid mechanics entangles the inertial and compliance aspects so thoroughly that it is difficult to construct a series of finite elements converging to the continuum description without using a large number of active bonds except in the acoustic approximation.

Consider the isentropic, 1-dimensional flow of fluid through a tube of unit cross-sectional area. Using s for the space coordinate, $v(s, t)$ for velocity, (s, t) for the density, and $p(s, t)$ for the pressure, the equations describing the system are:

$$\rho \frac{\partial v}{\partial t} + \rho v \frac{\partial v}{\partial s} = -\frac{\partial p}{\partial s}, \tag{9.121}$$

$$\frac{\partial \rho}{\partial t} + \frac{\partial \rho v}{\partial s} = 0, \tag{9.122}$$

$$p = p(\rho). \tag{9.123}$$

Equation (9.121) is just Newton's law for a length of fluids with the acceleration expressed in Eulerian form. Equation (9.122) is a statement of conservation of mass that, with the constitutive law for the gas, Eq. (9.123), can be used to define the compressibility effect. The density, $\rho(s, t)$, enters both the inertia law, Eq. (9.121), and the compressibility laws. In contrast, a Lagrangian description would express the inertia in terms of a constant density of fluid in a reference state (see Reference [8]).

It is probably not sensible to attempt to construct a bond graph for Eqs. (9.121)–(9.123) unless ρ in Eq. (9.121) is nearly constant [in Eqs. (9.122) and (9.123) ρ *must* be allowed to vary] and $v \, \partial v / \partial s$ is a second-order small quantity. Basically, Eq. (9.121) is Newton's law following the flow infinitesimally. At each succeeding instant of time, the law refers to *different* infinitesimal slices of fluid. Thus, the time-varying inertial effect will not conserve energy as normal bond-graph inertial elements do. Also, the term, $\partial v / \partial s$, implies that the velocity is different at the two ends of the differential control volume. Thus, the inertial effect and the compressional effect both appear in Eq. (9.121). Finally, the terms appearing in these equations (pressure and velocity for volume flow since we are considering unit area) do not multiply to give the true power flow at position, s, and time, t. As we have seen, the convected internal energy has been left out of the equations. One should not, therefore, expect that these (correct) equations can be represented by a bond graph in which power conservation and energy conservation are built in. It seems that the alternatives, if one wishes to represent fluid dynamic lines in terms of bond graphs are to (1) use a Lagrangian description; (2) use an Eulerian description, but with a restriction essentially to the acoustic approximation; (3) consider only the incompressible or noninertial cases, as was done above. Thus, the intuitively appealing scheme of Figure 9.26 appears to be more restrictive than has been generally appreciated even without consideration of shear stresses, boundary layers, and so forth, which also complicate the modeling of fluid transmission lines.

9.5.4 Conclusions

It has been demonstrated that many fluid mechanical systems can be described with the same sort of bond-graph elements that are familiar in electrical and solid-mechanical systems. In some cases, elements exist that are unusual in other types of systems. For example, the resistance relation that is required in Bernoulli's equation represents a loss or gain in power associated with static pressure as dynamic pressure increases or decreases. The resistance actually represents an inertial effect as seen from an Eulerian or control-volume standpoint. When both inertia and compressibility effects are present and one desires to use an Eulerian description of the fluid motion, the lumping process is more complex. Most familiar lumped-parameter systems are represented in Lagrangian form, so it is not surprising that unexpected effects can occur in open systems, through which matter and energy can flow. It is perhaps more surprising that, in some important cases (such as the incompressible case), models closely analogous to Lagrangian lumped systems can be found for Eulerian systems. This no doubt explains why many analysts have never even given thought to the difference between Eulerian and Lagrangian descriptions in making crude lumped models of distributed systems.

REFERENCES

1. G. Oster and D. M. Auslander, "Topological Representations of Thermodynamic Systems: I. Basic Concepts," *J. Franklin Institute* **292**, (1) (July 1971), 1–17; "II. Some Elemental Subunits for Irreversible Thermodynamics," *Ibid.*, **292**, (2) (Aug. 1971), 77–92.
2. G. Oster, A. Perelson, and A. Katchalsky, "Network Thermodynamics," *Nature*, **234**, (Dec. 17, 1971), 393–399.
3. G. F. Oster, A. S. Perelson, and A. Katchalsky, "Network Thermodynamics: The Analysis of Biological Systems," *Donner Laboratory and Lawrence Berkeley Laboratory Preprint*, LBL-588, (Feb. 1972), 245 pp.
4. D. M. Auslander, G. F. Oster, A. Perelson, and G. Clifford, "On Systems with Coupled Chemical Reaction and Diffusion," *Trans. ASME J. Dyn. Sys. Meas. and Control*, **94**, Ser. G., (3) (Sept. 1972), 239–248.
5. Sir William Thomson and P. G. Tait, *Principles of Mechanics and Dynamics*, N.Y.: Dover Publications, 1962.
6. C. Lanczos, *The Variational Principles of Mechanics*, Toronto, Canada: University of Toronto Press, 1957.
7. R. C. Rosenberg, "Multiport Models in Mechanics," *Trans. ASME, J. Dyn. Sys. Meas. Control*, **94**, Ser. G., 3 (Sept. 1972), 206–212.
8. S. Crandall, D. Karnopp, E. Kurtz, and D. Pridmore-Brown, *Dynamics of Mechanical and Electromechanical Systems*, N.Y.: McGraw-Hill, 1968.

9. D. Karnopp, "Power-conserving Transformations: Physical Interpretations and Applications using Bond Graphs," *J. Franklin Institute*, **288**, (3) (Sept. 1969), 175–201.

10. R. C. Rosenberg, "State-space Formulation for Bond Graph Models of Multiport Systems," *Trans. ASME, J. Dyn. Sys. Meas. Control*, **93**, Ser. G, (1) (March 1971), 35–40.

11. G. Martin, *Kinematics and Dynamics of Machines*, N.Y.: McGraw-Hill, 1969, p. 7.

12. E. Colin Cherry, "The Duality Between Interlinked Electric and Magnetic Circuits and the Formation of Transformer Equivalent Circuits," *Phys. Soc. London Proc.*, **62**, (Feb. 1949), 101–111.

13. R. W. Buntenbach, "Improved Circuit Models for Inductors Wound on Dissipative Magnetic Cores," 1968 Conference Record of Second Asilomar Conference on Circuits and Systems, 68 C 64-ASIL, pp. 229–236, IEEE, New York, 1969.

14. P. G. Stohler and H. R. Christy, "Simulation and Optimization of an Electromechanical Transducer," *Simulation*, Vol. **13**, (4) (Oct. 1969), 202–210.

15. F. H. Raven, *Automatic Control Engineering*, N.Y.: McGraw-Hill, 1961.

16. R. H. Cannon, *Dynamics of Physical Systems*, N.Y.: McGraw-Hill, 1967.

17. W. M. Swanson, *Fluid Mechanics*, N.Y.: Holt, Rinehard and Winston, 1970.

18. C. J. Radcliffe and D. Karnopp, "Simulation of Nonlinear Air Cushion Vehicle Dynamics Using Bond Graph Techniques," Proceedings 1971 Summer Computer Simulation Conference, Boston, Mass., July 19–21, 1971.

19. L. Tisza, *Generalized Thermodynamics*, Cambridge, Mass.: The M.I.T. Press, 1966.

20. P. M. Morse and K. U. Ingard, *Theoretical Acoustics*, N.Y.: McGraw-Hill, 1968.

21. D. C. Karnopp, "Bond Graph Models for Fluid Dynamic Systems," *Trans. ASME, J. Dyn. Sys. Meas. Control*, **94**, Ser. G, (3) (Sept. 1973), 222–229.

PROBLEMS

9-1

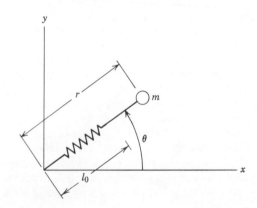

A particle can slide on a frictionless horizontal plane constrained by a linear spring pivoted at the origin of x, y and with spring constant, k, and free length, l_0.

Make a bond graph for this system by using r, θ for the $\mathbf{q}_C$ and $\mathbf{q}_k$ vectors and $\dot{x}$, $\dot{y}$ for the $\mathbf{v}_I$ vector. Can the system accept all integral causality? Make another bond graph by using x, y for $\mathbf{q}_k$, $\dot{x}$, $\dot{y}$ for $\mathbf{v}_I$, and r for $\mathbf{q}_C$.

9-2

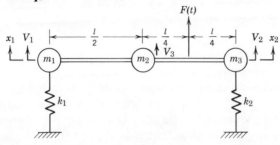

Three mass particles are attached to a spring-supported rigid massless bar that executes small vibratory motion. The $\mathbf{q}_C$ vector is x_1, x_2, the $\mathbf{v}_I$ vector is V_1, V_2, V_3, and several choices for $\mathbf{q}_k$ are possible. Show a bond graph for the system with the following choices for $\mathbf{q}_k$:

(a) $\mathbf{q}_k = x_1$, $x_2 = \mathbf{q}_C$;
(b) $\mathbf{q}_k = x_3$, θ, where x_3 is displacement of m_3 and θ is the angle of inclination of the bar;
(c) $\mathbf{q}_k = x_1$, x_3.

Pick one bond graph, and write the equations of motion for the system using the graph.

9-3

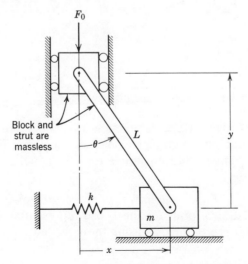

Set up bond-graph representations of the system using the three alternative *MTF* forms shown below

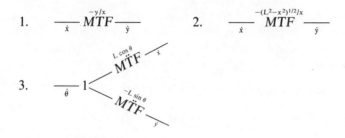

1. $\dfrac{}{\dot{x}}$ MTF $\overset{-y/x}{\underset{\dot{y}}{}}$ 2. $\overset{-(L^2-x^2)^{1/2}/x}{\dfrac{}{\dot{x}}\,MTF\,\dfrac{}{\dot{y}}}$

3. $\dfrac{}{\dot{\theta}}$ 1 $\overset{L\cos\theta}{\underset{\ddot{MTF}}{}}\,\overset{\dot{x}}{}$
                              $\overset{-L\sin\theta}{\underset{\ddot{MTF}}{}}\,\overset{}{\dot{y}}$

Write *complete* state equations in each case using integration-causality methods or Lagrange equations. Comment on any advantages or disadvantages you see in the alternative formulations.

9-4 Consider the model of an off-road vehicle shown below.

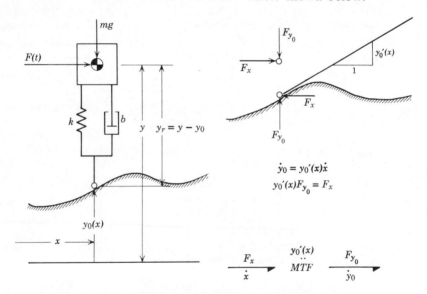

$$\dot{y}_0 = y_0'(x)\dot{x}$$
$$y_0'(x)F_{y_0} = F_x$$

Note that the surface interaction may be represented by the equations and bond graph element shown above. Develop a bond graph and state space equations for this model.

9-5 Construct a bond graph for this 2-degree-of-freedom vehicle model. Use two forms for the *I*-field representing the rigid body: one matrix representation using the $\dot{y}_1$, $\dot{y}_2$ port variables directly,

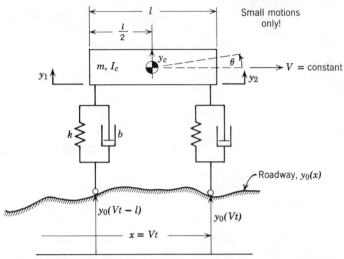

the other using the auxiliary variables, y_c and θ, and a multiport TF.

9-6 (In which the inherent nastiness of geometric nonlinearity is exhibited.)

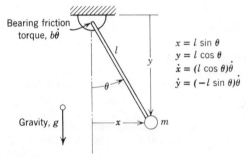

Consider the simple pendulum of length, l, mass, m, swinging through large angles under the influence of gravity and bearing friction.

(a) Set up a modulated transformer representation of the relations between $\dot{x}$, $\dot{y}$, and $\dot{\theta}$ and the corresponding forces, F_x, F_y, and torque, τ, that are enforced by the rigid bar. Demonstrate that the multiport displacement-modulated transformer relations are power conservative.

(b) Show a bond graph for the system using your MTF. Set up sign conventions and a causal pattern in which integration causality is applied to the x-motion $-I$.

(c) Write the equations corresponding to your graph. Be sure that your state space is complete and in the standard form.

(d) Show that your state space is consistent with $ml^2\ddot{\theta} = -b\dot{\theta} - mgl\sin\theta$, which would be the result of many standard analyses.

9-7 Consider the double pendulum shown (see problem 9.6).

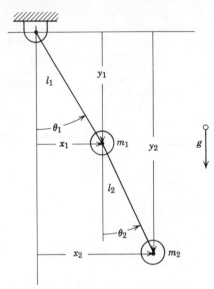

(a) Set up the multiport displacement-modulated transformer that relates $\dot{x}_1, \dot{y}_1, \dot{x}_2, \dot{y}_2$ to $\dot{\theta}_1, \theta_1, \theta_2, \theta_2$, and exhibit both the velocity and force relations in matrix form.

(b) Using the *MTF*, show a system bond graph, and predict one possible set of state variables using the integration-causality method.

(c) Use the junction structure of (b) to compute $T^*(\theta_1, \dot{\theta}_1, \theta_2, \dot{\theta}_2)$—kinetic coenergy—and generalized torques so that Lagrange equations can be utilized as an alternative procedure to find state equations. What are the state variables if Lagrange equations are used?

9-8

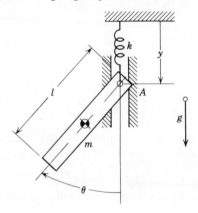

The uniform thin bar is pivoted at A. The pin slides freely in the vertical guideway. Represent the bar by three 1-port —Is representing the horizontal and vertical motion of the center of mass and rotation about the center of mass.

(a) Show that the junction structure of the multiport MTF involved may be rearranged to yield a bond graph for the system of the following form:

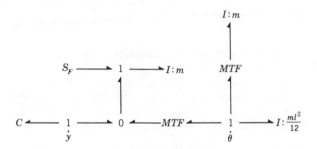

(b) Apply causality to the system, and write the equations of motion.

9-9

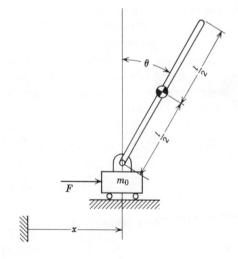

Mass, m
Moment of inertia, J
Assume $\theta \ll \pi$.

(a) Noting that the equilibrium point for the upside-down pendulum is $\theta = 0$, write expressions for kinetic co-energy and potential energy valid up to second order in θ and $\dot{\theta}$. Use these expressions to validate the linear bond graph shown below:

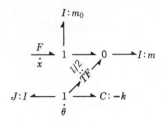

Find the negative spring constant for the $-C$ that models the gravity effect.

(b) Assuming F is an input, and that the $-I$s with parameters m_0 and J have integral causality, write state equations, and indicate the matrix that must be inverted because of the differentiation causality.

(c) Show that the system can be represented by

$$\xrightarrow[\dot{x}]{F=\dot{P}_x} I \xleftarrow[\dot{\theta}]{\dot{P}_\theta} C:+k,$$

and find the matrix representation of the I-field.

9-10 Following the pattern of analysis of problem 9.9, find a bond graph for the double-inverted pendulum shown. Write state equations for the system. (For illustration see p. 389.)

9-11 The German-speaking automobile model shown below uses the subscripts, v, h, and R, for vorn (forward), hinten (rear), and rad (wheel), respectively. Otherwise, it seems to be sufficiently international to be bond graphed.

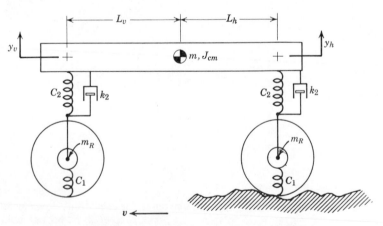

Dynamic model of a simple two-axle vehicle.

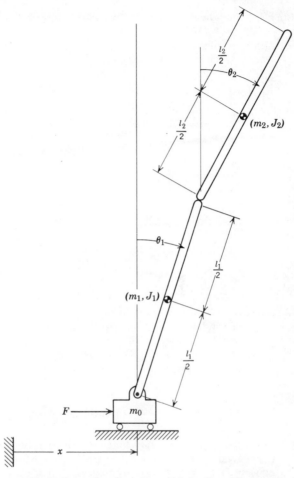

Consider small pitch angles for the vehicle. Recall that the sum of the vertical forces is equal to the mass times the acceleration of the center of mass and that the sum of the torques about the center of mass is the centroidal moment of inertia times the angular acceleration.

(a) Construct a bond graph for the model in which the vehicle mass, m, and centroidal moment of inertia appear as 1-port element parameters.

(b) Show that consistent causality may be assigned for your bond graph, and list the state variables. Discuss very briefly any difficulties you may find in assigning causality.

(c) Show that if the vehicle body is considered to be an I-field, the assignment of causality is simplified and the writing of state equations is easier.

(d) Either find the properties of the 2-port I-field describing the rigid body or outline a procedure to find them.

(e) Can you think of a way to represent the nonlinear behavior of the wheels when they leave the ground?

9-12

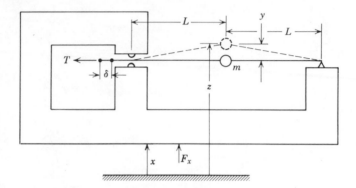

The mass, m, is supported on a thin inextensible wire stretched across the frame with tension, T.

(a) Show that if T=constant, and if $y/L \ll 1$, then the force on the mass is given approximately by $F = -(2y/L)T$, and a bond graph for the system (neglecting the inertia of the frame) is

$$k_{eq} = \frac{2T}{L} : C \longleftarrow 1 \underset{\dot{y}}{\overset{F_x}{\longleftarrow}} 0 \underset{\dot{x}}{\overset{F_y}{\longleftarrow}} .$$

with I and $F_z | \dot{z}$ above the 0-junction.

(b) When T is allowed to vary by connecting a force source to the wire, the equivalent spring seems to simply have a "variable constant." This concept is not very profound, however, since a $-C$ element must be conservative, and with variable T the relation between F_y and y is not.

Show that if $(y/L) \ll 1$, the distance the end of the wire moves against the force source, δ, is $\delta = y^2/L$. Use this relation to show a bond graph that represents the effect of the wire by means of a force source and a displacement-modulated transformer. Note that this model shows a causal restriction that was not evident in the previous bond graph.

9-13 Consider the spherical pendulum shown in part a of the figure. Verify the bond graph shown in part b by finding the constitutive laws for the MTF relating $\dot{\theta}_1$, $\dot{\theta}_2$ to $\dot{x}$, $\dot{y}$, $\dot{z}$. Write out the multiport MTF as a junction structure using 0- and 1-junctions and 2-port MTFs.

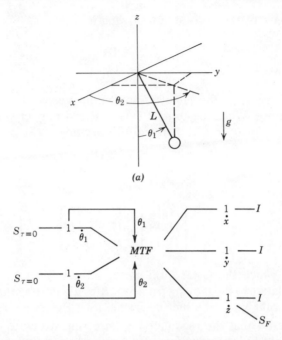

(a)

Now show that all three Is cannot accept integral causality because the MTF cannot accept flow input causality on all three of the $\dot{x}$, $\dot{y}$, $\dot{z}$ bonds. Show that consistent complete causality can be achieved if only two of the Is have integral causality.

9-14   A brief outline of the relation between applied torque and angular momentum for general motion of a rigid body using components in a coordinate frame moving with the body and aligned with the principle axes is given below:

Following the notation of Reference [8], Section 4.4, a rigid body may be defined by a relation between angular velocity, $[\omega]$, angular momentum, $[H]$, and an inertia matrix, $[I]$, as follows:

$$[I][\omega] = [H]. \tag{i}$$

In principal axis coordinates, this becomes

$$\begin{bmatrix} I_1 & 0 & 0 \\ 0 & I_2 & 0 \\ 0 & 0 & I_3 \end{bmatrix} \begin{bmatrix} \omega_1 \\ \omega_2 \\ \omega_3 \end{bmatrix} = \begin{bmatrix} H_1 \\ H_2 \\ H_3 \end{bmatrix}. \tag{ii}$$

As long as torques are computed about a fixed point or the center of mass of the body, the following is true:

$$\frac{d\mathbf{H}}{dt} = \tau, \tag{iii}$$

where $\mathbf{H}$ is the angular momentum vector and τ is the torque vector. When the components of $\mathbf{H}$ and τ are chosen in the moving coordinate frame, then Eq. (iii) becomes

$$\frac{\partial \mathbf{H}}{\partial t_{\mathbf{rel}}} + \boldsymbol{\omega} \times \mathbf{H} = \tau \tag{iv}$$

or

$$\begin{bmatrix} \dot{H}_1 \\ \dot{H}_2 \\ \dot{H}_3 \end{bmatrix} + \begin{bmatrix} \omega_2 H_3 - \omega_3 H_2 \\ \omega_3 H_1 - \omega_1 H_3 \\ \omega_1 H_2 - \omega_2 H_1 \end{bmatrix} = \begin{bmatrix} \tau_1 \\ \tau_2 \\ \tau_3 \end{bmatrix}, \tag{v}$$

in which the $\boldsymbol{\omega} \times \mathbf{H}$ term corrects the terms representing change relative to the moving frame to correctly portray the total change in $\mathbf{H}$ relative to the inertial frame. (See Reference [9].)

Verify that the modulated gyrator ring structure correctly represents Euler's equations, Eq. (v).

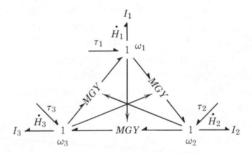

Select one of the MGYs, and write the equations that it represents. Verify power conservation in the form,

$$\dot{H}_1\omega_1 + \dot{H}_2\omega_2 + \dot{H}_3\omega_3 = \tau_1\omega_1 + \tau_2\omega_2 + \tau_3\omega_3.$$

9-15 The mechanical system shown in part a contains one rigid body that rotates about a fixed axis and one that moves in more general motion about a single fixed point. Using the results of problem 9.14, verify the bond-graph representation shown in part b. By computing ω_1, ω_2, ω_3 in terms of θ, ψ, find the constitutive laws for the multiport MTF.

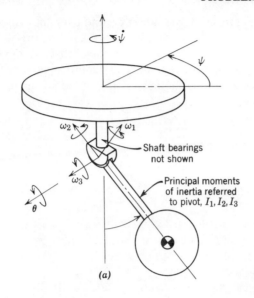

(a)

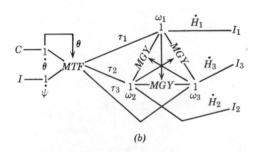

(b)

If θ and ψ were used as generalized coordinates, a nonlinear, fourth-order set of state equations could readily be found by applying Lagrange's equation to $T^* - V$. If all dynamic elements could accept integral causality, a fifth-order state space would result. This suggests that derivative causality may be necessary for some of this system. See if this is true by expanding the multiport MTF into a junction structure of 0- and 1-junctions and 2-port MTFs and applying causality.

9-16 Consider the device of Figure 9.17. Suppose you are given the permeability of the core material and of air, the pertinent physical dimensions of the core, and the number of turns, N. Estimate the inductance of the device if the core material remains in the linear range of the B-H curve.

9-17   In Figure 9.18, find an expression for the capacitance parameters for the sections of core material in terms of permeability and physical dimensions in the linear case. Combine the capacitance parameters into a single equivalent capacitance, and show how to estimate the inductance using this equivalent capacitance.

9-18

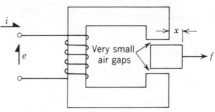

In the transducer sketched above, a slug of magnetic material slides partially in and out of the flux path. Make a simple bond-graph model of the device neglecting all loss effects and all leakage paths. By imagining how the reluctance at the slug would vary with displacement, x, discuss qualitatively the difference you might expect between the f versus x relation at constant current for this device as compared with the device of Figure 9.20.

9-19

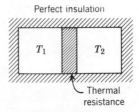

Two bodies that do not expand or contract are in thermal contact through a thermal resistance. Make two bond graphs for the system using both $T\dot{Q}$ and $T\dot{S}$ variables. Explain the different relations for the $-R-$ and $-C$ elements when the two different variable sets are used. Using the $T\dot{S}$ bond graph and Figure 9.21, show that if $T_2 \neq T_1$ initially, the entropy of the system can only increase.

9-20

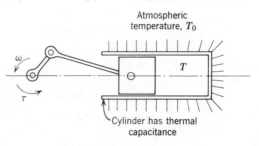

Consider the compression of a gas by means of a crank-piston arrangement. Let the cylinder have a single average temperature, and define thermal resistances between the gas temperature, T, and the cylinder temperature, and between the cylinder temperature and the atmospheric temperature, T_0. Make a bond graph that would allow you to predict the crank torque, τ, for relatively slow speeds of rotation, ω. (Note that you do not have enough information to evaluate all the system parameters.)

9-21

The dashpot in the suspension system has a force-velocity constitutive law that varies with the average temperature, T, of the dashpot since it utilizes oil that changes viscosity with temperature. Make a simple model of the system that would predict how the dashpot heats up when the input base velocity, $V(t)$, is given. Discuss your assumptions, and how you might estimate the system parameters you need for the thermal part of your model.

9-22

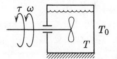

Suppose a shaft is connected to a paint stirrer such that, for rather slow angular rates, ω, all the power, $\tau\omega$, goes to heating up the paint. Make a bond graph relating the mechanical power to an entropy flow, $\dot{S}$, and including possible heat transfer to the atmosphere. Discuss the simplifying assumptions you have used.

9-23 In the acoustic approximation, the bulk modulus, B, is given by

$$B = \rho_0 c^2,$$

where ρ_0 is the mean mass density of the fluid and c is the speed of sound. If Δp represent a small increase in pressure over the mean pressure, p_0, then

$$\Delta p = \rho_0 c^2 \frac{\Delta \rho}{\rho_0},$$

where $\Delta \rho$ represents the change in density.

(a) Considering a fixed mass of fluid that occupies volume, V, when the pressure is p_0, show that

$$\frac{\Delta \rho}{\rho} = \frac{-\Delta V}{V}, \qquad \Delta p = -B \frac{\Delta V}{V},$$

where $-\Delta V$ represents a *decrease* in volume.

(b) Evaluate the inertia and capacitance parameters for the length of pipe shown in Figure 9.26.

(c) Using $\lambda = fc$ where λ = wave length, f = frequency, and c = sound speed, relate the length, l, for the pipe segment to the highest circular frequency, ω, of interest so that even the shortest wave length will span several "lumps" if one used Figure 9.26 to make a model of a long pipe by cascading many segments.

9-24

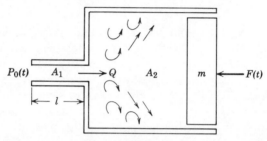

A high-speed hydraulic ram is forced by a pressure source, and we desire to predict how fast it can be stroked. Let the inlet pipe have length, l, and area, A, and consider only inflow, Q. The ram has mass, m, area, A_2, and has a force, $F(t)$, applied to it. To be conservative, assume that only the static pressure in the ram acts on the piston, that is, all the dynamic pressure is assumed to be lost. Make a bond graph for this system using a Bernoulli resistor to model the dynamic-pressure loss. Apply causality and write equations of motion for the system.

9-25

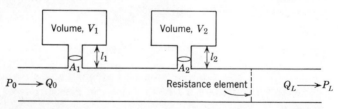

A schematic diagram of a muffler system employing two Helmholtz resonators and a resistive element is shown. Assume all dimensions are less than one wavelength of the highest frequency of interest. Make a bond graph for the system in which the inertias of the necks of the resonators of effective lengths, l_1, l_2, and areas, A_1, A_2, the capacitances of the volumes, V_1, V_2, and the resistance, R, are all represented. Using the results of problem 9.23, list the capacitance and inertia parameters in terms of the density, ρ_0, and the sound speed, c.

9-26

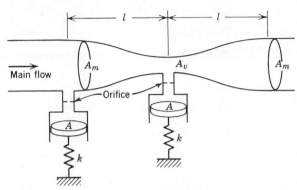

Part of an automatic flow-metering system is shown. The main flow goes through a smooth venturi that reduces the area from A_m to A_v and in which we may assume that all dynamic pressure is fully recovered. The spring-mounted pistons deflect due to the pressure in the main pipe and at the throat. Orifices in the pipes connecting the piston chambers to the main pipe restrict the flows to the pistons to low values.

Construct a bond graph for this subsystem, and indicate inertia elements and Bernoulli resistors for the two area change sections of length, l. Why should we expect the two pistons to deflect differing amounts? Would the difference between the two deflections serve to measure the main flow?

INDEX